AF505819

MANAGEMENT OF RADIOACTIVE MATERIALS AND WASTES:
Issues and Progress

The Pennsylvania Academy of Science Publications
Books and Proceedings

Book Editor: Shyamal K. Majumdar
Professor of Biology
Lafayette College
Easton, Pennsylvania 18042

1. *Energy, Environment, and the Economy,* 1981. ISBN: 0-9606670-0-8. Editor: Shyamal K. Majumdar.

2. *Pennsylvania Coal: Resources, Technology and Utilization,* 1983. ISBN: 0-9606670-1-6. Editors: Shyamal K. Majumdar and E. Willard Miller.

3. *Hazardous and Toxic Wastes: Technology, Management and Health Effects,* 1984. ISBN: 0-9606670-2-4. Editors: Shyamal K. Majumdar and E. Willard Miller.

4. *Solid and Liquid Wastes: Management, Methods and Socioeconomic Considerations,* 1984. ISBN: 0-9606670-3-2. Editors: Shyamal K. Majumdar and E. Willard Miller.

5. *Management of Radioactive Materials and Wastes: Issues and Progress,* 1985. ISBN: 0-9606670-4-0. Editors: Shyamal K. Majumdar and E. Willard Miller.

6. *Proceedings* of the Pennsylvania Academy of Science. Two issues per year; current volume (1985) is 59. ISSN: 0096-9222. Editor: Daniel Klem, Jr.

A volume on *Threatened, Rare and Endangered Species* is in preparation.

MANAGEMENT OF RADIOACTIVE MATERIALS AND WASTES:
Issues and Progress

EDITED BY

SHYAMAL K. MAJUMDAR, Ph.D.
Professor of Biology
Lafayette College
Easton, Pennsylvania 18042
                and
E. WILLARD MILLER, Ph.D.
Professor of Geography and
    Associate Dean for Resident
    Instruction (Emeritus)
The Pennsylvania State University
University Park, Pennsylvania 16802

Founded on April 18, 1924

A Publication of
The Pennsylvania Academy of Science

Library of Congress Cataloging in Publication Data

Management of Radioactive Materials and Wastes: Issues and Progress

Bibliography
Appendix
Includes Index

 I. Majumdar, Shyamal K., 1938-   , ed.
II. Miller, E. Willard, 1915-   , co-ed.

Library of Congress Catalog Card No.: 85-61443

ISBN 0-9606670-4-0

Printed in the United States of America by

Typehouse of Easton
Phillipsburg, New Jersey 08865

Honorable William W. Scranton, III
Lt. Governor, Commonwealth of Pennsylvania

FOREWORD

As the quality of life in Pennsylvania continues to improve and brings our state national recognition, all of us must look to the future and ways to preserve and enhance the environment and the economy of our Commonwealth.

The safe handling of all waste, and especially radioactive waste, is critical to our quality of life. Business, government, the research community and all citizens have a stake in the policies and planning to safely dispose of such waste.

The Pennsylvania Academy of Science advances our efforts through its work and study, and thereby makes an invaluable contribution to our state. I commend the work of the academy and encourage all Pennsylvanians to help our Commonwealth move forward in this area.

By working together, we can have a future that includes the development of new technologies to dispose of waste safely and new levels of cooperation among all communities to protect our environment and our economy.

Sincerely,

William W. Scranton, III
Lieutenant Governor
Commonwealth of Pennsylvania

COMMONWEALTH OF PENNSYLVANIA
DEPARTMENT OF ENVIRONMENTAL RESOURCES
P.O. Box 2063
Harrisburg, PA 17120

The Secretary

Nicholas DeBenedictis
Secretary, Department of Environmental Resources

REMARKS

Implementation of comprehensive federal and state laws to regulate solid, hazardous and radioactive wastes has made business, government and the scientific community examine our current waste disposal methods and evaluate alternative techniques for the future. The preservation of our environment, the health of our economy and the maintenance of our quality of life are dependent upon the safe and proper disposal of the waste produced by all members of our society.

Efforts to establish and develop proper disposal sites have been hampered by public opposition, much the result of problems associated with old, abandoned dumps. To gain the public's confidence, government, industry and community leaders must be willing to lead a constructive, open dialogue to site and manage new and innovative waste facilities.

Our immediate task is to identify and clean up those dumps that pose a threat to the public and the environment and to prevent a repeat of past mistakes. Only then can the confidence of the public be gained to develop new technologies and facilities.

With limited space remaining in many landfills across the Commonwealth and opposition to opening new sites, attention is being focused on resource recovery and waste-to-energy projects.

Approximately 300 permitted landfills are in operation in Pennsylvania; over 800 inadequate sites have been closed by the state since 1968. The inability to locate new landfills demonstrates the need for more energy-efficient solid waste disposal.

In 1983, Pennsylvania awarded a record $873,000 to 23 communities for solid waste planning grants to study long-range disposal alternatives and evaluate recycling possibilities. This year, more communities than ever before in the program's 14-year history applied for grants, which serve as pilot studies for other municipalities as well as guiding immediate help for solid waste problems.

The safe disposal of hazardous wastes will present a difficult challenge in the future, as public concerns over "midnight dumping" and groundwater contamination from illegal and unregulated disposal flourishes. Much of the scientific community is skeptical of the long-term safety of total dependence on landfills for hazardous waste disposal, creating another impetus for alternatives to land disposal.

The nation's industries could reduce the volume of hazardous wastes through reprocessing of used materials and the evaluation of manufacturing methods, but the development of new treatment and disposal methods cannot be overemphasized. Research into high-temperature incineration, biological treatment and detoxification of hazardous wastes may provide the best disposal alternatives available for the future.

The scientific problems presented by the disposal of radioactive wastes are only part of the social and political considerations surrounding the issue that must be faced in the coming years. For too long, band-aid solutions to the disposal of radioactive wastes have been accepted.

The nation, including Pennsylvania, is required by law to establish low-level radioactive waste disposal sites, either alone or in cooperation with other states, by January 1986. With the assistance of community leaders, scientists and elected officials, Pennsylvania can properly locate a suitable site for low-level radioactive wastes.

The Pennsylvania Academy of Science is to be commended for its efforts in the study of waste management, through the expertise of government, industry and the academic communities. Development of new techniques in waste disposal is dependent on new ideas, such as are included in these volumes. The academy will play an important role in the development of solutions essential to a clean environment and a sound economy.

Sincerely,

Nicholas DeBenedictis
Secretary
Pennsylvania Department
of Environmental
Resources

PREFACE

This is one of three books published by the Pennsylvania Academy of Science on Hazardous and Toxic Wastes, Solid and Liquid Wastes, and Radioactive Materials and Wastes. These volumes are comprehensive, authoritative source books describing and analyzing the technology and disposal of waste products, one of the most critical problems of our times. They consist of an assemblage of quality papers contributed by a group of leading experts from five countries and over fifteen states.

The problem of the disposal of the various types of wastes created by an industrial society is not a new one, but one that has reached a critical dimension in our times. The modern industrial wastes can not only devastate the natural environment, but have the potential to affect the health of millions of human beings. The seriousness of the problem becomes evident when waste materials have not been disposed of for many years and entire neighborhoods become contaminated, or when scores of abandoned hazardous waste sites are discovered in densely populated regions and the total environment is subject to contamination.

Each of the three volumes in this series considers a different type of waste material. This volume on Management of Radioactive Materials and Wastes is divided into five parts. Part one considers types, sources and management of radioactive wastes and covers such aspects as classification of radioactive wastes, low-level waste management, and shallow land-burial of radioactive wastes. The storage, transportation and disposal problems of radioactive materials are discussed in part two. Part three covers socio-political considerations in disposal of radioactive wastes. Part four considers emergency planning and preparations including radioactive spills, nuclear accidents and the preparation of hospitals for handling victims of radiation accidents, and the last part is devoted to radiation standards, environment and public health.

The volume on Hazardous and Toxic Wastes discusses treatment and disposal methods, waste incineration, selection and geological considerations of waste sites, emergency response and preparations needed in a toxic spill emergency, regulations, economic considerations and the environmental and health effects of hazardous wastes.

The book on Solid and Liquid Wastes classifies waste types, discusses management possibilities, describes pretreatment and treatment methods, evaluates environmental and health impacts, examines recyling and energy recovery potential from wastes, and considers laws, regulations and socio-economic problems.

These books will be of value to a wide audience. Such individuals as engineers, scientists, medical doctors, social scientists, and environmentalists will find them useful. They provide a wide perspective so that a specialist in one field can be informed about developments and trends in another branch of the subject. The volumes will also be of value to individuals who want to be informed about some of the most critical problems of the day.

We express our deep appreciation for the excellent cooperation and dedication of the contributors, who recognize the importance of solving the critical problems of waste disposal. For a task of this magnitude many individuals in addition to the authors made contributions, and we are most pleased to acknowledge them. The advice and guidance of the members of the editorial and advisory boards are gratefully acknowledged. Gratitude is extended to Dr. Robert S. Chase, Head, Department of Biology, Lafayette College, and to Dr. C. Gregory Knight, Head, Department of Geography, The Pennsylvania State University, for providing facilities for editorial work to the editors of the three volumes. Thanks are due to Caryn Golden of Lafayette College, and Nina McNeal and Joan Summers of The Pennsylvania State University for competent secretarial assistance. S. K. Majumdar and E. W. Miller extend heartfelt thanks to their wives Jhorna and Ruby, respectively, who graciously shared weekends and evenings with the preparation of the series and provided help and encouragement.

Shyamal K. Majumdar, Ph.D.
Lafayette College
Easton, Pennsylvania
and
E. Willard Miller, Ph.D.
The Pennsylvania State University
University Park, Pennsylvania

Editors, September 1985

Management of Radioactive Materials and Wastes: Issues and Progress

Table of Contents

CONTRIBUTORS

L. H. Barrett, (Chapter 9), Deputy Director, Three Mile Island Program Office, U.S. Nuclear Regulatory Commission, P.O. Box 311, Middletown, PA 17057.

Faith N. Brenneman, (Chapter 4), U.S. Nuclear Regulatory Commission, Office of Nuclear Reactor Regulation, Washington, D.C. 20555.

Charles R. Bolmgren, (Chapter 7), Westinghouse Electric Corporation, Advanced Power Systems Divisions, Nuclear Center, Box 355, Pittsburgh, PA 15230.

Richard J. Bord, Ph.D., (Chapter 14), Professor of Sociology, The Pennsylvania State University, 201 Liberal Arts Tower, University Park, PA 16802.

Edward F. Branagan, Ph.D., (Chapter 21), Radiological Assessment Branch, Division of Systems Integration, United States Nuclear Regulatory Commission, Washington, D.C. 20555.

David R. Brill, M.D., (Chapter 8), Chief, Nuclear Medicine, Geisinger Medical Center, Danville, PA 17822.

Joseph A. Coleman, Ph.D., (Chapter 2), Director, Division of Storage and Treatment Projects, U.S. Department of Energy, Washington, D.C. 20545.

Frank J. Congel, Ph.D., (Chapter 21), Radiological Assessment Branch, Division of Systems Integration, United States Nuclear Regulatory Commission, Washington, D.C. 20555.

Ralph DiSibio, Ed.D., (Chapter 10), Hittman Nuclear & Development Corporation, 9151 Ramsey Road, Columbia, Maryland, 21045.

William P. Dornsife, (Chapter 1), Nuclear Engineer, Pennsylvania Department of Environmental Resources, P.O. Box 2063, Harrisburg, PA 17120.

David G. Ebenhack, (Chapter 19), Vice President, Regulatory Affairs, Chem-Nuclear Systems, Inc., Columbia, S.C. 29210.

James E. Fairobent, (Chapter 21), Radiological Assessment Branch, Division of Systems Integration, United States Nuclear Regulatory Commission, Washington, D.C. 20555.

Robert S. Friedman, Ph.D., (Chapter 15), Professor of Political Science, Department of Political Science, The Pennsylvania State University, University Park, PA 16802.

P. J. Grant, (Chapter 9), Chief, Technical Support, Three Mile Island Program Office, U.S. Nuclear Regulatory Commission, P.O. Box 311, Middletown, PA 17057.

J. M. Harrison, Ph.D., (Chapter 16), Science Advisory Board of the Northwest Territories, 4 Kippewa Drive, Ottawa, Ontario, Canada K1S 3G4.

D. G. Jacobs, Ph.D., (Chapter 6), Technical Program Manager, HER Technical Associates, Inc., Oak Ridge, TN 37830.

A.R. Jarrett, Ph.D., (Chapter 3), Associate Professor of Agricultural Engineering, 249 Agricultural Building, The Pennsylvania State University, University Park, PA 16802.

William A. Jester, Ph.D., (Chapter 23), Associate Professor of Nuclear Engineering, The Pennsylvania State University, Breazeak Nuclear Reactor, University Park, PA 16802.

Shyamal K. Majumdar, Ph.D., (Chapter 26), Professor of Biology, Lafayette College, Easton, PA 18042.

Charles W. Mallory, (Chapter 11), Technical Advisor, Hittman Nuclear and Development Corporation, 9151 Ramsey Road, Columbia, MD 21045.

Stephen Marchetti, (Chapter 17), Manager, Environmental and Occupational Safety Department, West Valley Nuclear Services Co., (A Subsidiary of the Westinghouse Electric Corporation), West Valley, New York 14171.

E. Willard Miller, Ph.D., (Chapter 26), Professor of Geography and Associate Dean for Resident Instruction (Emeritus), The Pennsylvania State University, University Park, PA 16802.

Kenneth L. Miller, M.S., C.H.P., (Chapter 24), Associate Professor of Radiology, Director, Division of Health Physics, The Milton S. Hershey Medical Center, The Pennsylvania State University, Hershey, PA 17033.

E. Preston Rahe, (Chapter 18), Manager, Nuclear Safety Department, Westinghouse Electric Corporation, Box 355, Pittsburgh, PA 15230.

Richard R. Rawl, (Chapter 13), Chief, Radioactive Materials Branch, U.S. Department of Transportation, DMT-223, 400 Seventh St. S.W., Washington, D.C. 20590.

R. R. Rose, Ph.D., (Chapter 6), Manager, Evaluation Research Corporation, 800 Oak Ridge Turnpike, Oak Ridge, TN 37830.

Michael T. Ryan, Ph.D., (Chapter 19), Director, Environmental and Dosimetry Laboratory, Chem-Nuclear Systems, Inc., Barnwell, S.C. 29812.

A. T. Sabo, (Chapter 22), Director, NES Licensing, Safeguards and Safety, Westinghouse Electric Corporation, Box 355, Pittsburgh, PA 15230.

Anne D. Stubbs, (Chapter 5), Senior Research Associate, CONEG Policy Research Center, Hall of the States, 400 North Capitol Street, Washington, D.C. 20001.

George K. Tokuhata, Dr. P.H., Ph.D., (Chapter 25), Director, Division of Epidemiology Research, Pennsylvania Department of Health, P.O. Box 90, Harrisburg, PA 17108.

Michael A. Vince, M.Sc., (Chapters 12 & 20), Medical and Health Physicist and Radiation Safety Officer, St. Luke's Hospital, Bethlehem, PA 18015.

John B. Yasinsky, Ph.D., (Chapter 7), General Manager, Westinghouse Electric Corporation, Advanced Power Systems Divisions, Nuclear Center, Box 355, Pittsburgh, PA 15230.

Charley Yu, (Chapter 23), The Pennsylvania State University, Breazeak Nuclear Reactor, University Park, PA 16802.

SYMPOSIA

I *Hazardous and Radioactive Wastes.* Holiday Inn, Grantville, Pennsylvania, October 29, 1982.
Chairman: Donald Zappa, President, Vector Corporation, Pittsburgh, PA

II *Solid and Hazardous Wastes.* Host Corral, Lancaster, Pennsylvania, April 10, 1983.
Chairman: Justice John P. Flaherty, Supreme Court of Pennsylvania. Chairperson of The Pennsylvania Academy of Sciences' Advisory Council

III *Radioactive Materials and Wastes.* Marriott Hotel, Monroeville, Pennsylvania, October 27, 1983.
Chairman: Dr. George C. Shoffstall, President, Pennsylvania Academy of Science

ACKNOWLEDGMENTS

The Pennsylvania Academy of Science published this book in association with the Lt. Governors office and the Pennsylvania Department of Environmental Resources (DER). Any opinions, findings, conclusions, or recommendations expressed are those of the author(s) and do not necessarily reflect the views of the Lt. Governor's office, the DER, or The Pennsylvania Academy of Science.

The publication of this book was aided by contributions from The Pennsylvania Power and Light Company, Allentown, Pennsylvania, U.S. Ecology, Louisville, Kentucky and other companies.

Pictorial Coverage of the Pennsylvania Academy of Sciences' Presentation of *Hazardous and Toxic Wastes* and *Solid and Liquid Wastes* Books Ceremony in the Governors Reception Room, Harrisburg, PA—April 23, 1985

Dr. George C. Shoffstall, President, PAS, (left) greets Governor Thornburgh. Dr. S. K. Majumdar, President-Elect, PAS, is on right.

Governor Thornburgh discusses the books with Drs. Majumdar and Shoffstall.

Officers and several members of the Pennsylvania Academy of Science and representatives from the Department of Environmental Resources gather after the books' presentation to pose for a picture with Pennsylvania Governor Richard Thornburgh.

Message from

JOHN P. FLAHERTY
Justice
Supreme Court of Pennsylvania
Chairman, Advisory Board
Pennsylvania Academy of Science

A society advances, indeed survives, measured by its control and disposition of societal waste. As is quite evident by even a cursory look at history, nothing is more destructive to life than uncontrolled human, industrial and, a fortiori, radioactive waste. During the German blitz of London in 1941-42, for example, Winston Churchill's greatest fear was a breakdown of the sewer system! No amount of explosive then known could have caused the human destruction of which such an event was capable. If we are, thus, to accommodate an increasing population on our now highly urbanized and industrialized planet, it is essential that the scientific community devote itself *with priority* to neutralizing the inundating waste which, unabated, will cause catastrophe to mankind, unparalleled in history.

The Pennsylvania Academy of Science, recognizing the importance of the subject, has endeavored by this publication to stimulate the scientific reader to further innovation, as well as to provide an anthology of present methodology.

Highly industrialized, with a large urban population, Pennsylvania is particularly an appropriate situs for this work, as within its borders occurred an event of stark terror, presaging potential future disaster — *Three Mile Island!*

The President of the Pennsylvania Academy of Science

Dr. George C. Shoffstall, Jr.

Dr. George C. Shoffstall, Jr., President of the Pennsylvania Academy of Science (1976-78 and 1984-86) was the Director of Education and Organizational Development, The Western Pennsylvania Hospital, Pittsburgh, Pennsylvania and Assistant to the Dean in the College of Science (1966-1980) at The Pennsylvania State University. Dr. Shoffstall served as President (1981-82) of The National Association of Academies of Science and was also named a Fellow in the American Association for the Advancement of Science. He is currently serving on the AAAS Council (1984-86).

Active at the national and state levels, Dr. Shoffstall holds a doctorate in radiation biology from The Pennsylvania State University. Named an Outstanding Educator in 1975, Dr. Shoffstall is listed in *American Men and Women in Science, Who's Who in the East, Who's Who in Training and Development, Community Leaders of America, 11th ed.,* and the *1981 Directory of Distinguished Americans.*

Along with the American Association for the Advancement of Science and the Pennsylvania Academy of Science, his memberships include the Association for the Advancement of Medical Instrumentation, American Society for Training and Development, National League for Nursing, Pennsylvania League for Nursing, Pennsylvania Association for Medical Education and Smithsonian Associates.

MESSAGE FROM THE PRESIDENT OF THE PENNSYLVANIA ACADEMY OF SCIENCE

Dr. George C. Shoffstall, Jr.

With the advent of the world's first electric generating nuclear power plant in 1957 located at Shippingport, Pennsylvania, the role of nuclear power was hailed as a cheap, clean, efficient and safe energy source for the world's populace. However, public support for nuclear energy began to dwindle over the years and with the unfortunate accident at Three Mile Island on March 28, 1979, near Middletown, Pennsylvania, support declined radically.

Reasons proffered for this decline in public support are manifold—nuclear power's latent potential for disaster; media bias against science and technology; public assumptions that a significant proportion of scientists oppose further development of nuclear energy; and a dubious legacy related to the disposal of radioactive wastes.

Uniquely, Pennsylvania has played key roles in the development of nuclear energy for peaceful as well as military application. Unfortunately, the Commonwealth of Pennsylvania is labeled as a major producer of radioactive wastes. This is quite inevitable with the state's many nuclear reactors, hundreds of hospitals and clinics utilizing radioisotopes in the protocols of diagnosis and treatment of disease, numerous research laboratories using nuclear tracers, radiopharmaceutical manufacturing, industrial, institutional and other corporate users of nuclear materials adding to the stockpile.

Currently, much of Pennsylvania's radioactive wastes comes from the utilization of nuclear materials in the generation of electricity. However, a significant quantity of radioactive wastes has nothing to do with the generation of power by nuclear energy. Virtually every type of radioactive waste is produced in the state with the exception being the high level waste that results in the reprocessing of fissionable materials used in the production of plutonium for nuclear weapons.

In general, the types of radioactive wastes produced or stored in Pennsylvania are: high-level wastes, products of used or spent fuel from nuclear power plants and reactors; low-level wastes from contaminated filters, sludges, experimental animal carcasses used in medical and biological research, floor sweepings, gloves, clothing, rags, glassware, tools and plastics; transuranic wastes from materials contaminated with radioactive elements other than uranium; and mill tailings which are earthen radioactive residues remaining from the processing of uranium at former mill sites.

Concomitant with the accrual of nuclear wastes is a need to safely transport the material to a disposal site. Compounding this dilemma is a deep apprehension in countless communities across America regarding the transportation of the various types of radioactive material through their communities. Coincidentally, over the past several years, large sectors of the public and some members of the scientific community have become concerned that a major population will suffer exposure from radioactive leaks due to improperly packaged nuclear waste products during shipment or, worse, an accident during the actual transport of the nuclear waste storage casks will cause widespread havoc.

This scenario has been amplified with the negative treatment of nuclear issues by the national media. Accordingly, the public-at-large is badly served by facile misrepresentations of national needs. Worldwide, an overwhelming majority of scientists have remained strong supporters of nuclear energy, a counterpoint to the escalation of public opposition fueled by media criticism, political activists, and a small but extremely vocal minority of anti-nuclear scientists.

Paralleling increased public concern regarding the nuclear industry, more stringent policies have been implemented by governmental and regulatory commissions. Coupled with licensing reform, the nuclear industry has responded by dramatically upgrading its operational practices, monitoring and safety systems.

Moreover, with large quantities of radioactive wastes accruing via a diverse industry, a need exists for additional disposal sites and an expanded nuclear and hazardous wastes transportation system. However, antipodal to the Federal Department of Transportation and the Nuclear Regulatory Commission agreement that current methods of transporting radioactive materials are adequate and have an excellent record of safety, are the unfavorable non-scientific allegations by activists who refute this technology. Ergo, wherein lies the problem?

It would appear that nuclear energy and its related technologies are presently caught in a world convulsed by social, political, cultural and religious reform. The role of nuclear power in these United States of America is still evolving and can ill-afford the unscientific, hysterical opposition that is thwarting the economic growth and the enhancement of the quality of life of this nation and its citizenry.

Paramount for the nuclear industry to meet its potential is a mandate from the public sector for improved science education in our schools and a commitment from the American public-at-large to think in terms of rational scientific principles rather than ideologies based upon fear, ignorance, and misinformation.

This textbook importunes the reader to assay the writings of the authors and the quality of the scientific information presented. Moreover, in addition to the topics being intrinsically heuristic and scientifically factual, the Pennsylvania Academy of Science is pleased to present to the reader this publication—*Management of Radioactive Materials and Wastes: Issues and Progress.*

INTRODUCTION

Dr. HEINZ G. PFEIFFER
Manager, Technology and Energy Assessment
Pennsylvania Power and Light Company
Allentown, Pennsylvania

One characteristic of all living processes is that they produce waste. Throughout most of history the wastes and the environment reached a balance of growth and decay. Since the start of the industrial revolution and the attendant population growth, this balance has been upset both by the increased quantity of traditional wastes produced and by the introduction of new classes of wastes. At the same time, increased affluence and attention to public health have increased expectations for life, both quality and duration. Industrial wastes are inevitable; and we will see more and more that industries, as they expand or relocate, will have as a major consideration the safe, economic, and convenient disposal of these wastes.

Compared to the real progress on air and water problems, solid waste disposal is in the most confused state. The tendency to solve the problem by sending them "elsewhere" is short-sighted. By and large, unnecessary transportation increases the problems associated with the waste.

There are three alternatives for the disposal of the solid wastes: reuse in some form, destruction, or landfill. Of these, reuse is by far the most desirable; but many regulations that have appeared in the last few years have put barriers to the ability of an industry to reprocess its byproducts to useful ends. Most organic wastes can be successfully destroyed by properly controlled combustion and are being so destroyed in many places. The public fears of combustion are based on some examples where adequate attention was not paid to the process. This should not remove combustion as an alternative—it should enliven public interest in supporting careful combustion as a permanent solution. Landfill should be considered as the last resort because the material remains and can cause later problems.

Another regulatory trend has been a tendency to confuse the boundary between hazardous wastes and residual wastes that pose no health threat, to the detriment of our ability to deal with either. Overkill on the wastes that pose no health problems raises the costs of everything we produce and reduces the effort we can put on those wastes that really pose a threat.

Radioactive wastes are an entirely separate category; if one excludes transmutation, they will gradually destroy themselves in a time determined by their own internal clock. These times for some of the high level wastes are extremely long, depending on the particular form of the radioactive waste. It would take one to ten thousand years before the toxicity of the wastes (as measured by their impact on the ground water) has fallen to the level of the original ore from which the radioactive material was made. This level is a good measure because at that time, if the wastes were in place of the original ore, no net gain or loss in radioactive hazard would be present. Particularly in the case of spent nuclear fuel as opposed to reprocessed fuel, the further radioactivity would then proceed much as the original ore would have. Many of the statements that talk about millions of years for radioactive materials to decay to "safe" levels forget that natural uranium and thorium in the ground take four billion and fourteen billion years, respectively, to lose half their radioactivity.

The disposal of a material that has to be isolated for at least one thousand years poses a new set of problems and controls that can only be solved by cooperative effort. The times involved, long in human lifetimes but short for most geological events, have led to the selection of a series of burial alternatives—granite, salt, clay, etc.—that are now being compared as alternatives throughout the world. Even though the nuclear community answers the questions and doubts concerning the disposal, a large scale project is necessary to start to rebuild public confidence in the ability and will to do the job right.

This volume along with two previously published books *(Hazardous and Toxic Wastes: Technology, Management and Health Effects* and *Solid and Liquid Wastes: Management, Methods and Socioeconomic Considerations*, by the Pennsylvania Academy of Science) will serve to describe our alternatives and point out the directions that our efforts have to take to enable us to live in health, comfort and beauty with the inevitable byproducts of our industrial society.

PART 1
Types, Sources and Management

The problems of the disposal of radioactive wastes may be the most critical issue facing modern society. Part one provides fundamental information on types, sources and management of these waste products. The initial chapter presents a classification scheme identifying major types of radioactive materials and wastes with a discussion of the characteristics of low-level waste, and comparisons of the toxicity of radioactive wastes.

Because of the complexity of handling radioactive wastes, management considerations are paramount. Chapters two and three consider waste management technology and management of water problems at low-level radioactive waste repositories.

As the amount of radioactive wastes increased in recent times, there has been a growing recognition of the need for regional and national co-operation in securing disposal sites. Chapters four and five discuss the compacts that are evolving for low-level waste management. The state and federal activities are summarized. The problem of locating a satisfactory site and the length of time needed to construct a facility are emphasized.

The Low-Level Radioactive Waste Policy Act of 1980 gave each state the responsibility to provide for disposal of its low-level radioactive wastes generated, but clearly indicated that low-level waste sites should not be generated in each state. As a consequence, regional management systems are evolving. The regional compacts are based on the five major principles of good faith, equity, state sovereignty, minimal cost to states and consistency with federal authority. In the development of radioactive waste disposal sites the most fundamental criteria are the health, safety, and welfare of the local citizens. The compacts are in an initial stage of development and will face many problems in the immediate future as facilities are needed and ultimately developed.

A technology to handle radioactive wastes is gradually evolving. The shallow land burial of radioactive wastes is discussed as to the design of the facility, radionuclide migration, biological intrusion and translocation and space utilization. The final chapter reviews the technology of handling high-level waste, spent fuel, transuranic waste and low-level waste.

Chapter One

CLASSIFICATION OF RADIO-ACTIVE MATERIALS AND WASTES

William P. Dornsife, Nuclear Engineer

Pennsylvania Department of Environmental Resources
Bureau of Radiation Protection
P.O. Box 2063
Harrisburg, Pennsylvania 17120

Radioactive materials find a wide variety of uses in today's highly technical society. These materials are most frequently used or generated in producing electricity, various industrial processes, medical applications, various research activities, defense and research and development activities of the Federal Government. As is generally the case, various wastes are produced as a by-product of the use of this material, and these wastes must be safely managed and disposed of in order for the use of material to continue to benefit mankind.

MAJOR TYPES OF RADIOACTIVE WASTE

Radioactive waste is legally defined as belonging to one of four categories; high level and spent nuclear fuel, transuranic, mill tailings and low level. These four waste categories are defined by different statutes as follows:

1. High level waste is defined[1] as "the highly radioactive material resulting from the reprocessing of spent nuclear fuel, including liquid waste produced directly in reprocessing and any solid material derived from such liquid waste that contains fission products in sufficient concentrations; and other highly radioactive material that the Nuclear Regulatory Commission by rule requires permanent isolation." Spent nuclear fuel is defined[1] as "fuel that has been withdrawn from a nuclear reactor following irradiation, the constituent elements of which have not been separated by reprocessing." These wastes are exclusively generated by commercial power reactors,

research reactors, or reactors used for defense activities of the Federal government. The waste is typically produced in relatively small volumes, but contain very high concentrations of fission products and transuranics, and thus requires massive shielding during handling. The transuranics and some of the fission products have a very long toxic lifetime (greater than 1,000 years), and therefore this waste requires long-term isolation from the environment using very secure techniques, such as geological repositories. The Nuclear Waste Policy Act of 1982[1] has given the responsibility for ultimate disposal of this waste to the Federal Government.

2. Transuranic (TRU) waste is defined[2] as "waste material containing radionuclides with an atomic number greater than 92, which are excluded from shallow land burial by the Federal Government." In the past, waste containing these elements with a concentration greater than 10 nano (10^{-9}) curies per gram of waste was considered TRU waste. If the concentration was lower, it was considered low level waste. However, recent NRC rules[3] governing land disposal of radioactive waste has raised this limit by an order of magnitude to 100 nanocuries per gram of waste. TRU waste is primarily generated by defense activities of the Federal Government, and therefore, responsibility for permanent disposal of this waste mostly rests with the Federal Government. Since most of the TRU radionuclides have a very long toxic halflife, the most suitable method for disposal will be isolation in geological repositories with the high level waste.

3. Mill tailings are defined[4] as "the remaining portion of the metal bearing ore after some or all of the material, such as uranium, has been extracted, or other wastes produced by the extraction or concentration of uranium or thorium from any ore processed primarily for its source material content." Mill tailings are principally generated as a by-product of the extraction of uranium for nuclear fuel. These wastes are produced in very large volumes but contain very low concentrations of the daughter products of uranium and thorium (such as radium and radon), some of which may have very long lifetimes. Special consideration for disposal must be given to radon, since it is a noble gas and, therefore, very difficult to contain. Because of the large volumes that are generated, disposal of mill tailings usually occurs very near the source of generation.

4. Low level wastes (LLW) are defined[5] as "radioactive waste not classified as high level radioactive waste, transuranic waste, spent nuclear fuel or mill tailings." The Low Level Radioactive Waste Policy Act[5] has made disposal of this category of waste a state responsibility, and has allowed states to enter into compacts for the development of regional disposal facilities as the most suitable way to manage this waste. Regional compacts may exclude LLW not generated in those regions after January 1, 1986. Pennsylvania has been negotiating with other northeastern states to form a regional compact. After initially rejecting a compact negotiated with all

northeastern states in January, 1983, Pennsylvania agreed to a smaller compact with several states contiguous to Pennsylvania. It is low-level waste that is produced by the bulk of the large variety of industries and institutions that use radioactive material. Because of this fact, and since LLW is produced in so many different forms, its general characteristics and classification will be the subject of the remainder of this Chapter.

For detailed information on the technical and social political issues that must be addressed in siting a disposal facility for LLW, it is recommended that the recent Penn State study entitled "Low Level Radioactive Waste Disposal Siting: A Social and Technical Plan for Pennsylvania"[7] be consulted.

Information on estimates of the annual volumes that are generated and constituents of the various categories of radioactive waste is presented in Table 1.

TABLE 1

Data on Various Types of Radioactive Wastes[6,7]

Waste Type	Annual Generation Rate in 1982 (m^3)	Currently in Storage Awaiting Disposal (m^3)	Major Isotopes and representative concentrations
High level waste (defense activities)	1500	290,000	Cs^{137} or Sr^{90}-800 Ci/m^3 Pu^{238} - 3 Ci/m^3
Spent fuel	240	3,300	Cs^{137} or Sr^{90} - 4000 Ci/m^3 Pu^{239} - 13 Ci/m^3
Mill tailings	13,000,000	—	Ra^{226} - 4.9 x 10^{-4} Ci/m^3
Transuranics	4,000	55,000	—
Low Level Waste	90,000	—	H^3 - 7 Ci/m^3 Cs^{137} - 0.3 Ci/m^3 I^{129} - 2 x 10^{-5} Ci/m^3

Classification of Low-Level Waste

Since LLW is legally defined so broadly, there may be certain waste streams which are defined as LLW but contain high concentrations of various radionuclides which may not be suitable for disposal using near surface techniques. To resolve this problem, NRC has developed a LLW classification system which is included as part of the recently developed LLW disposal regulation, 10CFR 61[3]. This waste classification system separates the LLW into three classes, A, B, and C. The class into which the LLW falls is determined by the concentration of the various contained radionuclides, as shown in Table 2.

Class A waste contains the lowest concentration of radionuclides and must meet minimum requirements for stability, such as: 1) cardboard or fiberboard packages are not acceptable; 2) solid waste should contain as little freestanding liquid as possible; 3) waste cannot be pyrophoric, explosive or highly reactive, or capable of generating toxic gases; and 4) if hazardous material is present, waste must be treated to remove the hazard to the maximum extent possible.

TABLE 2

Maximum Radionuclide Concentrations in Curies/m^3 for the Various Waste Classes from NRC Waste Classification System

Radionuclide	Class A	Class B	Class C
H-3	40	MC	—
Co-60	700	MC	—
Nc-63	3.5	70	700
Ni-63 (in activated metal)	35	700	7000
Sr-90	0.04	150	7000
Cs-137	1	44	4600
C-14	0.8	—	8
C-14 (in activated metal)	8	—	80
Ni-59 (in activated metal)	22	—	220
Nb-94 (in activated metal)	0.02	—	0.2
Tc-99	0.3	—	3
I-129	0.008	—	0.08
Total all radionuclides with less than 5 year half life	700	MC	—
Alpha emitting TRU with half life greater than 5 years	10*		100*
Pu-241	350*		3500*
Cm-242	2000*		20,000*

MC - maximum concentration—all waste above class A limit is class B
— - no limit is applicable for this class
* - units are nanocuries/gram

Class A waste is considered relatively unstable, compared to the other classes and, therefore, may partially be decompose and cause slumping of the disposal cell cover, which could lead to the eventual failure of the integrity of the disposal cell. Since this class by itself contains radionuclides which have a relatively low concentration and short halflife, significant exposure would not occur to the public as a result of this failure. However, because of this factor, Class A waste must be segregated in scparate disposal cells from the Class B and C waste, if it cannot meet the stability requirements of these higher classcs.

Classes B and C wastes must meet the same minimum requirements for Class A waste, but in addition, must meet more rigorous requirements concerning waste form to insure that the wastes will remain stable after disposal. These additional requirements include: 1) providing structural stability to the waste under conditions which can be expected after disposal—This structural stability can be provided by the waste form or the disposal container; 2) conversion of liquid waste or waste containing liquid to a waste form that contains as little freestanding liquid as possible; and 3) the filling of void spaces within the waste package and between waste packages after disposal to the maximum extent possible. In addition, the Class C waste must be disposed of with consideration given to protection of inadvertent intruders that may eventually contact the waste.

These additional measures may include providing engineered or natural intrusion barriers or placing the waste at the bottom of the disposal cell.
If the waste exceeds the Class C limits, it is generally considered not suitable for near surface disposal. NRC regulations do allow for specific case-by-case consideration for disposal using near surface techniques. However, since these very high activity wastes require a greater degree of isolation from the environment, special facilities which allow deeper than near surface disposal may be required. In the Northeast and other regions that are humid, this option may not be feasible due to a generally shallow watertable. For this reason, special arrangements, such as agreements between regions or transfer to the Federal Government, may be necessary to provide disposal for this category of waste.

CHARACTERISTICS OF LLW

Low level radioactive wastes generally consist of various materials which have become contaminated by radionuclides in the various processes which use radioactive material. The radioactive material users that generate LLW can be divided into three broad categories: power reactor, industrial and institutional. The various major types of generators in each category and the types of waste material that they produce is summarized in Table 3[7].

It should be noted that all materials which are generated as low level radioactive waste is not necessarily disposed of by shipping to a commercial shallow land disposal site. NRC regulations allow the release of waste liquids and gases which contain low concentrations of radionuclides which are less than specified limits to the environment. In addition, if the waste contains radionuclides with very short halflifes, it may be temporarily stored until the isotopes decay to insignificant levels and then disposed of as normal trash. This controlled release of radioactive waste results in insignificant radiation exposure to the public.

The LLW which is shipped for disposal represents a special category that requires special handling and precautions to isolate this waste from the environment until it has decayed to insignificant levels.This waste is usually identified by the volume and the activity of the contained radioactive material. An estimate of the average annual volume and activity which was produced by the various types of generators during the period from 1982-1983 in Pennsylvania, the Northeast and the entire U.S. is shown in Table 3. As can be seen from this table, Pennsylvania currently generates about 25% of the LLW that is generated in the Northeast and about 9% of the LLW in the country. Primarily because of the fact that several nuclear power plants are scheduled for operation in the next few years in Pennsylvania, the quantity of LLW that is projected to be generated in the future will significantly increase. It has been estimated[8] that by 1990, Pennsylvania will be generating about 9,900 m³ of LLW per year, which will be about 33% of all the waste generated in the Northeast.

TABLE 3

Average Annual Estimated Volume and Activity of Low Level Radioactive Waste Generated in Pennsylvania, the Northeast, and entire U.S. from 1982-1983.

		Volume (m³)	Activity[c] (curies)
PENNSYLVANIA[a]			
	Reactor	4,807 (70%)	6,400.8 (33%)
	Industrial	1,358 (20%)	13,068.0 (67%)
	Institutional	669 (10%)	27.4 (<1%)
	Total	6,834	19,496.2
NORTHEAST[a,b]			
	Reactor	15,554 (57%)	108,503.5 (58%)
	Institutional/Industrial	11,543 (43%)	80,008.0 (42%)
	Total	27,097	188,511.5
U.S.[c]			
	Reactor	48,067 (63%)	289,563.4 (63%)
	Institutional/Industrial	28,229 (37%)	170,055.2 (37%)
	Total	76,296	459,618.6

[a] From Reference 8

[b] Includes all New England states, New York, Pennsylvania, New Jersey, Maryland and Delaware

[c] From 1982 and 1983 State by State assessment of LLRW to Commercial Disposal Sites prepared by the National LLW Management Program EG&G Idaho, Inc.

A COMPARISON OF THE TOXICITY OF VARIOUS RADIOACTIVE WASTES

Primarily because of the way in which the media has been presenting information on radioactive waste, the public does not generally distinguish between the various categories of radioactive waste, much less recognize that each category has a very different potential toxicity as a function of time. Various studies have been performed[7,9] which provide some perspective on the relative toxicity of the various categories of radioactive wastes compared to other common material which contain various toxic substances. The relative toxicity of a cubic meter of the various types of radioactive waste, assuming that all the contained toxic material is released via the groundwater pathway, is presented in Figure 1 and compared with the toxicity of soil which can be considered as a background toxicity. This comparison shows that an equal volume of spent fuel is many orders of magnitude more potentially toxic than typical LLW or background over its lifetime. Defense HLW has a potential toxicity which is a couple of orders of magnitude less than spent fuel because it is generally irradiated for a much shorter period of time and the plutonium-239 and uranium are removed by reprocessing. LLW decays to within the range of background after about a few hundred years, but mill tailings will not decay to the range of background until after many thousands of years.

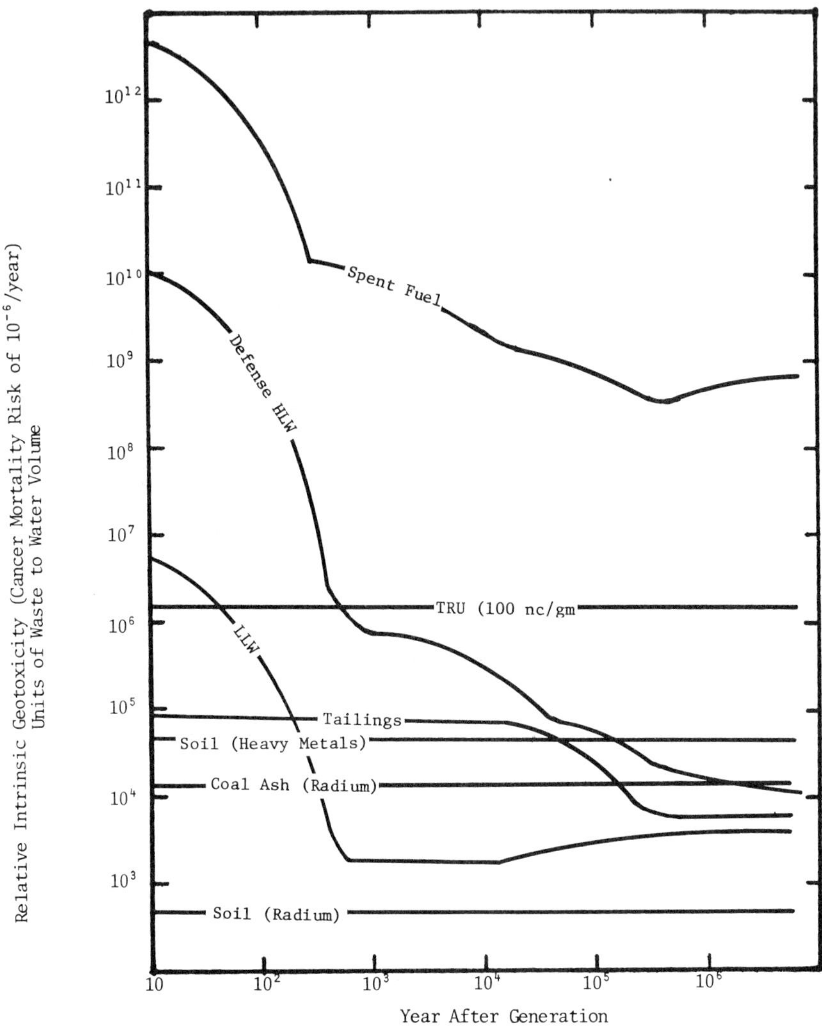

FIGURE 1. Relative Intrinsic Geotoxicity of a Cubic Meter of Various Types of Radioactive Waste Compared to Background.

CONCLUSIONS

Radioactive waste is generated by a wide variety of industries and institutions that provide vital services which are of great benefit to mankind. Since the states are now responsible for assuring adequate disposal of LLW and Pennsylvania is a major generator of this waste, more attention needs to be given to this problem. Possible solutions are either joining a compact with other northeastern states to manage the waste on a regional basis, or establishing an in-state disposal capacity for Pennsylvania only generators. The general public does not typically distinguish between the various types of radioactive waste. This is unfortunate because the potential toxicity of spent fuel and HLW are many orders of magnitude greater than LLW or mill tailings, and therefore these high activity wastes must be managed and disposed of with much greater care.

REFERENCES

1. "Nuclear Waste Policy Act of 1982," P.L. 97-425, 96 Stat. 2201, December 1982.
2. "Northeast Interstate Low-Level Radioactive Waste Management Compact," Official Draft, February 1983, Report of the CONEGLLRW Policy Working Group, March 1983.
3. USNRC, No. 248, December 1982, "10 CFR 61-Licensing Requirements for Land Disposal of Radioactive Waste," *Federal Register,* 47: 57446-57482.
4. "Uranium Mill tailings Radiation Control Act of 1978," P.L. 95-604, 92 Stat. 3021, November 1978.
5. "Low-Level Waste Policy Act," P.L. 96-573, 94 Stat. 3347, December 1980.
6. USDOE, January 1981, "The National Plan For Radioactive Waste Management."
7. Penn State University, August 1983, "Low-Level Radioactive Waste Disposal Siting: A Social and Technical Plan for Pennsylvania."
8. CONEG, June 1984, "Low-Level Radioactive Waste in the Northeast: Revised Waste Volume Projections," Report of the LLW Policy Working Group Technical Subcommittee.
9. Dornsife, W.P., May 1979, "A Prospective on the Relative Hazard of Low-Level Radioactive Waste Disposal," Proceedings of the Health Physics Society, *12th Major Topical Symposium,* EPA 520/3-79-002.

Chapter Two

LOW-LEVEL RADIOACTIVE WASTE MANAGEMENT TECHNOLOGY DEVELOPMENT*

Joseph A. Coleman, Ph.D., Director
Division of Storage and Treatment Projects
U.S. Department of Energy
Washington, DC 20545

Significant amounts of low-level radioactive waste have been generated in this country since the government nuclear weapons program began in the 1940's. The controlled disposal of low-level waste has resulted from the sizable amounts of wastes generated by a wide variety of government and commercial activities. Government low-level waste results primarily from production, research, and development activities of the Atomic Energy Commission and its successor agencies, the Energy Research and Development Administration, and the Department of Energy (DOE). Commercial low-level waste results primarily from commercial nuclear power production, manufacturing activities, and use of radioactive materials at institutions such as hospitals and universities. The amount of waste disposed of currently is divided fairly evenly from government and commercial sources.

The primary means of disposal of low-level waste has been the accepted and regulated practice of shallow land disposal, i.e., placement of low-level waste in trenches 5 to 10 meters deep with several meters of special soil cover[1]. Department of Energy waste is primarily disposed at six major shallow land disposal sites[2]. Commercial waste is currently disposed of at three major sites in the nation—Barnwell, South Carolina; Richland, Washington; and Beatty, Nevada.

In the late 1970's public concern arose regarding the management practices of sites operated by the civilian sector and by the Department of Energy.

---

*Presented at the Low-Level Radioactive Waste Symposium, Harrisburg, Pennsylvania, October 27, 1983

Although reviews of disposal practices and site performance indicated that there were no releases to the environment that would affect public health and safety, it became clear that: (a) several burial grounds were not performing as expected; (b) long-term maintenance of closed trenches could be a costly problem, and (c) more cost-effective methods could be developed for the treatment, packaging, and disposal of low-level waste. As a result of these reviews, the Department of Energy developed the Low-Level Waste Management Program to seek improvements in existing practices, correct obvious deficiencies, and develop site closure techniques that would avoid expensive long-term maintenance and monitoring. Such technology developments provide a better understanding of the physical and technical mechanisms governing low-level waste treatment and disposal and lead to improvements in the performance of disposal sites.

Technology Development

The Technology Development portion of the program supported by the Department of Energy addresses four major technical areas: (a) waste treatment, packaging, and handling; (b) shallow land disposal; (c) correction of inadequate performance of existing disposal sites, and (d) greater confinement disposal. Technology development activities in the areas of shallow land disposal practices and waste treatment are aimed at improving the management of about 90 percent of the low-level waste generated today. Improvements in greater confinement disposal relate to the remaining 10 percent of the low-level waste. Corrective measures are aimed at improving performance of closed sites.

In these four technology development areas, major emphasis is being placed on developing improved shallow land disposal practices. In addition to protecting the health and safety of the public in the short-term by providing more effective confinement, these improvements would avoid costly recurring maintenance and provide for long-term protection of the biosphere. Reducing the volume of waste produced would also extend disposal site lifetime without requiring additional capacity. Improved waste treatment technologies generally will reduce the volume of waste requiring disposal, result in improved waste forms, and provide increased stability of the entire disposal system. Corrective measures efforts focus on activities that will solve such problems as subsidence. These technology improvements will provide both interim and long-term solutions for both active and inactive disposal sites.

Waste Treatment

Waste treatment technology developments are designed to resolve issues related to the handling and disposal of low-level waste by developing treatment processes and improved waste forms. For example, incineration of low-level waste is one practice that could greatly reduce the quantity of waste requiring disposal. For institutional waste, the Department of Energy is co-sponsoring the demonstration of a large (300-500 lbs/h) controlled air incinerator at the University of Maryland at Baltimore. The incinerator has been constructed and test burns were conducted this past spring. With completion of the licensing and permitting process, incineration of radioactive waste commenced in June. Design and installation information is currently available, performance data will be available later this year.

DOE has been sponsoring the demonstration of an incineration process for defense low-level radioactive wastes at the Savannah River Laboratory. The processing steps of the facility include waste feeding, incineration, dry off-gas cleanup, and ash residue packaging. The incinerator is a two-stage controlled air unit capable of incinerating about 390 lbs of solids/h or about 242 lbs of liquids/h. The dry off-gas system consists of a quench tank with air-atomized spray nozzles to cool the off-gases, a baghouse for coarse filtration, and HEPA filters for final particulate removal. The incinerator was tested using nonradioactive solid waste from October 1981 through September 1982. Emissions of off-gas components NO_x, SO_2, CO and particulates were well below South Carolina State standards.

As a result of these tests, the process has been upgraded by the Savannah River Laboratory to accept low-level beta-gamma combustible waste. A 2-year demonstration is scheduled to begin in early 1984.

Another technology development project addresses treatment of reactor wastes. The Department of Energy will sponsor the demonstration of advanced waste treatment systems for more problematic wastes typical of nuclear power reactors, such as filter sludges and ion exchange resins. The purpose of the project is to demonstrate a system that can produce a superior waste form, has acceptable operating characteristics, and most importantly, has cost advantages over present waste treatment practices.

Shallow Land Disposal

Specific tasks for improving shallow land disposal practices are designed to resolve issues related to water management, radionuclide migration, environmental monitoring, modeling or performance prediction, and site operation. Several technology development activities are being conducted at humid sites—Oak Ridge National Laboratory (ORNL), Savannah River Laboratory, and Maxey Flats, Kentucky. Arid site studies are being conducted principally at Los Alamos National Laboratory in New Mexico, and at the Pacific Northwest Laboratory in Washington State.

Of primary importance to management of disposed waste is the control of ground and surface water infiltration. Radionuclide migration studies investigate ground water transport mechanisms, factoring in chemical and physical effects on the behavior of specific radionuclides. Site performance monitoring studies include development of monitoring techniques and modeling methodologies. Modeling provides estimates of comprehensive site performance that include all significant pathways of radionuclide transport. Operational implementation studies include review of waste inventory records management, waste characterization, and facility operational practices.

The Department of Energy is sponsoring the design of a passively drained low-level waste disposal site at the Georgia Institute of Technology. Such a design will allow for any moisture entering the disposal trench to pass by the emplaced waste through a permeable, but stable, backfill material. This moisture will then be removed from the trench by passive means. The drained site designs will minimize the contact between the emplaced waste and any moisture entering the trench, thereby minimizing the leaching of radionuclides. This improved site design, coupled with the ongoing trench cover and waste form research, will improve the containment capabilities of shallow land waste disposal sites.

Corrective Measures

Corrective measures improvements are designed to upgrade site performance and reduce the high degree of maintenance to control the release of radionuclides from the site. Research activities include efforts to control water infiltration, reduce subsidence, control erosion, and prevent biological intrusion.

Water infiltration may be controlled by using various types of trench caps employing passive subsurface drainage systems, and/or certain treatment of wastes within the trench. Subsidence is normally caused by the settling and compaction of backfilled material, the filling of void spaces between waste packages, and the physical and chemical breakdown and decomposition of the package and wastes. Improvement projects aim at understanding the fundamental mechanisms and rates of subsidence, determining the effects of subsidence on the overall disposal system; and developing methods to avoid subsidence by using improved waste forms, engineered barriers, or trench treatments to increase the structural stability of the disposal system. Control of erosion by wind and water is applicable to both open and closed trenches. Biological intrusion that may result in the transport of radionuclides from the disposal site can be minimized by developing physical barriers to prevent plant roots and animals from reaching the waste.

The application of improved engineered practices to the design of land disposal facilities in humid regions is being pursued at the Oak Ridge National Laboratory in Tennessee. ORNL has designed and constructed an Engineered Test Facility to focus on waste leaching and contaminant transport problems associated with disposal in a humid climate. Two improved disposal techniques are being evaluated at nine demonstration trenches at ORNL. These are the use of a cement-bentonite grout applied as a waste backfill prior to closure of the trench and a complete lining of the trench with a synthetic impermeable lining material. In the year since the trenches were filled with waste and closed, sampling at 36 monitoring wells has indicated no leachate movement from any of the nine trenches. To date, the research has demonstrated that trench lining and grouting can successfully be used to isolate waste from contact with water.

An economic analysis of the trench grouting and lining has shown that lining was considerably cheaper than grouting the trenches ($1055/demonstration trench for lining compared to $1585/demonstration trench for grouting.) In addition, lining becomes relatively less expensive as trench size increases.

Greater Confinement

In this area, activities include the evaluation of low-level wastes that require greater confinement for disposal, identification of options for disposal, and development of a proof-of-concept facility to demonstrate specific technologies (i.e., the large diameter borehole concept).

The Greater Confinement Disposal Test (GCDT) at the Nevada Test Site will demonstrate the use of large diameter boreholes for the disposal of low-level wastes considered unsuitable for shallow land burial. A 3-m diameter, 37-m deep shaft has been constructed in alluvial sediment and nine monitoring holes have been drilled around the shaft and instrumented. During 1984, wastes will be loaded into the GCDT. At the beginning of this experimental phase, nonradioactive gaseous tracers will be released and monitored to gather information on the migration of tracers and radionuclides, and on properties of the alluvial sediment. Predictive computer models will also be validated in this phase.

SUMMARY

The Department of Energy low-level waste technology development program is currently investigating many ways to improve low-level waste management and disposal. Reports on specific studies and results are provided at our annual program participants meeting in August of each year[3]. The results derived from these demonstrations are available for public review and significant results are provided through comprehensive documentation in both proceedings from the annual meeting and in technical reports. Handbooks for each of the technology development areas are currently in preparation.

REFERENCES

1. Nuclear Regulatory Commission, "Licensing Requirements for Land Disposal of Radioactive Waste," 10 CFR 61, *Code of Federal Regulations* (January 1, 1983).
2. U.S. Department of Energy *Spent Fuel and Radioactive Waste Inventories, Projections, and Characteristics,* DOE/NE-0017-2 (September 1983).
3. Oak Ridge National Laboratory, *Proceedings of the Fourth Annual Participants' Information Meeting — DOE Low-Level Waste Management Program,* ORNL/NFW-82/18 (October 1982).

Management of Radioactive Materials and Wastes: Issues and Progress. Edited by S. K. Majumdar and E. Willard Miller. © 1985, The Pennsylvania Academy of Science.

Chapter Three

EFFECTIVE WATER MANAGEMENT FOR LOW-LEVEL RADIOACTIVE WASTE REPOSITORIES

A. R. Jarrett, Ph. D.
Associate Professor of Agricultural Engineering
249 Agricultural Building
The Pennsylvania State University
University Park, Pa. 16802

Though radionuclides can be transported by air or with plant, animal or soil material, water is the primary transporter of radionuclides from low-level radioactive waste (LLRW) repositories. Therefore, if radionuclides placed in a repository are to remain entombed until radioactive decay reduces the activity to background levels—usually about 7 to 10 half lives—all water falling on the site as precipitation, flowing under the site in the groundwater system and flowing over the site surface as runoff must be properly managed. Proper management may mean collecting, monitoring and releasing surface runoff water. It may also mean keeping contaminated materials away from the water as in the case of groundwater.

Planning for effective water management begins during site selection, long before the first trench is opened or the first waste buried, by choosing a site where surface water can be managed and groundwater can be avoided. This total water management plan must be initiated during site selection, implemented during site design and facility operation, and concluded during site closure. The result of a carefully conceived and executed water management plan will be waste which is properly stored and confined in a leachate free environment.

Many of the LLRW repository closings, which have occurred in recent years, can be traced to inadequate water management. Maxey Flats, Kentucky stopped accepting LLRW in 1976 when it became evident that inadequate waste consolidation and stabilization plus biological decomposition in the filled trenches was creating voids under the trench caps resulting in subsidence and cracking of the caps. The surface runoff water which accumulated in these depressed areas easily entered the trenches creating leachate[1]. The LLRW site at West Valley, NY, was closed in 1975, when water leaking through a cracked and subsiding trench cap filled a trench with leachate causing it to spill onto the surface[2]. In both cases, good sites, by todays standards, were closed because a comprehensive and effective surface water management plan was not implemented. The Sheffield, Illinois site was filled in 1978 and was closed having apparently performed as expected. A 1982 report revealed, however, that a 30-meter wide tritium plume has been located 76 meters from the nearest trench[3]. The Sheffield site is underlain by pebbly sand and fine grained sediments through which the tritium is moving at 8 meters/yr. There is still debate about whether the site could be licensed from a hydrogeological perspective under todays 10 CFR 61[4] guidelines.

PRINCIPLES OF RADIONUCLIDE TRANSPORT

For purposes of this chapter, there are two phenomena which can result in radionuclide transport related to water management. They are advection-dispersion, and erosion of soil particles containing absorbed nuclides.

Advection-Dispersion

The physical processes that control the movement of radionuclides in a solution are advection and hydrodynamic dispersion. Radionuclides can be added to or removed from the solution by chemical or biochemical reactions or removed by radioactive decay[5].

Advection is the movement of a fluid resulting from energy gradients. The rate of advective transport is equal to the flow velocity. In a porous media flow system, where the flowstream is partially filled with solid obstructions causing the fluid to assume a tortuous path, the flow velocity, $\bar{v}$ is equal to v/θ where v is the Darcy velocity defined by

$$v = K\frac{dH}{dx}$$

where K is the hydraulic conductivity, H is the hydraulic head and x is the direction of flow. The Darcy velocity assumes flow through an unobstructed channel. Since soil fills a part of the flow channel, the Darcy velocity must be divided by the cross-sectional area of flow, or porosity, to obtain the actual flow velocity. In cases of unsaturated flow, where the total pore space is not filled with water, but a combination of air and water, the Darcy velocity must be divided by the volumetric moisture content, θ, to obtain the actual flow velocity.

Hydrodynamic dispersion is caused by the mechanical mixing which occurs as the solute flows by advection and by molecular diffusion caused by the thermal-kinetic energy of the solute particles[5]. The cumulative effect of advection and hydrodynamic dispersion can be represented by the solute transport equation for nonreactive constituents in saturated, homogeneous, isotropic, steady-state, uniform flow as

$$\frac{\delta C}{\delta t} = D \frac{\delta^2 C}{\delta x^2} - \bar{v}\frac{\delta C}{\delta x}$$

where x is the direction of flow, D is the coefficient of hydrodynamic dispersion in the direction of flow, C is the solute concentration and t is time. For a more detailed discussion of this relationship and its derivation see Freeze and Cherry[5].

Solute Absorption

Radionuclides are also influenced in the soil by chemical processes which tend to slow or retain nuclide movement called adsorption. Colloidal sized clay soil particles are large compared to individual molecules in solution but small enough so that interfacial forces control their behavior producing a large electrical charge relative to their surface area. Freeze and Cherry[5] give two reasons for this surface charge; (1) imperfections or ionic substitutions within the crystal lattice and (2) chemical dissociation reactions at the particle surface. Ionic substitutions cause a net positive or negative charge on the crystal lattice creating a charge imbalance which is compensated for by a surface accumulation of positively charged ions. The ions in this adsorbed layer are readily exchanged for other ions seeking to neutralize the overall charge of the clay particle. Most radionuclides found in LLRW leachate are readily attracted to these ion exchange sites. This mechanism effectively removes radionuclides and retains them on the colloidal particles.

The two most notable radionuclide which are not retained in the soil by ion exchange are tritium and cobalt-60. Tritium usually occurs in the form of tritiated water, generally HTO. In this form it readily becomes part of the soil water flow where it is not slowed by ion exchange, hence it moves with and is diluted by the soil water at the waste site[6]. Cobalt appears as a divalent cation which would readily be tied up by the ion exchange capacity of the soil except that it is readily complexed by inorganic and organic constituents in the soil water flow system forming neutral or anionic complexes for which soils have poor sorptive properties. Complexing is especially serious in waters having pH's below 6. As the soil pH increases, it tends to precipitate out as the insoluble hydroxide[7,8,9,10].

Ion exchange adsorption also causes soil, once exposed to radionuclides, to remain contaminated until radioactive decay diminishes the hazard. Therefore, if the contaminated soil particles are transported from the disposal site by erosion, the radionuclides are carried along.

Typical Shallow Land Burial Site and Operating Procedure

To date, LLRW is packaged and placed in shipping containers which meet NRC requirements set forth in 10 CFR Part 61[11] and Department of Transpor-

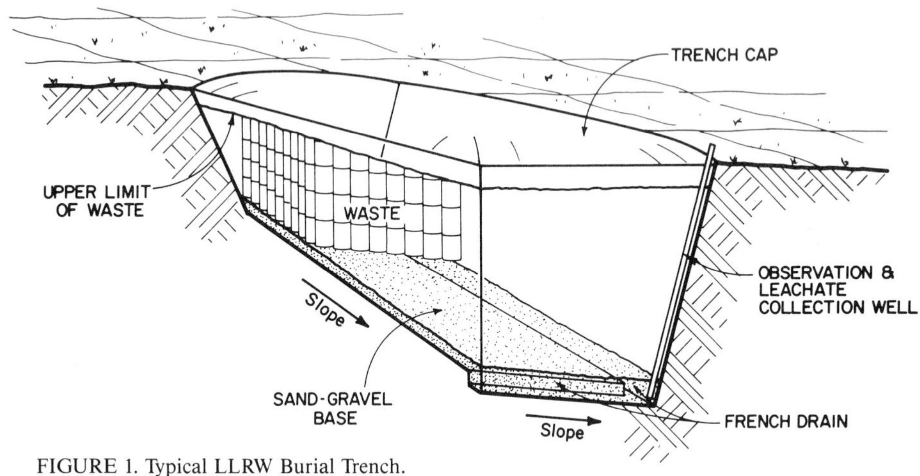

FIGURE 1. Typical LLRW Burial Trench.

tation requirements set forth in 49 CFR Parts 171-179 and trucked to a licensed waste repository. Typically waste arrives at the repository in steel drums, wooden boxes or some type of high integrity containers. At the repository, the shipments are monitored for compliance with the above cited requirements, off loaded and stacked into large shallow land burial trenches by crane or forklift. The trenches, which average 600 feet in length by 100 feet wide by 25 feet deep, have sloping floors (floors slope away from the end where filling starts and to one side) covered with a sand-gravel base which provides drainage and stability for equipment working in the trench, Figure 1. The low side of the floor contains a french drain which is monitored and sampled through a well to the surface. Any leachate which accumulates in the trench is periodically pumped, treated and solidified before reburial. As a trench is filled to within 3 feet of the trench top, excavated soil, dry sand or cement grout is backfilled into the waste, filling (as well as possible) the intercontainer voids. When the whole trench has been filled with waste and backfilled, the backfill is compacted to help fill any voids. When backfilling and compacting has been completed the trench will have been filled to within three feet of the trench top. A plastic, cohesive material—often the material excavated to form the trench—is then placed over the backfilled waste and highly compacted to completely fill the trench to the level of the sur-rounding soil. Extra material is added near the center of the trench to provide a small mound over the filled trench, which diverts surface runoff water. The area is then seeded to short-rooted grasses. The grass is mowed as needed and the trench cap is inspected periodically for evidence of cracking, subsidence or burrowing animals. If any are found, remedial action is taken immediately.

When the site has been filled to capacity (all trenches filled and capped), some site closure plans call for the addition of a 3- to 15-foot thick site cap placed over the entire site. This site cap is constructed of plastic, cohesive material, compacted and graded to divert surface runoff water. Usually contained within the site cap, and at some sites within the trench cap, is a course rock layer about one to two feet below the surface to serve as a burrowing animal resistant layer. The site cap is then seeded to short-rooted grasses. In all cases, periodic site inspections continue throughout the 100-year institutional control period.

Water management is often not a primary problem for sites in arid regions and water management was not a priority concern at several of the early LLRW burial sites. Problems resulting from improper water management has resulted in increased awareness and improved water management techniques, especially at sites located in humid regions. Proper water management requires continued awareness of potential problems resulting from improper land shaping, steep natural slopes, permeable soils and natural precipitation. These concerns must be addressed throughout the planning and implementation phases of any LLRW repository.

SITE SELECTION

During the site screening and site selection process, there are several water management considerations which must be addressed if problems are to be avoided during the operational and post closure phases. These include the depth of unconsolidated soil material between the trench bottom and the watertable, the possibility of site inundation from tides or floods, the permeability and degree of homogeneity of the soils, and the land slopes at the site.

The objective of any LLRW shallow land burial site is to isolate the radiation produced by the waste from the uncontrolled environment. Most radionuclides are carried from the burial site in leachate, produced when the waste comes in contact with water. All shallow land burial trenches for LLRW have a leachate collection system, but with proper site selection and water management, the collection system should never be used except as part of the monitoring system.

The process by which water is kept from the waste is closely related to the soil moisture and the rate of soil moisture movement. First, it is not possible to keep moisture from contacting buried waste since all soil contains some water. At relatively dry moisture contents this water is held to the soil very tightly in very thin films. Due to the minute cross sectional areas of flow, this water moves very slowly (10^{-10} cm/s or slower). When the moisture content increases, the films become thicker and the water flows faster. The flow velocity continues to increase until the moisture content reaches the point where all the pores are filled with water or saturated. At moisture contents between saturation and a point known as field capacity, the gravitational energy is sufficient to remove soil water from the soil. This gravitational water is the water which is usually referred to as leachate. The goal of site selection and waste management is to bury the waste in an environment where the moisture content will never increase above field capacity—about 80 percent of saturation in most soils. This implies that quantities of water large enough to produce leachate must be excluded from the waste cell. These large volumes of water may come from surface runoff water flowing into the trench, from precipitation which infiltrated into the trench cap and subsequently percolated through the trench cap into the waste cell on its way to the groundwater system, or from groundwater influenced by hydraulic heads in the region, rising into the waste cell from below. Water can also enter the waste cell through zones or layers of high permeability which are often present in the regolith. Probably the last source of leachate is from liquids buried in the trenches or produced when biogradable wastes, such as animal carcasses, decay. Therefore, in order to minimize the leachate produced in the trenches it is necessary to select and manage the site such that surface and groundwater will not easily enter the waste cells.

To provide for water management during the operational and closure phases of a LLRW site, it is necessary to consider each of the following parameters during site selection. Specific criteria are given to provide limits on physical parameters which may be present. These criteria were adapted from a study to determine site criteria for a Pennsylvania LLRW site[12].

Flood Plains

No LLRW disposal site shall be located within the horizontal or vertical limits of the "maximum probable flood" adjacent to any river or stream. Flooding of a LLRW disposal site will interrupt site operation, cause surface erosion, alter ground and surface water flow, breach the burial units and possibly expose hazardous waste materials. Moreover, by saturating the waste, flooding will accelerate the generation of leachate which can increase the rate of contaminant transfer to the groundwater table. The alternate saturation and drying of the waste containers may facilitate chemical reactions and biological decay.

In light of the 500-year anticipated life of a LLRW disposal site, prudence dictates the use of the calculated "maximum probable flood" in determining the vertical and horizontal limits of the prohibited floodplain areas. It also requires that no site be located within the historic limits or predicted maximum probable flood level of any surface lake or other water body, swamp or marsh.

Tidal Zones

No LLRW disposal site shall be located in a coastal area within the horizontal or vertical extent of flooding and/or erosion caused by action of the maximum recorded storm wave and/or the highest-high spring tides adjacent to any coast or tide water. Flooding and erosion by lake or ocean waters will have effects similar to those of river or estuarine flooding, and must be avoided.

Cavernous Geology

No LLRW disposal site shall be located in an area where soil and regolith rest on limestone, dolomite or carbonate-cemented bedrock. Cavernous bedrock beneath the soil cover leads to vertical instability of the soil and may allow subsidence or collapse of the soil cover. Such collapse will jeopardize site performance by disrupting the site itself, the structures associated with it, the intrusion barriers surrounding it and the access to it. It may destroy the integrity of the burial units, carrying waste materials directly into the fractures or caverns beneath, where their subsequent dispersal cannot be controlled.

Both vertical and lateral movement of groundwater in cavernous bedrock is localized and rapid; in such terrains groundwater movement and contaminant dispersal cannot be reliably predicted. These terrains are consequently unsuited for LLRW waste disposal.

Soil Thickness and Permeability

A LLRW disposal site must have unconsolidated soil or weathered rock mantle of not less than 10 feet thickness between the waste and the permanent regional watertable (or saturated zone) including the capillary fringe, if the soil permeability is 10^{-7} cm/s or less. The soil thickness must be increased in proportion to the permeability so as to assure that leachate will not reach the groundwater reservoir or site boundary.

Practical operational considerations usually require shallow disposal trenches of twenty feet depth or more and the provisions of 10 CFR 61 mandate that the upper surface of the groundwater-saturated soil or rock, including the capillary fringe, be at sufficient depth beneath the base of the disposal unit that, groundwater intrusion, perennial or otherwise, into the waste will not occur[11]. This provision would require a minimum separation of 10 feet between the base of the disposal unit and the upper surface of the permanent water table. Allowing for the 20-foot depth of the burial trench itself, then, the minimum acceptable soil thickness above the groundwater table will be greater than 30 feet. If a mound or partial mound is used this 30-foot depth could be reduced proportionately. This requirement presumes that the top of the trench cap will be built entirely above original ground level to facilitate drainage, and will thus stand as a low mound above the surrounding land surface.

Slopes

Slopes steeper than 1:3 (vertical:horizontal) will generally be unacceptable for disposal site location. Stability of slopes in unconsolidated soil or regolith varies substantially and is significantly affected by moisture content, mineralogy, particle size and shape, particle sorting, and soil fabric. Site-specific determination of slope stability will generally be required except for slopes of one on three or less. Where topographic slope parallels the dip of underlying bedrock, extra precautions are required, and similar precautions will be necessary where there may exist danger of oversteepening by erosion or human activity at the foot of the slope.

Overly steep slopes are subject to mass failure by slumping, landslide or debris or mud-slide. Such mass failure will alter the flow regimen of surface and ground-water, may interrupt site operation, may damage site facilities including buildings, barriers and the burial units themselves, and may expose or even transport waste from the controlled area of the site.

OPERATIONAL CONSIDERATIONS

Once a site has been selected, licensed and opened for disposal, the major concern should be surface water management. Proper site selection should have eliminated the need to be concerned about subsurface or groundwater. Therefore, the only two sources of water available to the waste as it is placed in the trench prior to closure are precipitation and surface runoff. It is important to use procedures which keep these waters from the waste, thus eliminating the possibility of leachate production.

Surface Runoff

Managing surface runoff generally requires the construction of a storm water conveyance system which will intercept, collect and carry all runoff, resulting from precipitation, from the disposal area at non-erosive velocities to some type of detention structure where the water can be checked for contamination before being released to the uncontrolled environment. These stormwater conveyance systems are usually constructed by land shaping around the edges of the trenches so water flows away from the lip of the trench into nearby channels rather than into the trench where it may become contaminated. These channels are

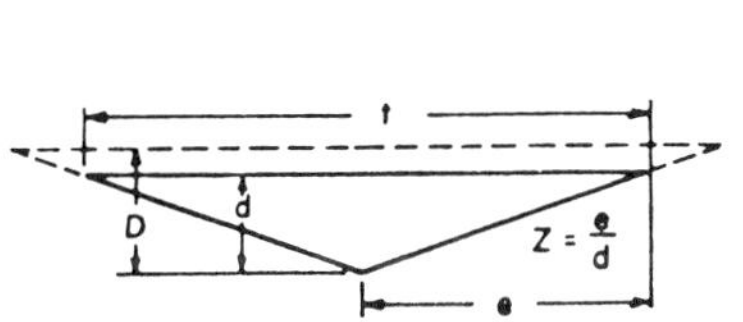

Area, A	Wetted Perimeter, WP	Hydraulic Radius, R	Top Width
Zd^2	$2d\sqrt{Z^2+1}$ or $2dZ$ approx.	$\dfrac{Zd}{2\sqrt{Z^2+1}}$ or $d/2$ approx.	$t = 2dZ$

TRIANGULAR CROSS-SECTION

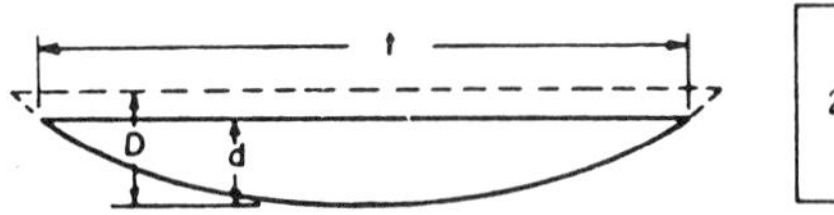

$2/3\,td$	$t + \dfrac{8d^2}{3t}$	$\dfrac{t^2 d}{1.5^2 + 4d^2}$ or $2d/3$ approx.	$t = \dfrac{a}{0.67d}$

PARABOLIC CROSS-SECTION

FIGURE 2. Channel Cross-Section notation and formulas. (Adapted from Schwab et al., 1981)

usually located between trenches and convey the water to the detention structure. Extreme care must be taken to keep radioactive waste materials from all surface areas at the site and especially from the drainage channels. These channels are usually parabolic or triangular shaped with side slopes in the 1:6 to 1:10 (vertical:horizontal) range, Figure 2. To provide protection during all events, the channels must be designed using the peak runoff rate expected during the site operation period—probably a 500-year return period event. The channel dimensions should be determined from Mannings Equation[13,14]

$$Q = \frac{1.486}{n} AR^{2/3} S^{1/2}$$

where Q = peak rate of discharge in ft³/sec
$\quad$ n = Manning's roughness coefficient
$\quad$ A = cross-sectional area of the channel in ft²
$\quad$ R = hydraulic radius, defined as the cross-sectional area divided by the wetted perimeter (A/WP) in ft., and
$\quad$ S = slope of the channel in the direction of flow in ft/ft.

Manning's Equation can be solved by trial and error using the relationships in Figure 2. During the active burial period, these drainage channels are usually constructed with heavy equipment in the undisturbed earth between the trenches. They usually remain as earthen channels because vegetation cannot be maintained when heavy equipment and trucks are using the channel areas for off loading. Lining these channels with more durable material such as concrete or asphalt is usually considered unnecessary. After the trenches, on both sides of each drainage channel, have been filled and capped, the drainage channel is usually seeded to grass and periodically mowed as part of the maintenance program for the site. These channels still serve as surface water drainage channels carrying runoff originating from the capped trench area.

For design purposes it is necessary to size the channels so they will not overflow into the trenches—either when they are earthlined during operation or sod lined after capping. The key parameter in the Manning Equation is the roughness coefficient which can vary greatly depending on whether the lining is grass, soil or other. Typical Manning n values are given in Table 1. Maximum permissible velocities are also given.

It is important to note that in order to keep the design flow velocities slow enough to prevent erosion it will be necessary to find an appropriate trade-off between channel depth and channel slope. The net result is usually to keep the depth small and to keep the slope very flat. Slopes exceeding one percent will usually cause excessive velocities unless a shallow, wide channel is used. This need to control slope, will influence greatly the layout of the trenches on the site, requiring that the long dimension of the trenches generally follow the natural contour. In most cases, the cost of such a drainage network is low compared to other costs of burial so there is no excuse for surface flooding caused by an under-designed system.

TABLE 1

Manning's Roughness Coefficient and Maximum Permissible
Velocities for Several Channel Linings.
(Adapted from Schwab et al., 1981).

Lining	Manning's n	Maximum Velocity
		ft/sec.
Concrete	0.014	—
Asphalt	0.015	—
Dense Grass Cut	0.030	4.0
Dense Grass Long	0.040	4.0
Bare Soil	0.030	1.5

Direct Precipitation

Managing the precipitation which falls directly on the open trenches is difficult. Normally burial operations cease during precipitation events. If the trench is partially filled at the time of the precipitation event, all of the water falling on the waste will become contaminated and subsequently require removal and treatment. Therefore, the trench cap is usually placed over any portion of the trench which has been filled. The geometry of the trench cap should direct all runoff to the surface drainage network described above.

In regions where precipitation is infrequent, the management scheme given above is usually sufficient. In humid and cold climates, additional measures may be required. It was felt by Aron et al.[12] that some means must be employed to keep precipitation from the partially filled trenches. This can be done with plastic which is difficult to move, keep placed and otherwise manage or by building a more substantial structure which can be placed over the trench during the waste placement period. The Environmental Impact Statement on 10 CFR 61 estimates that this structural shielding from weather will cost about $4.20 per cubic meter of waste buried[11]. Such a structural shield can be effectively used during winter weather when waste burial operations should be stopped especially if there is snow or ice on the ground. This weather shield could be placed on tracks and moved aside facilitating the use of a crane for placement operations.

BACKFILLING AND CAPPING

Next to selecting a site which will insure that the burial cell is always above the watertable, backfilling is the most important aspect of water management at LLRW sites. Failure to implement a backfilling procedure which will provide structural stability for the cap placed over the filled trench will result in cracking, slumping, and other forms of subsidence leading to entrance of surface water and the creation of leachate. Therefore, it is imperative that a backfilling program be implemented which will fill all interpackage voids with material which will provide the necessary structural support for the cap. This has been the weakest part of the overall LLRW site management to date and as alluded to earlier, it was at least partially responsible for causing the closing of the Maxey Flats and West Valley sites.

These early sites simply backfilled the trenches with the soil material which had been excavated to form the trench. Since trenches are usually constructed in unconsolidated clayey material, the backfill material was cohesive and plastic causing it to easily wedge in the upper layers of the waste leaving unfilled voids in the lower layers of the waste. After an indefinite period, the natural soil moisture and the weight of the backfill material caused settling, filling, or partially filling voids in the lower layers, opening new voids under the caps. The weight of the trench cap caused cracking and subsidence of the cap which provided open channels for the entrance of surface water. The problem caused by poorly placed backfill material was enhanced by the presence of biological material buried in the trenches which degraded leaving additional voids.

Based on earlier experiences, the quality of the backfill material was upgraded to improve the void filling with clay by using a high energy vibrator compactor. When this additional procedure was only partially effective, some sites, such as Barnwell,began to backfill with air dried sand vibrated into the voids. The dry sand flows into and more easily fills the interpackage voids creating a structurally sound, filled trench on which the cap can be constructed. Though Barnwell has not had any problems with subsidence, the compacted sand still presents the potential for settlement if biologically degradable material is buried. One means of overcoming the voids created by biological degradation is to backfill all voids with cement grout[11,12]. The cured concrete gives the structural support and voids which may develop after capping will not compromise the structural stability of the filled trench.

If the waste is placed and the backfilling completed to provide long term structural stability, the trench cap can be constructed by compacting the clayey, cohesive, plastic soil material over the trench. The cap must be densely packed and shaped to facilitate surface runoff without causing erosion. This generally means that slopes should not exceed 5 percent. The whole area is then seeded to short-rooted grasses. As long as the support under the trench cap remains sound, the cap will direct precipitation away from the waste in the trench as surface runoff with only a very small portion infiltrating into the trench cap. Any compromise in the cap's structural support leading to subsidence will render the cap ineffective permitting the runoff from subsequent precipitation events to enter the waste cell producing leachate.

A final water management concern related to the trench cap has to do with the need to provide some type of barrier over the waste cell which will serve to keep plant roots, rodents and even inadvertent humans from reaching the material buried in the waste cell. Plant roots and percolating water can be repelled by using some type of impermeable layer such as plastic within the trench cap.

Such a layer will not, however, reduce the ability of rodents or humans to reach the waste from above. One of the most effective methods of resisting intrusion is to include a layer of coarse rock in the cap. This coarse layer, placed one to three feet under the top of the compacted cap will generally retard the passage of rodents and slow any humans digging in the site. This layer will also retard the vertical movement of water through the cap since water will tend to be retained above the clay-stone interface rather than freely flowing into the stones. Only if the moisture content in the cap above the stone layer exceeds field capacity, will water pass into the coarse stone layer. If water does enter the stone layer, it will quickly flow laterally through the stones rather than penetrate into the waste. Such a stone layer would not provide much protection against root penetration but this can be controlled by selecting shallow rooted species. Such coarse rock layers are more commonly included in the 3 to 10-foot thick site cap rather than in the thinner trench cap.

SITE CLOSURE

After a LLRW repository has been filled and each trench properly backfilled and capped, the disposal facility is closed (decommissioned) and a period of institutional control established. The closure procedure involves dismantling buildings and equipment and decontaminating or burying them. After all waste has been buried, the closure may or may not involve stripping the organic surface layer (grass etc.) of the trench cap and placing a thick site cap over the total burial area. This practice has not been practiced in the past, but due to trench cap subsidence observed at several burial sites, it is felt by many that the additional depth of compacted, relatively impermeable material will aid in retarding the entrance of surface water. In general, it will not cause more water to enter the waste cells, but the additional cost will not eliminate the entrance of some surface water if the structural support under the trench cap is not sound. Whether or not the site cap is added during closure does not reduce the concern for proper shaping of the surface. The site must be shaped to facilitate removal of surface runoff without relying on manmade structures which may deteriorate and malfunction without proper maintenance. Simple surface drainage ditches with very flat side slopes (1:6 to 1:10) sloped at about 1% in the direction of flow will serve to direct water away from the trenches and the site. It is assumed that new drainage channels will exist in the site cap and that they will lie just above the original channels constructed during site operation.

Another site closure consideration is monitoring. During the institutional control period, the site must be surveyed regularly to determine if leachate is accumulating in any of the trenches. If regular sampling does not detect water in the trench drains, there should be little cause for concern. If leachate is detected, remedial measures must be undertaken to determine the source of the water and stop it. The leachate must then be removed, treated, and disposed of at another site. If the site was properly located, any leachate should be originating from water passing the trench and site cap. Leachate left in trenches will percolate vertically to the watertable below and subsequently follow the gradient of the watertable to its discharge point. It should be remembered that most radionuclides present in the leachate will be removed as it percolates through the soil below the trenches by adsorption onto the clay soil particles. The only nuclides which would not be removed from this leachate stream would be tritium, which does not interact with any soil, and possibly cobalt-60 which may have chelated with organics present thus becoming non-adsorptive. Any hazard from cobalt-60 should be eliminated within 25 to 35 years after burial due to its short half life of 5.2 years.

REFERENCES

1. Grant, J. L. 1982. Geotechnical Measurements at the MAXEY Flats, Kentucky Low-Level Radioactive Waste Disposal Site-Lessons Learned. Proc. of the Symposium on Low-Level Waste Disposal—Site Characterization and Monitoring. Arlington, Virginia p. 151-162.
2. Clancy, J. 1981. Status of Low-Level Radioactive Waste Management Regulations in *Low-Level Waste Management* (Tutorial) ANS/ED/TP-4, ANS Annual Meeting, Miami, FL.
3. Foster, J.B. 1982. Lessons Learned in a Hydrogeological Case at Sheffield, Illinois. Proc. of the Symposium on Low-Level Waste Disposal—Site Characterization and Monitoring. Arlington, Virginia. p. 237-244.
4. NUREG-0782. 1981. Draft Environmental Impact Statement on 10 CFR Part 61 "Licensing Requirements for Land Disposal of Radioactive Waste." Summary.
5. Freeze, R. A. and J. A. Cherry. 1979. *Groundwater*. Prentice-Hall, Inc.
6. Stanley, C. B., et al., 1977. Radionuclide Mirgration from Low-Level Waste: A Generic Overview. Paper presented at the Symposium: Management of Low-Level Radioactive Waste, Atlanta, Georgia, May 23-27.
7. Straub, C. P. 1964. Low-Level Radioactive Wastes: Their Handling, Treatment, and Disposal. chapter 3, pp. 107-112, USAEC.
8. Dugid, J. O., 1976. Annual Progress Report of Burial Ground Studies at Oak Ridge National Laboratory: Period Ending September 30, 1975. Oak Ridge National Laboratory, ORNL-5141.
9. Robertson, J. B. 1977. Numerical Modeling of Subsurface Radioactive Solute Transport from Waste-Seepage Ponds at the Waste Seepage Ponds at the Idaho National Engineering Laboratory. USGS, Open-File Report 76-717, IDO-22057.

10. Schultz, R. K. 1965. Soil Chemistry of Radionuclides. Health Physics, Vol. II, pp. 1317-1324.

11. NUREG—0945. 1982. Final Environmental Impact Statement on 10 CFR 61 Licensing Requirement for Land Disposal of Radioactive Waste.

12. Aron, G., R. J. Bord, F. A. Clemente, W. P. Dornsife, A. R. Jarrett, W. A. Jester, R. F. Schmulz, and W. F. Witzig. 1983. Low-level Radioactive Waste Disposal Siting: A Social and Technical Plan for Pennsylvania. Volume III. Final Completion Report for Subcontract No. C29-007909 to EG & G Idaho, Inc., Idaho Falls, Idaho.

13. Schwab, G. O., R. K. Frevert, T. W., Edminster and K. K. Barnes. 1981. Soil and Water Conservation Engineering, Third Ed. John Wiley & Sons, New York.

14. NRC. 1983. Trench Design and Construction Techniques for Low-Level Radioactive Waste Disposal. NUREG/CR-3144.

Management of Radioactive Materials and Wastes: Issues and Progress. Edited by S. K. Majumdar and E. Willard Miller. © 1985, The Pennsylvania Academy of Science.

Chapter Four

A REVIEW OF LOW-LEVEL RADIOACTIVE WASTE COMPACTS ON A NATIONAL LEVEL*

Faith N. Brenneman
U. S. Nuclear Regulatory Commission
Office of Nuclear Reactor Regulation
Washington, D. C. 20555

Since the 1950s, increased quantities of low-level radioactive waste (LLW)[a] have been produced in the United States as a result of the use of radioactive materials in medical diagnoses and treatment, research, industrial processes, and electrical power generation by nuclear plants. During the 1950's, LLW was disposed of by burying it or by dumping it into the ocean. However, in 1960 the Atomic Energy Commission (AEC) placed a prohibition on the issuance of new licenses for sea disposal, but permitted existing licenses to remain in effect. In addition, the AEC authorized licensees to use, on an interim basis, AEC burial facilities.[1]

With increasing volumes of commercially generated waste, the private sector was encouraged to develop LLW disposal facilities, to be licensed by the AEC or by AEC Agreement states.[b] In 1962, the commercially operated Beatty, Nevada low-level waste facility was opened. During the ensuing nine years, five additional low-level waste disposal facilities opened, resulting, although not planned, in a regional distribution of such facilities (Figure 1). They were: Maxey Flats, Kentucky (1963); West Valley, New York (1963); Richland, Washington (1965); Sheffield, Illinois (1967); and Barnwell, South Carolina (1971). During the 1970s, for several reasons, including environmental, three of the facilities closed, namely: West Valley (1975), Maxey Flats (1977), and Sheffield (1978). Thus, by 1979, only three were operating, two west and one east of the Rocky Mountains.

Given the preponderance of LLW generated in the East, the possibility that LLW generated in the cleanup of the accident at Three Mile Island might be disposed of at Barnwell, and a number of violations of packaging and transportation regulations (including leaky shipments and faulty vehicles), the Governors of Nevada, South Carolina, and Washington were no longer willing to have

---

*Status as of December 31, 1983.

their states be the "dumping grounds" of the country and recognized the need for a national LLW program for developing additional facilities. In 1979 and 1980, the temporary closing of the Nevada and Washington facilities, the announcement by South Carolina of a phased 50% reduction in the volume of LLW received at Barnwell, the Washington State Initiative 383 to exclude out-of-state non-medical LLW by July 1981, and the recommendation by the Federal Interagency Review Group[2] that, at a state's option, either states or the federal government should control LLW sites contributed to the growing national consensus that new facilities were needed, that the responsibility for such development belonged to the states, and that the states should develop the needed LLW facilities on a regional basis.

The National Governors' Association, the State Planning Council on Radioactive Waste Management, and the National Conference of State Legislatures were particularly instrumental in the formulation and subsequent implementation of national LLW policy. The U. S. Nuclear Regulatory Commission (NRC), the U. S. Departments of Energy (DOE) and Transportation (DOT), and the U. S. Environmental Protection Agency were recognized as those federal agencies with responsibilities in the development of regulations and/or in the provision of technical and financial resources for state efforts. The activities of these organizations and agencies and the Congress culminated in the passage of the Low-Level Radioactive Waste Policy Act (PL 96-573) in December 1980[3]. The Act incorporated the key features of the national consensus and defined national low-level waste policy, so that: "each state is responsible for providing for the availability of capacity either within or outside the state for the disposal of low-level radioactive waste generated within its borders except for waste generated as a result of defense activities of the Secretary (Department of Energy) or federal research and development activities"; "low-level radioactive waste can be most safely and efficiently managed on a regional basis;" states

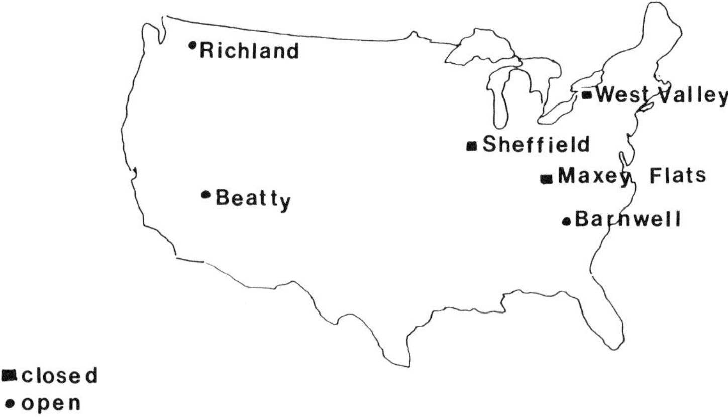

FIGURE 1. Commercial LLW Burial Sites.

"may enter into such compacts as may be necessary to provide for the establishment and operation of regional disposal facilities"; after January 1, 1986, compacts may restrict the use of the facilities to waste generated within the region; and Congress must consent to such compacts.

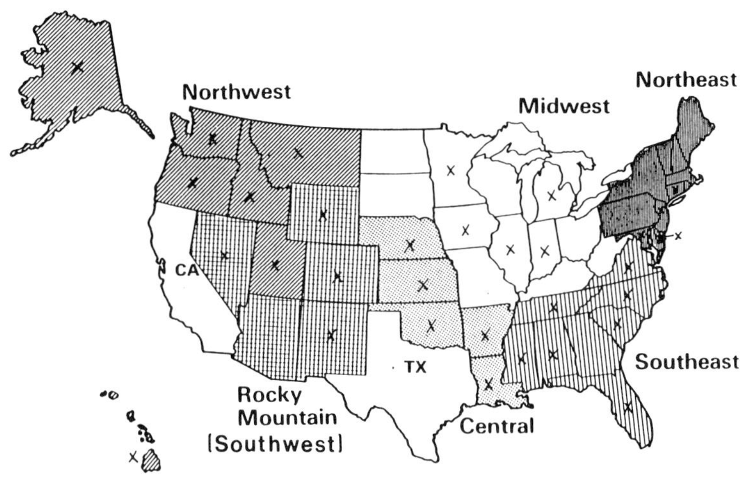

FIGURE 2. Compact Regions.

REGIONAL ACTIVITIES

Although some states had already taken the initiative to form compacts,[c] the Act provided the incentive for others to begin regional negotiations. It was clear from the beginning that not only the presence of operating LLW disposal facilities, but also historical patterns, political ties, and public perception would influence the manner in which the regional efforts would proceed and would define the regional boundaries (Figure 2). The Northwest and Southeast, where there are operating LLW facilities, took an early lead in drafting compact language. The Southwest, having some political problems associated with the use of the Beatty facility, proceeded more cautiously. The Central, Midwest, and Northeast regions, to varying degrees, deliberated the issues before drafting compact language. Several states around the country chose not to be clearly aligned initially, either to keep their options open or to develop a state facility solely for waste generated within their state. Six compacts, encompassing

TABLE 1

Interstate Low-Level Radioactive Waste Compacts
Legislative Status as of December 20, 1983

Northwest	Central States	Midwest	Southwest	Northeast
(S. 247	(S. 1581		(S. 1749	
H. R. 1012)	H. R. 3002)		H. R. 3777)	
Alaska—E	Arkansas—E	Illinois—A	Alabama—E	Connecticut—E
Hawaii—E	Kansas—E	Indiana—E	Florida—E	Delaware—E
Idaho—E	Louisiana—E	Iowa—E	Georgia	Maine
Montana—E	Nebraska—E	Kentucky	Mississippi—E	Maryland—E
Oregon—E	Oklahoma—E	Michigan—E	N. Carolina—E	Massachusetts-I
Utah—E		Minnesota—E	S. Carolina—E	New Hampshire-I
Washington—E		Missouri—I	Tennessee—E	New Jersey—E
		N. Dakota—D	Virginia—E	New York
		Ohio—I		Pennsylvania
Rocky Mountain		S. Dakota—T		Rhode Island—I
(S. 1991		W. Virginia—P		Vermont—I
H. R. 4388)		Wisconsin—I		
Arizona				
Colorado—E				*Unaffiliated*
Nevada—E				California
New Mexico—E				Texas
Wyoming—E				Washington, DC
				Puerto Rico
				Virgin Islands

KEY   Office of State Programs, U.S.N.R.C.
E—Enacted; I—Introduced; D—Introduced and Defeated
T—Introduced and Tabled; P—Petition to Join; A—Enacted Amended Version

most of the states, were enacted to varying extents by the eligible states (Table
1). Provisions in essentially all the compacts define the responsibilities of all
member states, of states which host facilities, and of the compact commission
or board[4,5]. Some address issues of particular concern, such as public participation, liability, financial responsibilities, and judicial review.

Northwest Region

The Northwest Interstate Compact on Low-Level Radioactive Waste Management, which applies to the use of the LLW facility at Richland, Washington,
was enacted by Alaska, Hawaii, Idaho, Montana, Oregon, Utah, and
Washington, and was introduced into both Houses of the Congress. The Compact Committee, established under the compact and consisting of all member
states, held public meetings on the implementation of the compact, procedures
for citizen interface with the committee, and criteria for access to the regional
facility by non-compact members after the 1986 date for exclusion of out-of-
region waste and before new sites are developed in other regions.

Southwest Region

The Rocky Mountain Low-Level Radioactive Waste Compact was enacted by Colorado, Nevada, New Mexico, and Wyoming, but not by Arizona, which is also eligible to join. The compact was introduced into both Houses of the Congress. The Regional Board met to discuss operational issues and receipt of federal waste, and adopted proposed congressional consent language and two resolutions in order to clarify or modify the compact. The consent language directed the NRC to require its licensees to comply with the terms of the compact, and reserved the right of Congress to alter, amend, or repeal the compact. Resolution I restricted the receipt of waste containing transuranic elements[d] and waste from federal facilities. Resolution II clarified the regulatory role of the Regional Board, restrictions on exports, and consistency with federal law and regulations. Colorado offered to host the first regional LLW disposal facility. An economic study of small-scale sites (1,300,000 ft³—3,600,000 ft³ disposed over 20 years) prepared by the State and based on data of Ford, Bacon, and Davis, Inc., conciuded that disposal costs could be competitive with those at the Barnwell LLW facility[6].

Early in the compact negotiation process in the West, California was excluded from membership in both the Northwest and Rocky Mountain compacts because of questions concerning its unwillingness to host a facility not withstanding the fact that more LLW was generated in California than in all of the other states in the western regions combined. However, with a change in the political climate, a state statute was enacted which required California to identify areas within the state for a site, develop LLW regulations, select an operator, and negotiate and enter into a compact.

Central Region

The Central Interstate Low-Level Radioactive Waste Compact was enacted by Arkansas, Kansas, Louisiana, Nebraska, and Oklahoma, and was introduced to both Houses of the Congress. The Regional Commission selected officers, drafted by-laws, and developed plans for an exclusionary siting study (whereby areas unsuitable for a site would be eliminated).

In Texas, the Low-Level Radioactive Waste Authority was established by statute to develop and operate a facility to accept waste generated solely in Texas. The Authority has options on land in two areas of the State, and will make a final choice in mid-1984.

Southeast Region

The Southeast Interstate Low-Level Radioactive Waste Management Compact was enacted by Alabama, Florida, Mississippi, North Carolina, South Carolina, Tennessee, and Virginia, and was introduced to both Houses of the Congress. This version of the Compact is the second one to go through the state legislatures for enactment. The first was sufficiently amended by South Carolina to require re-enactment by the other member states. Georgia enacted the compact prior to enactment of the amended version by South Carolina, and intends to enact the amended version. The Compact Commission elected officers and established committees to develop administrative procedures and by-laws and to investigate alternatives to shallow land burial.

Midwest Region

The Midwest Interstate Low-Level Radioactive Waste Compact was enacted by Indiana, Iowa, Michigan, and Minnesota, but has not yet been introduced to Congress. In order to implement the compact, the Midwest states held public meetings, prepared draft by-laws to be distributed for public comment, and explored staffing and financial needs.

The Missouri House and Senate passed the compact with companion legislation, which awaits the Governor's signature. The legislature in North Dakota defeated the compact. In Ohio, the compact passed in the House and was referred to the Senate. In South Dakota, a joint House and Senate committee is deliberating options for the state to join the Midwest Compact, to petition to join the Rocky Mountain Compact, to form a compact with North Dakota, or to develop a facility for state waste only. The Wisconsin Senate held hearings to consider joining the compact or developing its own facility for state waste. Action is expected in early 1984.

In Illinois, an amended version of the Midwest Compact was passed by the House and Senate, and signed by the Governor. Before legislative action, the amendments were discussed with all eligible states. The states which had already enacted the unamended version were unwilling to go back to their legislatures for consideration of the amended compact. The amendments address selection of host states, requirements for long-term care and liability funds, and third-party liability insurance.

Kentucky, although still interested in the Midwest Compact, initiated compact discussions with West Virginia and Puerto Rico, and prepared a draft compact for consideration by the three entities. The use of a part of the Maxey Flats facility was suggested for consideration.

Mid-Atlantic Region

The Mid-Atlantic states held preliminary compact discussions in early 1981, and also attended other regional meetings. A draft compact with Delaware, Kentucky, Maryland, North Carolina, Virginia, West Virginia, and Washington, D.C., was ratified by Virginia. However, the compact effort died as the various states aligned themselves with other regions.

Northeast Region

The Northeast Interstate Low-Level Radioactive Waste Management Compact was enacted by Connecticut, Delaware, Maryland, and New Jersey, but has not yet been introduced to Congress, since a majority of the states, as discussed below, have not yet acted on it. For the same reason, the Compact Commission has not been organized. However, all of the eligible states, as members of the Policy and Implementation Working Group, will focus on the implementation of the compact. This will include preparation of a draft regional management plan and draft administrative rules and procedures to be used by the Compact Commission when it is established.

In Maine, the Low-Level Radioactive Waste Siting Commission was mandated to develop recommendations for the Governor and the Legislature on the best option for Maine. Studies on site screening, facility design, and economic feasibility are under way.

In Massachusetts, the statutorily established Special Low-Level Waste Commission has focused in depth on transportation, liability, economic, legal, and technical issues. Until recently, it considered the Northeast Compact as its only option. However, its focus has broadened to consider the options of joining with a few states or developing a facility solely for waste generated in the state. The Massachusetts Department of Health independently developed a proposed plan for LLW management..

The compact is in committee in New Hampshire for interim study. The committee is considering a Task Force report and recommendations for modifications to the Northeast Compact. In New York, the State Energy Office in conjunction with an appointed advisory committee was directed by statute to study, report, and make recommendations in March 1984 on the potential economic, environmental, and social advantages and disadvantages of the Northeast Compact, another compact, or a facility for waste generated only in New York State.

Pennsylvania has made no decision on the compact, and is considering its options—essentially the same options available to New York. A study by Pennsylvania State University on LLW disposal siting reviewed the options and alternatives for Pennsylvania and laid out a social and technical plan for the state[7]. This plan could be implemented under any one of the options available to Pennsylvania.

In Rhode Island, the compact was passed in the House and is on the Senate calendar for 1984. In Vermont, the compact was referred to committee for interim study.

At the invitation of a Maine legislator, Vermont and New Hampshire legislators and state agency representatives held several public meetings to explore the potential of a three-state compact. Several of the participants are concerned about the potential for their states of being designated as a host state for an eleven-state compact with large generating states as members.

Unaffiliated Entities

The Virgin Islands and Washington, D.C. are not affiliated with any compact, nor has any action been taken by their governments to address the disposal of LLW.

CONGRESSIONAL ACTION AND ISSUES

Congressional hearings have been held and will continue into early 1984 when final committee action is expected. The Senate Judiciary Committee held oversight hearings on the Northwest Compact in Seattle, Washington in November 1982 and in Washington, D.C. in March 1983. A field hearing on the Southeast Compact was held in South Carolina in September 1983, and a hearing on the Central Compact in Washington, D.C. in October 1983. The Energy and Environment Subcommittee of the Committee on Interior and Insular Affairs held a general oversight hearing in October 1983. In addition, in November 1983, the House Energy, Conservation, and Power Subcommittee held an oversight hearing on the compacts.

What are the issues that Congress, states, and federal agencies have been concerned about; and what has been done to address these issues? Sensitive or unresolved issues have been identified for some time and were the focus of two national low-level waste meetings (in Seattle, November 1982[8], and Baltimore, June 1983[9]); meetings of federal agencies alone and with congressional staff; meetings of representatives of the compact groups alone and with congressional staff; and meetings of all three groups—federal agencies, states, and congressional staff.

Post 1986

Early in the negotiating process, one of the concerns recognized was the time required to develop new LLW disposal sites and the January 1, 1986 date for exclusionary authority under PL 96-573. Estimates are that it will take at least five years to select, license, and construct a facility. Therefore, as of January 1, 1986, it is likely that the Central, Midwest, and Northeast regions will be excluded from the facilities now operating and will not have new operating facilities in those regions. The U.S. Government Accounting Office report on LLW[10] focused on this issue and recognized as options to solve the problems: (1) extension of the 1986 date; (2) use of U.S. Department of Energy (DOE) sites on an interim basis; (3) development of central storage facilities in the regions, and (4) interregional agreements for use of the three presently operating facilities. Those solutions which did not require federal action were identified as the most desirable. Congressional staff have made it clear that Congress does not intend to extend the date. As for the second option, the DOE argued against the use of its facilities on legal, economic, political, and technical grounds. Because states recognize that siting regional or interim storage facilities may entail the same processes and time constraints as siting disposal facilities, the third option has not yet been seriously pursued. It is, therefore, on the fourth option that states are focusing.

It was at the national LLW meeting in Baltimore, June 1983, that the State of Washington floated a trial balloon on criteria by which states with operating facilities might evaluate requests by states or regions without facilities for interim use of operating facilities. These criteria included: significant volume reduction of the LLW proposed to be sent to the facility; on-site storage by generators; interim storage within the region; membership in a compact of a state seeking use of the facility; and congressional consent to the compact by January 1, 1986. The states would want evidence that the applicant (a state) had a plan with target dates and alternatives, and was moving forward accordingly to develop a new LLW facility. Discussions between states with operating facilities and those from regions without facilities have further explored the conditions and terms for interregional agreements. Congressional staff have indicated that the states must come to Congress with a consensus on such agreements and should not expect congressional committees to resolve the problem.

Federal Waste

An issue of particular concern for both federal agencies and states is that of the disposal of federal waste and the intent of PL 96-573, which falls short in the definition of federal waste. States have sought more precise characterization of federal waste, and have emphasized the need for federal LLW generators utilizing regional facilities to be subject to the same regulatory system, inspections, and fees as would apply to commercial LLW generators. The Act states that the compacts shall not be applicable to the disposal of LLW from federal research and development activities. In order to clarify "federal research and development," states are considering consent language which limits federal waste to that resulting from activities at atomic energy defense or research installations or institutions owned by or operated for the DOE, its agents, successors or assigns.

The U.S. Department of Defense (DOD) has expressed concerns with the requirements that states plan to impose equally on military and commercial waste, namely: (1) inspections of LLW packaging areas—DOD wants to assure protection of "secure" areas; (2) assessment of inspection fees—DOD would seek congressional approval to pay such fees; (3) interim measures arranged by the regional commission if the regional site were unavailable—DOD would seek to be exempt from such measures and would have free access to other regional facilities. Congressional consent language is in preparation by both states and federal agencies.

Going-it-alone

The issue of a state developing an LLW facility for waste generated solely within the state is of particular interest not only to Texas, but to a number of other states considering this option. PL 96-573 did not address this, nor does the legislative history of the Act. However, those who choose to go it alone read the Act as supporting their position. That is, the policy statement in PL 96-573 gave to states the responsibility for disposal either within or outside the state. The issue of whether a state can develop a facility that is state-owned and operated and restrict its use to residents or businesses in the state may have precedence in case law[11,12]. However, there is no case law pertaining to the LLW disposal issue and of a state going it alone. The Supreme Court decision in *Reeves* v. *Stake* 48 LW 4746, among others, has been cited in support of the concept. The Court ruled that the State of South Dakota could restrict the sale of cement from a state-owned and operated plant to residents of the State. On the other hand, those who argue against a state going alone cite PL 96-573 wherein it stated that LLW may best be managed on a regional level, and that compacts can exclude out-of-region waste. The Supreme Court case of *City of Philadelphia* v. *New Jersey* 437 U.S. 617, among others, is cited to deny the ability of a state to go it alone. The Court ruled that New Jersey could not prohibit the import of out-of-state waste (garbage) in order to conserve landfill space—i.e., economic protectionism was declared invalid. The issue of a state going it alone has been brought to the attention of congressional staff, but it is not clear if Congress will address it.

Other Issues

A number of technical and regulatory issues were raised over a period of two years by states and federal agencies, and, for the most part, were resolved. The NRC, DOE, and DOT reviewed and commented on the compacts throughout the drafting of compact language. Comments addressed the scope of the compacts ("management" vs. "disposal"), inspection of NRC licensees, regulatory roles of compact commissions, and regulatory requirements inconsistent with federal regulations, to name a few[5].

Among those unresolved issues in some compacts is the definition of LLW, which varies among the compacts. Those in PL 96-573 and the Nuclear Waste Policy Act of 1982 were both used, with and without variations. The definition of transuranic waste and those concentrations allowable at the disposal facilities are not uniform in the compacts.

LOOKING AHEAD

In its deliberations on the LLW compacts, Congress will be looking for a national system for LLW disposal—where are the inconsistencies; will all LLW be taken care of; has any state been left out; are there gaps in the system; what issues are resolved, and which ones remain?

With the consent of Congress to the compacts, states will have reached their first major milestone in their assumption, under the Low-Level Radioactive Waste Policy Act, of their responsibility for LLW disposal.

Since it is unlikely that there will be new facilities developed by the January 1, 1986 date in PL 96-573 for exclusion of out-of-region waste, states and regions without disposal facilities will continue to explore options and implement measures to bridge the gap between the 1986 date and the opening of new facilities. These measures include interregional agreements for interim use of existing LLW facilities, volume reduction and interim storage by LLW generators, and possibly centralized interim storage facilities.

The next milestone for those regions without LLW disposal facilities will be the designation of a state to host a regional LLW facility. The final milestone will be reached when the host state sites the facility and ensures its development in order to receive the LLW generated in the region.

GLOSSARY

a. Low-level radioactive waste means radioactive material that: (A) is not high-level radioactive waste, spent nuclear fuel, transuranic waste, or tailings or wastes produced by the extraction or concentration of uranium or thorium from any ore processed primarily for its source material content; and (B) the NRC, consistent with existing law, classifies as low-level radioactive waste. Typical low-level radioactive waste is low activity paper and plastic trash, tools, resins, sludges, evaporator bottoms, liquids, vials, animal carcasses, tissue cultures, labware, syringes, and sealed sources.

b. An Agreement State is a state which has assumed regulatory authority from the AEC or NRC over radioactive by-products, source materials, and small quantities of special nuclear materials. The Agreement States are: Alabama, Arizona, Arkansas, California, Colorado, Florida, Georgia, Idaho, Kansas, Kentucky, Louisiana, Maryland, Mississippi, Nebraska, Nevada, New Hampshire, New Mexico, New York, North Carolina, North Dakota, Oregon, Rhode Island, South Carolina, Tennessee, Texas, and Washington.

c. A compact is a legally binding interstate agreement, which requires enactment by the respective state legislature and governors, and, in the case of LLW compacts with authority to exclude out-of-region waste, the consent of Congress.

d. Transuranic elements mean radionuclides with an atomic number greater than 92.

REFERENCES

1. Brenneman, Faith N. 1982. Status of Activities on Low-Level Radioactive

Waste Compacts. Presented at Atomic Industrial Forum, Inc. Fuel Cycle Conference '82, New York, N.Y. March 24, 1982.

2. Report to the President of the Interagency Review Group on Nuclear Waste Management. March 1979. TID-29442, Department of Energy, Washington, D.C.

3. Low-Level Radioactive Waste Policy Act, PL 96-573, 94 STAT. 3347, 42 USC 202 lb.

4. Regional Low-Level Radioactive Waste Compacts: A Progress Report. Proceedings of a Conference on State Initiatives to Implement the Low-Level Radioactive Waste Policy Act. May 1982. National Governors' Association, Washington, D.C.

5. Status-Report—Low-Level Radioactive Waste Compacts as of July 1982. Prepared for Senator James A. McClure, Chairman, Committee on Energy and Natural Resources, United States Senate. DOE/NE 0045, U.S. Department of Energy, Washington, D.C.

6. Whitman, Mary, Leonard Slosky, Susan Mizner, and Colleen Murphy. June 1983. Economics of a Low-Level Radioactive Waste Management Facility for the Rocky Mountain Region. Office of the Governor, State of Colorado.

7. Low-Level Radioactive Waste Disposal Siting: A Social and Technical Plan for Pennsylvania. 1983. LW8303-1. Warren F. Witzig, William P. Dornsife, Frank A. Clemente, Ed., Institute for Research on Land and Water Resources, the Pennsylvania State University, University Park, PA.

8. Resource Notebook on Establishing Regional Compacts for Low-Level Waste Management, A National Conference. November 30-December 1, 1982. Seattle, Washington. Sponsored by the Northwest Interstate Compact Committee on Low-Level Radioactive Waste Management and the State of Washington.

9. A Resource Notebook on State and Regional Initiatives in Low-Level Radioactive Waste Management. Prepared for the National Meeting on Low-Level Radioactive Waste, June 20-21, 1983, Baltimore, Maryland. Sponsored by Southern States Energy Board in conjunction with the National Governors' Association and the U.S. Department of Energy.

10. Report by the U.S. General Accounting Office, Regional Low-Level Radioactive Waste Disposal Sites—Progress Being Made But New Sites Will Probably Not Be Ready By 1986. GAO/RCED-83-48, April 11, 1983. U.S. Government Accounting Office, Washington, D.C.

11. Shapar, Howard K., July 15, 1980. State Ability to Prohibit Disposal Within Its Boundaries of Low-Level Radioactive Waste Generated in Other States. Memorandum for William J. Dircks, Director, Office of Nuclear Materials Safety and Safeguards, U.S. NRC, Washington, D.C.
Also: May 3, 1980. Answers to Legal Questions Regarding Low-Level Waste: Do States Have Authority to Prohibit the Disposal Within Their Boundaries of Low-Level Radioactive Waste Generated in Other States?

12. Doub and Muntzing, Chartered (Perry B. Seiffert and James R. Shoemaker). October 12, 1983. Authority of Single State to Prohibit Disposal Within Its Borders of Out-of-State Low-Level Nuclear Waste. Memorandum to Argonne National Laboratory, Doub and Muntzing, Chartered, Washington, D.C.

Management of Radioactive Materials and Wastes: Issues and Progress. Edited by S. K. Majumdar and E. Willard Miller. © 1985, The Pennsylvania Academy of Science.

Chapter Five

THE NORTHEAST LOW-LEVEL WASTE COMPACT: REGIONAL COOPERATION IN LOW-LEVEL WASTE MANAGEMENT[1]

Anne D. Stubbs

Executive Director
CONEG Policy Research Center
Hall of the States
400 North Capitol Street
Washington, D.C. 20001

When Congress enacted the Low-Level Radioactive Waste Policy Act in 1980 (P.L. 96-573), each state was given responsibility to provide for disposal of low-level radioactive waste generated within its borders. In assigning this new responsibility, the Congress indicated its clear preference that separate low-level waste disposal facilities not be developed in each state. It encouraged interstate cooperation in this new undertaking by allowing states which join together in Congressionally approved regional compacts to prohibit waste generated in non-member states from a regional disposal site, effective January 1, 1986.

This paper reviews the major provisions of the Northeast Interstate Low-Level Radioactive Waste Management Compact and discusses the prospects for emergence of a new regional management system. A more lengthy discussion of the Compact is provided in a separate "Report to the States".[1]

BACKGROUND

The new responsibility given to Northeast states poses major economic, legal and political challenges. The region accounts for one-third of the total low-level waste generated in the nation, yet there is not an operating commercial

---

[1]The views expressed in this paper are those of the author only and do not reflect the views of the Coalition of Northeastern Governors.

disposal facility in the region. The states vary widely in the volumes of waste generated by their industries, utilities, and medical and research institutions. States such as Delaware, New Hampshire and Rhode Island currently generate less than 3000 cubic feet (ft.[3]) annually; yet the region also contains six of the top twenty generating states: Massachusetts, New York, New Jersey, Pennsylvania, Connecticut and Maryland[2]. Given this disparity in waste volumes and the economic importance of the waste-generating industry to the states, hammering out an equitable, politically acceptable legal agreement would not be easy.

Following passage of the federal Act, the Northeastern Governors agreed to explore development of a regional compact or compacts as the solution to the disposal problem. Acting under the auspices of the Coalition of Northeastern Governors (CONEG), the Governors established the organizational framework for discussions leading to a Northeast regional compact. The six New England states plus New Jersey, New York and Pennsylvania formed the original nucleus, with Maryland and Delaware joining upon invitation. A Policy Working Group (PWG) of credentialled executive and legislative representatives from the eleven states met regularly over an eighteen month period, in open sessions throughout the region, to develop the compact. They were assisted by a Technical Subcommittee, composed of senior radiological health officials from each state; representatives of the U.S. Nuclear Regulatory Commission (NRC), advisory groups and task forces in the states, and by citizen, industry and other interest groups throughout the region.

In February 1983, a draft Compact was forwarded to the Governors for their consideration and submittal to the legislatures. That document reflects the fact that a compact is both state law and a legally binding contract which must withstand the tests of time and diverse administrative and legal systems. The Northeast Low-Level Radioactive Waste Compact ("Compact") is designed to strike a balance between the states' collective interest in an equitable regional management system and their individual self-interest. Each provision of the Compact is based upon a set of principles which guided the PWG. Collectively, these provisions establish a sound, workable legal system for states as they cooperate in managing low-level waste.

MAJOR PRINCIPLES

Several basic principles were the touchstone for decisions on what provisions and language to include on the Compact. The PWG consciously chose not to anticipate every specific administrative, management or technical problem which might arise during the decades that the Compact would be in effect. The Compact sets forth the major rights and responsibilities of each member state with these rights and responsibilities to be elaborated through a defined rule-making process.

The Compact, a charter of interstate and state-federal relationships for low-level waste management, is built on five major principles:

Good Faith: While "good faith" is a subjective term, each state commits to carry out its responsibilities under the Compact in a timely, responsible manner. Each state has the right to expect every other party state to act in a similar manner.

Equity: Like "good faith", the principle of equity is often subjective. Equity does not imply numerical equality. The terms of the Compact, and the process by which they are carried out, are designed to ensure that responsibilities are to be shared in a fair and reasonable manner.

State Sovereignty: The fundamental principle that state sovereignty is to be maintained pervades the Compact and is the underlying explanation for many of the criticisms leveled against it. The Compact sets forth only those principles, rights, responsibilities, institutions and procedures which are essential for a regional management system to work. All regulatory authority over waste generators or regional facilities remains with the states or federal government. Each state determines, under its legal and administrative system, the specific actions and procedures it will take to carry out its responsibilities.

Minimal Costs to States: The regional management system established by the Compact is to be self-supporting, with state tax dollars used only as initial seed money.

Consistency with Federal Authority: The Compact does not attempt to carve out new state roles in the management and regulation of nuclear materials. The Compact is designed to be consistent with the federal government's primary responsibility for radioactive materials, established by the Atomic Energy Act, as amended. While the Compact does not require basic federal policy on state and federal roles, it does provide states with full authority to carry out their responsibilities under the federal act.

COMPACT PROVISIONS

The Compact does four basic things. First, it sets forth the major rights, responsibilities and obligations of the party states (i.e., signatory states), the host states where facilities are located, and the regional commission. Second, it establishes the Northeast Interstate Low-Level Radioactive Waste Commission as the body to administer the Compact. Third, it established a process for selecting a state to host a regional facility. It does not specify how a host state, once selected, would site, develop and oversee management of a regional facility. Finally, it sets forth the terms and conditions under which a state joins or withdraws from the Compact and, reflecting its contractual nature, provides for penalties for states which fail to meet their agreed-upon obligations.

Equally important as these four major actions is what the Compact does *not* do. It does not select the state nor the site of a regional facility. It does not specify what type of facility must be developed, other than stipulating that any facility must be capable of being licensed and must be adequate to meet the region's needs. The Compact does not specify how a regional facility is to be developed—that decision is left to the host state. It does not regulate generators of low-level waste nor the owners/operators of a regional facility. Finally, it does not apply to facilities which are not designed by the Commission as "regional facilities." It has no applicability to storage or treatment facilities on-site of generators.

Policy, Purpose and Definitions

Article I sets forth the basic purpose and policies of the party states. In enacting the Compact, each state acknowledges its responsibilities under P.L. 96-573 for providing for disposal capacity for its waste generators and agrees to a coordinated regional management approach as the most efficient means to provide for disposal capacity.

The Compact identifies several policy objectives which are to guide the party states in carrying out their responsibilities. These include limiting the number of facilities, consistent with effective and efficient management; encouraging reduction of waste generated; equitable sharing of the costs, benefits and obligations of waste management; and ensuring environmentally sound and economical management of low-level waste.

Definitions of terms and concepts (Article II) are intentionally not linked to current statutes, regulations, technologies and organizational structures. The omission of specific disposal, treatment or storage technologies caused concern on the part of some citizens and the NRC, but it minimizes the need for future statutory revisions as technologies change.

Several terms warrant comment.

Disposal includes but is not limited to land disposal. This more flexible definition allows consideration of other technologies for the isolation of waste.

Generator includes those who produce or process waste within the region, including a broker who handles waste generated in a party state. However, waste generated outside the region is not eligible for access to a facility, even if it is handled by a broker located in a party state.

High-Level, Low-Level and Transuranic Wastes are defined consistent with federal law and NRC regulations, but specific technical criteria which may change are omitted. The definition of low-level waste is consistent with the Nuclear Waste Policy Act of 1982, but it excludes federal atomic energy defense wastes and federal research waste as stipulated by P.L. 96-573. If, as expected, Congress clarifies the definition of federal research waste to include only atomic energy defense research waste, this change would apply to the Compact.

Under the Compact, states accept responsibility for waste greater than Class

C, a category generally excluded from shallow land burial by current NRC regulations. Disposal capacity for this waste can be provided on a case-by-case basis or through agreements with other regional compacts or the federal government.

Post-Closure Observation and Maintenance is the final phase of activities required by NRC regulations to be undertaken by a licensed operator at a disposal facility. Since only disposal facilities are currently licensed, it does not apply to treatment or storage facilities under federal regulations. The post-closure activities include continued observation and corrections of any problems at a properly closed facility. Legal, management and financial responsibility remains with the operator.

Institutional Control is a management phase involving continued observation, monitoring and care of a disposal facility once control and responsibility have been transferred to the state or federal government as required by federal regulations. NRC regulations 10CFR(61) state that, in the design and licensing of a disposal facility, institutional controls may not be relied upon for more than 100 years to ensure adequate isolation of the waste.

State Rights and Obligations

The rights and obligations of the party states and host states are set out in Articles III and V. The fundamental collective responsibility is provision of sufficient capacity to manage all low-level waste generated in the party states at one or more regional facilities located within the region or at other appropriate facilities made available through agreement with the federal government or other states and regions.

If states are responsible for providing adequate management capacity, they must ensure that any regional facilities are economically feasible, with sufficient resources for sound management practices. Decisions by generators to reduce waste volumes generated or shipped to the regional facility will affect its planned capacity and economic feasibility. While waste minimization is encouraged by the Compact, shipping waste outside the region for disposal is not. The Compact addresses the potential instability in the regional management system by providing that regional facilities are entitled to that waste generated in the region which is handled off the site of generation. This control over export is an addition to the ban on import provided by the Congress. However, in the interest of sound management, neither set of controls is absolute. The Commission, with concurrence of the affected host states, may approve both the export and importation of waste.

The rights and responsibilities of the party states are clearly delineated in Article III.

Access: Each party state which complies with its obligations under the Compact has the right of access to regional facilities made available by the Commission. This right of access extends to individual generators so long as they comply with applicable federal and state laws and regulations and no restrictions are placed upon the state's access.

Enforcement: Each party state must enforce federal and host state packaging and transportation regulations on generators, shippers and carriers operating within or transporting through the state. This responsibility for "shared enforcement" reduces the potential that improperly packaged or labeled waste will be shipped to the regional facility. Each state has a primary responsibility for enforcing violations which occur within its borders (Article VIII).

Fees: Each party state has the right to recover the costs of compact-related regulatory and oversight activities by imposing reasonable fees on generators, shippers and carriers. However, these fees may not create an unreasonable burden to the safe, economic regional management of the waste.

Host State Capability: Each party state must be capable of hosting a regional facility in a timely manner and to ensure the post-closure observation and maintenance and institutional control of any facility within its borders. In a major demonstration of its good faith, each state indicates its willingness and its legal and administrative ability to serve as a host state if selected. It must ensure that a licensable facility can be sited in a timely manner under federal and state laws and it must have adequate authority to oversee management of the facility.

Liability: A party state which is not host for a regional facility is not liable for personal or property injury resulting from operation of a regional facility or transportation to it.

Waste Minimization: Each state must encourage its generators to minimize the waste requiring disposal.

Information: Each state agrees to provide the Commission with the information needed to carry out its responsibilities.

Host states have all the rights and responsibilities of party states, as well as those additional responsibilities which are unique to its role as host. These activities are spelled out in Articles III and V.

Facility Development: A host state must carry out its capability of hosting a regional facility by ensuring that a facility is sited and developed in a timely manner. Any proposed site and facility must be capable of meeting applicable federal and state geologic, environmental and economic criteria, and is expected to be compatible with the regional management plan. The host state, according to its own procedures, may select a site and operator, respond to private sector initiatives, or develop and operate the facility under public authority. It must consult with each party state and the Commission in carrying out this responsibility.

Facility Oversight: The host state may choose to regulate and license the facility under agreement with the NRC, but it is not required by the Compact to do so. However, through its relationship to the NRC or contractual agreements with the owner/operator, it must ensure the safe operation, closure, post-closure observation and maintenance and, as appropriate, institutional control of any regional facility. Included within its management plan must be adequate provisions for third-party liability compensation and cleanup during the operational and post-closure periods (Article IX). This responsibility, which reinforces NRC licensing requirements, can be provided by requiring financial assurances of the operator or the state. The host state must consult with each party state and the Commission on its management oversight plans and must periodically report on the condition of the facility, its remaining useful life and the status of the post-closure and institutional control funds.

Facility Closure: A host state must report to the Commission any situation which requires emergency, temporary or permanent closure of a facility. The Compact specifies general notice procedures for each type of closure (Article V).

Facility Fees: The host state has the right and responsibility to ensure that fees charged by the facility operator are reasonable and adequate to cover all costs related to the facility. All charges must be imposed equitably upon all users, with charges allowed to differ for various types of waste but not for similar users in the various party states. The host state may also impose a surcharge on waste received at a facility within its borders to cover all current reasonable costs incurred by the state in carrying out its regulatory and management responsibilities. Both the host state and its local subdivisions retain authority to impose surcharges for host community incentives or compensation. However, the Commission must be allowed to review and comment on proposed fee and surcharge structures prior to their approval or imposition by the host state.

The Commission

The legal status and functions, duties and powers of the Commission are spelled out in Article IV. Each state is represented by a voting Commissioner (with two voting members for an active host state) appointed by the governor according to each state's procedures. A limited staff is authorized, with staff resources to be complemented by state expertise and resources, consultants under contract, and technical and advisory committees.

The Compact provides direction and principles to guide the Commission in administrative procedures to ensure its accountability. Detailed procedures are to be adopted by rulemaking. The Compact requires all binding actions to be by majority vote of all party states, with certain major actions (e.g., host state designation, waste importation) requiring a two-thirds vote. All meetings must be open to the public with reasonable prior notice, except that those dealing with sensitive personnel or legal matters may be closed. Following public notice and comment, the Commission must adopt operating rules and regulations which protect due process and ensure its orderly operation.

The Commission's responsibilities are of two types: coordination among the party states to ensure their collective interests are represented, and administration of the Compact.

Coordination: The Commission addresses the collective interest in developing adequate regional capacity through development, adoption and maintenance of a regional management plan which sets out the type and number of regional facilities needed. It designates the host state and accepts a facility proposed by the host state as a regional facility. The Commission can enter into agreements for access to facilities outside the region and must approve any waste export or importation. It reviews proposed fees and surcharges and must provide for long-term third party liability compensation for properly closed disposal facilities (Article IX). It can assist in resolving disputes between party states; provide guidelines and rules for efficient, consistent and reasonable implementation of the Compact; and can intervene in court or administrative actions on behalf of the party states.

Administration: As administrator of the Compact, the Commission can admit new members, review whether party states are fulfilling their obligations under the Compact, and impose sanctions and penalties. (Violations specifically identified in the Compact include waste importation without Commission approval and, with limited exceptions, waste disposal within the region at other than a regional facility Article VIII). It can obtain from the party states information needed to carry out its duties, and must provide them with information on its activities. It can adopt a budget; accept state appropriations for its initial support; accept grants, loans or donations so long as no conflict is involved; and levy a Commission surcharge on waste received at a regional facility sufficient to cover its approved costs and expenses.

Legal Status: As a coordinative body, the Commission is a legal entity separate from the states and liable only for its own actions. It has no legal or financial responsibility for any aspect of the regional facility. The Compact does not alter the incidence of liability of generators, shippers or carriers under applicable state and federal law. The Compact does provide guidance on standing, jurisdiction and venue for actions related to the Commission as a means to minimize confusion and prevent unreasonable delays. Jurisdiction lies with the federal courts, with any aggrieved party having standing. In the effort to minimize delays, the Compact specifies statutory time frames for filling actions and for selected court rulings (60 days to file a petition; 90 days for ruling on a challenge to host state selection).

Host State Selection

The heart of the Compact is selection of the host state. Procedures and criteria to be followed by the Commission are spelled out in Article V. The first step in this process is development of a regional management which provides the regionwide data base and analysis which allows the Commission to assess the number and type of regional facilities needed. The plan must include a current inventory of generators and all waste facilities in the region; a generic assessment of the type and number of regional facilities needed, consistent with applicable health and safety considerations; and reference guidelines to assist party states in its site selections.

The Compact provides for a host state to either volunteer or be designated by the Commission, with general procedures and standards specified in the Compact. A two-thirds vote of the Commission is required in either instance.

Volunteer: In developing criteria and procedures for reviewing a state's offer to serve as host, the Commission must seek public comment. At a minimum, the criteria must include consideration of that state's capability to host a facility in a timely manner and ensure its proper management, as well as the economic feasibility of the proposed facility. An offer which fails to meet the region's need as determined by the regional management plan could be rejected.

Designation: If no state volunteers, the Commission must develop and adopt procedures for designating a host state. It must conduct hearings and studies as required by its procedures and, upon request, hold public hearings in candidate states. A current host state may request that it continue to serve beyond its planned period.

The Compact limits the criteria which guide the Commission in its selection process to six and only six statutory criteria.

1) The first and fundamental criterion which cannot be subordinate to any other factors is the health, safety and welfare of citizens in the party states, as stipulated by the appropriate regulatory authorities (e.g., NRC, the U.S. Environmental Protection Agency, and state). Any state selected as host must be capable of selecting a site which, alone or in combination with engineered design factors, meets this criterion.

2) The second criterion—environmental, economic and social effects of a regional facility—is closely related to public health and safety. This would include effects on all the party states, not just the host state.

3) The third criterion is economic benefits and costs, including economic feasibility of a facility and its ability to generate sufficient revenue for sound management and long-term care; the costs and benefits derived by the host state; and the economic benefits associated with waste generating activities in each state.

4) The amounts and characteristics of waste generated in each state must be considered as an approximate measure of each state's use of and benefit derived from the facility.

5) Waste transportation must be minimized as a means to protect public health and safety and to minimize overall costs.

6) The existence of regional facilities, either currently operating or under

institutional control, within a party state must be considered in the selection process. In the effort to share responsibilities under the Compact, the burden already assumed by a host state must be considered a reason to select another state as host.

The Compact does not specify these criteria in detail nor assign weights to them, since that undertaking requires detailed information made available only through the regional management plan or detailed assessments of waste generation and management technologies.

Eligibility, Withdrawal, Revocation, Entry Into Force, Termination

The administrative section, Article VII, sets out criteria and procedures governing admission to and withdrawal from the Compact, its coming into existence and termination.

Membership: The eleven states which negotiated the Compact are initially elegible members, but do not have an automatic right of membership. Other states not members of other compacts may be granted eligible status if they meet conditions imposed by the Commission. Three conditions must be met to the satisfaction of the Commission: (1) the Compact enacted into law, (2) the entry fee of $70,000 paid, and (3) all statutes and statutory provisions inconsistent with the Compact must be repealed as they apply to the Compact. (Article VI provides laws not inconsistent remain in full force. Laws and regulations of the state or its subdivisions which are inconsistent are repealed by enactment of the Compact, including provisions which impede agencies in the designation, siting or licensing of a regional facility. Each state agrees not to enact or enforce inconsistent measures). Initial eligibility expires Junes 30, 1984; a cutoff date which allows the region to be defined and the regional planning and host state selection process to begin with a modest lead time before the 1986 exclusion date.

Entry Into Force: The Compact is effective as state law upon enactment, but is effective as an enforceable interstate compact only upon consent by the Congress. Three states must enact the Compact before the Commission can be formed and the Compact submitted to Congress.

Revocation: The Compact specifies procedures which must be followed before the rights granted by the Compact can be revoked or suspended. Since this is a significant action, opportunities for hearings and comment, a two-thirds vote and one year notice are required. Failure of a party state to comply with the Compact and carry out its obligations are grounds for suspension or revocation. The Compact specifically identifies creation of unreasonable barriers to siting or refusal to accept host state responsibility as grounds for revocation (Article IV(i)(14)).

Withdrawal: Any state may voluntarily withdraw from the Compact by providing a five-year written notice and repealing the Compact. A regional facility in a withdrawing state must remain available for five years following written notice to the Commission. The lengthy notice period is both an indicator of good faith and a necessary step for an orderly transition to a different configuration of members and waste management needs.

Termination: The Compact can be terminated only by the repeal of all authorizing laws in each party state or affirmative action by Congress. Withdrawal by one or more states does not affect the Compact's applicability in the remaining states.

NEXT STEPS

As expected, the Compact has been given close scrutiny by elected officials and citizens. Many of the concerns voiced—the lack of specific detail on the regional management plan, administrative procedures for public participation in Commission activities, and liability mechanisms— reflect the fact that a Compact is a statutory framework governing relations among states. Many of the specific requirements governing development and management of the regional facility would be developed by the host state under its laws and administrative process.

Four states moved promptly to ratify the Compact—Maryland, Delaware, Connecticut and New Jersey. In late 1984, the Northeast Low-Level Radioactive Waste Commission was formally organized. In Massachusetts, the Special Legislative Commission developed and forwarded to the Governor a separate compact which addresses that state's concerns with the lack of specificy with the Northeast Compact. As of spring 1985, that compact had not been introduced. After extensive study, a New York Advisory Committee recommended to the Governor and Legislature that New York develop an interim site and explore, as necessary, a compact arrangement with smaller generating states. Legislation to establish an above-ground interim storage site and a disposal siting process was forwarded to the legislature by the Governor in early 1985. Pennsylvania decided against joining the Northeast Compact and drafted an Appalachian Compact with membership limited to Pennsylvania and the contiguous states of West Virginia, Delaware and Maryland. The legislation, which calls for Pennsylvania to serve as the initial host state, was introduced in early 1985.

After initial legislative review, representatives of Maine, Vermont and New Hampshire joined together in 1983 to explore the feasibility of a smaller, more informal three state Northern New England Compact. After additional study, the Maine Low-Level Radioactive Waste Special Commission is examining in-state storage and a contractual arrangement with a compact region. The Rhode Island Legislature withheld action until a special study commission could examine the Compact and other options.

Two other issues compete for attention as state officials consider the Northeast Compact: Congressional ratification and post-1985 options. As the 98th Congress reconvened in January, the Northwest, Southeast, Central and Rocky Mountain Compacts were before it for approval. Representatives of these regions, anxious to invoke the January 1986 exclusion date, were pressing hard for immediate approval. If they are successful in this effort, the Northeast States and their generators face the prospect of being shut out of the three operating commercial sites.

The immediate issue facing states and generators in the Northeast is how they will provide for alternative management and disposal options after 1985. This problem faces all states, whether or not they join the Northeast Compact. Interim storage and/or negotiated access to existing disposal sites will likely be required, since a regional disposal facility is not expected to be operational before the end of the decade.

The Governors recognized that the need for regional cooperation and discussion did not end with completion of formal compact negotiations. They agreed to continue the Policy Working Group and expand its role. In addition to continued efforts to clarify states' remaining concerns with the Compact, the PWG represents the region's interest before the Congress and in discussions with other regions for access to existing sites. These discussions continue into 1985, as Congressional committees complete oversight hearings and move to markups, and as the Northeast determines its need for continued access to disposal sites. As the target date of January 1, 1986 comes closer, state, Congressional and inter-regional discussions are expected to take on a new sense of urgency.

FOOTNOTES

1. *Low-Level Radioactive Waste in the Northeast: Report to the States,* Report of the CONEG Low-Level Radioactive Waste Policy Working Group, CONEG Policy Research Center, Washington, D.C., March 1983.
2. *Low-Level Radioactive Waste in the Northeast: Disposal Volume Projections,* Report of the Technical Subcommittee, CONEG Policy Research Center, Washington, D.C. October 1982.

Management of Radioactive Materials and Wastes: Issues and Progress. Edited by S. K. Majumdar and E. Willard Miller. © 1985, The Pennsylvania Academy of Science.

Chapter Six

SHALLOW LAND BURIAL OF RADIOACTIVE WASTES

D.G. Jacobs,[1] Ph.D. and R. R. Rose,[2] Ph.D.

[1]Technical Program Manager
575 Oak Ridge Turnpike
HER Technical Associates, Inc.
Oak Ridge, TN 37830
and
[2]Technology and Engineering Services Manager
Evaluation Research Corporation
800 Oak Ridge Turnpike
Oak Ridge, TN 37830

Low-level, solid radioactive wastes have been buried in the ground since the startup of nuclear operations by the Manhattan Engineer District in the early 1940's. These operations were originally intended to be temporary so the primary consideration in locating land burial sites was their accessibility from the source of waste production. Early land-burial facilities were located on large reservations owned by the U.S. Atomic Energy Commission (AEC) and operated by their prime contractors.

Shallow land burial consists of excavating a trench or vault, emplacing the waste, minimizing void space within the disposal unit, and covering the waste with earth to control access to the waste.

Problems encountered in the land-burial of radioactive wastes can be classified into areas which are related to the environmental characteristics of the sites, waste characteristics, operational practices and control, and predictive capability.

The most serious environmentally related problems involve water management. Water provides a primary vehicle for both erosional processes, which affect the structural integrity of the waste trenches, and for the migration of radionuclides. Although there is consensus that the current level of off-site movement of radionuclides from operating burial grounds does not constitute an immediate health hazard, there is less certainty with respect to the ability of the facilities to provide long-term containment and isolation.

INTRODUCTION

Radioactive wastes have been generated and disposed of since the early use of radioactive materials in the late nineteenth and early twentieth century. However, when the Manhattan Project began during World War II the production of radioactive waste expanded significantly. Oak Ridge National Laboratory (ORNL) began operation as a part of the Manhattan Project and is one of the oldest nuclear centers in the United States. Shallow land burial has been used for the disposal of solid low-level radioactive wastes from the beginning of waste management operations at ORNL.[1,2,3]

Shallow land burial consists of excavating a trench or vault, emplacing the waste, minimizing void space within the disposal unit, and covering the waste with earth to control access to the waste. It is a simple operation and does not require highly skilled labor.

It is likely that the initial suggestion for burying waste was a memorandum dated January 5, 1943,* by Dr. S.T. Cantril of the Medical Department of the Central Safety Committee shortly after the graphite reactor attained criticality in late 1943 and plutonium separations began.[4]

Dr. Cantril stated:

> "It would seem that provision has not been made for the disposal of actively contaminated-broken glassware or materials not sufficiently clean to be used in other work. I am suggesting that a metal trash can with cover, with red lettering on the can, be provided for the disposal of such material as can be placed in the trash can. Mr. Schwertfeger has suggested that a suitable location for the burying of this material could be provided over on the burning ground. A suitable pit with enclosed fence could be made."

This suggestion was accepted by the committee and solid waste disposal by shallow land burial was begun in early 1944 as an extension of the disposal of municipal waste by burial in a sanitary landfill.[4,5]

The Manhattan Project was originally intended to be temporary so the primary consideration in locating early waste burial grounds was their accessibility to the source of waste production. Following World War II it became apparent that national laboratories would continue as a permanent part of government-supported activities. Generation of low level waste increased throughout the country as non-governmental groups such as hospitals, pharmaceutical firms, institutional and industrial research and development groups moved into the nuclear field and began to generate significant amounts of radioactive wastes. This resulted in an increased variety of waste materials and heightened the need for adequate disposal management.

Prior to 1960 the waste generated both at AEC facilities and in commercial

---

*The date of this memorandum appears to have been a "turn-of-the-year" typographical error. It seems probable that the correct date of the document is January 5, 1944, because the laboratory did not become operational until the late fall of 1943.

activities were buried at AEC sites. When it was evident that commercial low-level wastes would be generated in significant quantities, the AEC announced in 1960 that its land burial sites at the Idaho National Engineering Laboratory and the Oak Ridge National Laboratory would be used to dispose of low-level commercial wastes from AEC licensees pending the establishment of commercial waste sites. In 1962, the first commercial site, at Beatty, Nevada, was licensed by AEC; it is now licensed by the State of Nevada as an Agreement State. Five additional commercial sites were licensed over the next nine years: Maxey Flats, KY (1963); West Valley, NY, (1963); Richland, WA (1965); Sheffield, IL (1967); and Barnwell, SC (1971). AEC ceased its interim burial operations for commercial waste in 1963.

Commercial wastes result from nuclear power plants and their associated uranium conversion and fuel fabrication activities, the industrial use of radioisotopes and radiation sources, radiopharmaceutical manufacture, hospitals and medical schools, universities, and government and private research and development organizations.[6] Solid radioactive wastes arise at all operations in the nuclear fuel cycle and may contain a wide variety of materials, such as ion exchange resins, sludges from treatment of liquid wastes, and contaminated equipment and clothing.

Defense wastes are produced at facilities associated with the production of nuclear weapons and the research and development programs that support this effort. Defense waste is mainly generated and disposed at U.S. Department of Energy (DOE) facilities, but some low-level waste generated by non-DOE organizations (such as the Navy) is disposed at commercial burial grounds. There are six "principal" DOE burial sites: Los Alamos National Laboratory, Idaho National Engineering Laboratory, Nevada Test Site, Oak Ridge National Laboratory, Hanford, and the Savannah River Plant. Other smaller burial grounds exist but dispose of less than 10% of the DOE low-level waste.

During 1979 a volume of 8.0×10^4 m^3 of low-level solid waste containing 4.8×10^5 curies of radioactivity was buried at commercial burial grounds.[7] Approximately 86 percent was generated in states east of the Mississippi River and 14 percent in states to the west, indicating that nuclear-related activities in the eastern portion of the country are sources of most of the low-level waste.[7] Volumes disposed at commercial facilities increased to 10.7×10^4 m^3 in 1980 and decreased to 8.8×10^4 m^3 and 7.6×10^4 m^3 in 1981 and 1982, respectively.[8] Based on 1981 volumes, roughly 54 percent of the waste comes from power reactors, 44 percent from industrial and institutional sources, and 2 percent from government and military activities.[8] Projections based on growth in the nuclear power industry estimate 8-12 million cubic meters of waste will be buried by the year 2000[9] compared to the previous generation of roughly 2 million cubic meters.[10]

SHALLOW LAND BURIAL DESIGN

A near-surface disposal facility includes all of the land and buildings necessary to dispose of radioactive waste.[6] Disposal units have historically been earthen trenches, but a number of disposal unit designs similar to those used for sanitary landfills are possible.

Site design features and operating practices are selected to provide long-term isolation of the waste and avoid the need for continuing active maintenance of the site after its closure. Thus, the disposal site design should complement and improve the natural site features, where appropriate. It is especially important that the design minimize the potential for contact of water with the waste during storage to ensure long-term stability of the site. This is achieved by directing incident rainfall and surface runoff away from the disposal units, but with gradients and resultant velocities which will not result in excessive rates of erosion.

The disposal area should be laid out to permit efficient land utilization. Factors such as shape, size and orientation of disposal units; buffer zone; topography; and variations in soil types should be considered in laying out the disposal area.

A disposal facility is divided into two basic areas: a "restricted area" and an "administrative area." Access to the restricted area is controlled to protect individuals from exposure to radiation and radioactive materials and includes a "disposal area," in which disposal of radioactive waste takes place, as well as an "operational area" and a buffer zone between the disposal units and the restricted area fence. The operational area is used as a borrow area, for cask storage, and for other miscellaneous functions, such as a decontamination facility and a garage. The administration area is not controlled with respect to radiation protection. It includes support facilities, such as office space and parking areas.

Considerably less than the total site area is used for waste disposal. For example, specific areas of a particular disposal site may not be suitable for waste disposal due to geohydrological or topographical reasons, e.g., parts of a particular site might have excessively steep slopes or high water tables.

There must be a suitable buffer zone between the trenches and the site boundary to provide an area where monitoring can be conducted to detect radionuclide migration from the units and to allow for corrective actions to intercept such migration, if it occurs. Waste disposal is not allowed in the buffer zone, but this area may be used for office space, temporary waste storage or other uses, as long as they do not interfere with monitoring.

The buffer zone should be large enough to insure that any radionuclides released from the buried waste will not migrate outside the site boundary in excess of acceptable limits. Thus, the size of the buffer zone will depend on specific site conditions. However, under any conditions the minimum buffer distance should be 30 m and it should be wider in the direction of ground water flow. There should also be a buffer zone beneath the disposed waste.

Disposal units should generally be oriented with their longitudinal axes parallel to topographic contours of the site and there should be no significant differences in elevation between sidewalls of a disposal unit. In addition, the elevation difference of the ground surface between one end of a disposal unit and the other end should be less than the combined thickness of the backfill overlying the waste and the disposal unit cover. This will minimize the probability that erosion will expose waste and that water will collect in one end of a disposal unit and then flow out the top of the unit.

Spacing between disposal units at the ground surface should be sufficient to assure disposal unit integrity and provide for appropriate water drainage systems. The distance between units should be great enough so that equipment used at a newly excavated unit will not adversely affect the stability of the unit's sidewalls and will not disturb the closure and stabilization of a completed unit.

There is no unique optimum disposal unit size. The length and width of trenches are not controlled by federal regulations, but depend upon the physical size and topography of the disposal site, specific characteristics of soil and location of ground water table, the type and volume of waste, and the dimensions of the waste containers to be buried. The need for equipment access and maneuvering space, and surface water drainage before, during, and after waste emplacement also influence the size of the disposal unit.

The amount of waste to be disposed can be used to estimate the land requirements for the facility for a given time period. The depth of trenches or disposal units is site-specific and depends primarily upon the depth to the ground water table and stability of the sidewalls. The bottom of the trench should be well above the ground water table and slope to allow leachate to flow to a collection and monitoring sump. The bottom of the trench should be well above maximum seasonal fluctuations in the water table and the zone of capillary rise. Waste does not have to be disposed above the water table if radionuclide movement is controlled by molecular diffusion.

If trench depth is controlled by distance above the maximum level of the water table, greater depths can be obtained in the arid regions of the United States. The depth of trenches at various low-level waste disposal facilities in the United States range in depth from 8 to 16 m at arid sites, while trenches in more humid regions are generally less than 8 m deep.[11]

Stability of the sidewalls, and thus the safety of the operating crew, may limit the depth of trenches in arid regions. In certain soils vertical trench walls are not self supporting, so the sidewalls of the trench must be excavated at an angle to prevent sliding of the sidewall. If the sidewalls must be sloped for stability, deep trenches will be much wider at the top than at the bottom. In some cases the depth may be limited by the depth to which available equipment can operate cost effectively.

The sidewalls of the disposal unit should be stable during the period of waste emplacement and closure as well as over the long term. At sites with soils which require low slope angles, reinforcing or lining materials could be added if they are temporary or will not adversely effect the long-term stability of the site.

Once the designer has determined the depth of the trench for a given facility and the sidewall slope required for the trench, he can then estimate the filling time for a given length of trench and select the appropriate trench width. If trenches are excavated by an outside contractor, a wide trench is advantageous since the contractor would only have to be at the site at infrequent intervals; if the operator excavates the trenches himself, he may select a narrower trench so that a new trench can be excavated at the same time the open trench is being filled.

Trench length is mainly limited by the dimensions of the site after allowance is made for the site buffer zone. In areas of high rainfall, precipitation may collect in long, wide trenches prior to covering. This can be handled by design of the leachate collection system and should not be a major limitation. In areas with complex geology or where the surface topography is not flat, it may be better to have relatively short and narrow trenches so as to not intercept ground water and still make maximum use of trench depth.

Erosion of the site by wind and water should be minimized. Water is likely to be the primary cause of erosion in humid regions of the United States, whereas wind erosion is the major concern in arid sites. Some minor site grading may be required on portions of the site to divert surface water runoff, redirect previous drainage patterns on the site, and grade areas where roads, parking areas, and buildings will be located.

Water is a major transport vehicle for both erosion and subsurface radionuclide migration. Thus, the design of the burial facility must provide for removal of runoff resulting from precipitation and take into account the potential for ground water flow into and out of the trenches.

Water can enter the trench by infiltration of surface water through the cover material over the buried waste or by lateral intrusion of ground water. Although, the bottom of the trench should be above the water table, there may be localized areas where ground water can enter the trench. Since transport of contaminants can only occur if there is a moving medium, minimizing the entrance of water into and out of the trench is a primary environmental control. Infiltration cannot be eliminated, but it can be reduced by diverting surface water from the trenches with dikes, dams, diversion ditches, or regrading the area to allow better drainage, capping the trenches with impermeable covers, or by installing liners to prevent leachate from exiting the trench. Any or all of these actions can be used at a given site, or in specific areas of the site. The degree of control provided and its complexity depends on the local geology and hydrology and the nature of the waste material.

In areas where ground water flow may be a problem, some form of ground water interception system should be provided. Similar options can be used for ground water as for surface water; underdrains instead of ditches, grout curtain wall instead of dams. The direction and amount of ground water flow determines if, and by what means, it can be diverted. If drain tiles are used they should be located below the bottom of the landfill or the lowest depth at which significant ground water flow occurs. They should drain by gravity flow to a collecting sump where the water can be pumped, or to some point at which it can be released to surface streams. Although this type of system should be installed between the direction of flow and the trench, and should not intercept contaminated water, the collected water should be monitored.

Grout curtain walls have been used in construction where stability of the excavation and ground water have been problems. A thin trench is excavated and filled with grout or other materials impervious to water to create a diversion wall. The lower portion of the wall contains a permeable material to provide an underdrain system and remove the water.

Another alternative for reducing ground water intrusion is to lower the water table by pumping. This is a normal practice for construction in areas with a high water table, but is expensive and the water table will return to its normal elevation when pumping ceases.

Low permeability liners can be placed on the sides and bottom of the trench prior to waste placement to prevent or limit ground water access. Liner materials can be local soils mixed with bentonite, clays, asphalts, asphaltic concrete, or synthetic polymers. The lower 30 cm of the trench should be covered with sand or small gravel to intercept water and allow it to flow to a sump for removal. Installation of monitoring wells and sumps will then allow the presence of water to be detected and removed as necessary.

Facilities in arid climates generally do not need such an elaborate surface water management system. However, even at arid sites cloudburst storms of short duration and high intensity can result in local flooding and erosion of a site.[11] Therefore, the surface drainage system for an arid site could consider the effects of local flooding on disposal units (including debris flow).

In designing a surface water management system and in post-closure site grading, slopes in the disposal site should be selected so water runoff quantities and velocities will not cause significant erosion.

Large-scale engineering modifications of the upstream drainage area or near-site surface water system are not anticipated, but some modification of surface water flow may be feasible. Culverts or pipes can be used to divert surface water, but they are not permanent because of the limited life of most culvert materials and because the pipes may become clogged with debris. Modifications to the land surface may be acceptable if the area is properly revegetated or stabilized to prevent slope failures or excessive erosion. Periodic inspection of modified areas is necessary to ensure that the stabilization efforts had been successful.

Cover is placed on the waste as the trench is being filled or after filling is completed to serve as a barrier against surface water entering the trench. The cover also provides physical isolation of the waste and radiation shielding. Some burial ground operators use temporary covers over the trench during filling operations to allow operations in wet weather and to exclude rainfall from the trench prior to covering with soil. Local soils should be used whenever possible, but it may be necessary to use materials other than those obtained during trench excavation for cover material.[12] The key is to produce a final cover having low permeability, reasonably good strength, and a slope (2-10 percent) sufficient to prevent water accumulation on the trench.

The cover should be mounded to facilitate drainage and should be several feet thick at its thinnest point. A thick cover reduces the probability of the waste being exposed in future years by erosion or inadvertent excavation. Man-made cover materials such as concrete, soil-cement, asphalt, geotextiles and geomembranes can be used as cover materials, but these materials may not be feasible due to cost or insufficient design life. However, if the disposal site is designed in anticipation of the eventual breakdown of these materials, they can be used beneficially.

The cover should extend beyond the side walls of the unit and be integrated with the surface drainage system at the original or modified grade to assure surface runoff is not directed along the sidewalls down into the trench. Finally, the cover should be stabilized to assure that it is not significantly affected by wind or water erosion. In humid or moderate climates, covers can be stabilized by planting a vegetative cover, in which case the upper 15 cm of the cover material should be topsoil capable of supporting vegetation. The ground cover should be shallow-rooted and indigenous to the area. Ground cover also intercepts rainfall and transpires much of it to the atmosphere before it can infiltrate the cover material. In arid areas where it is difficult to maintain a vegetative cover, a layer of gravel or rock armoring may provide stability.

The site should be surrounded by a well constructed, well maintained security barrier. In most cases this is a chain link (cyclone) type fence at least 1.8 m, with barbed wire on the top to restrict access to the site by unauthorized personnel. In certain cases natural barriers such as cliffs or other geographic features can be used instead of a fence.

OPERATING PRACTICES

The design of a solid waste disposal facility and operating procedures are closely interrelated. For example, the waste form (packaged or loose) and the rate at which wastes are received dictate methods of initial handling, waste placement in trenches, and types of operating equipment needed. Trench design influences the frequency of cover and revegetation as well as documentation of waste location and environmental monitoring.

When waste is received at the burial site it is unloaded from the transporting vehicle and transferred to the trench or to an interim holding or storage area. If the disposal site operator specifies certain packaging requirements (size, shape, type of package, etc.), automatic handling equipment can be used for the transfer operations. (U.S. Department of Transportation (DOT) shipping regulations cover packaging requirements for interstate shipments and U.S. Nuclear Regulatory Commission (NRC) regulations also apply to shipments of radioactive materials; however, container size and shape are not specified).

The principal method of transporting low-level radioactive waste to the disposal site is by truck. Shipments may also be made via railroad, boat, or aircraft, but these are less convenient and seldom used. Trucks can pick up waste from, and deliver waste to, any part of the country with no enroute transfer of waste. Trucks also can be conveniently driven to the edge of the disposal trench where the waste can be removed and placed directly in the trench with a minimum of handling.

Waste receipts should be weighed prior to being placed in the trench to verify shipping papers and be certain that all of the waste shipped is actually received. In addition, each truck entering the disposal facility should be monitored for radioactivity to verify that the waste has been shipped in compliance with regulations of DOT (49 CFR 173), and NRC (10 CFR 71) and any applicable state regulations governing packaging and transportation of low-level wastes. Radiation monitoring will also identify any waste packages which may have been damaged during transport, or which for some other reason may require special handling throughout the disposal operation.

At times it may be impossible to bury waste at the normal or projected rate due to poor weather conditions (snow, ice, or heavy rains), heavy equipment malfunctions. labor force strikes, or accidents serious enough to close down an open trench. During such times, if waste is being received at the disposal site as scheduled, the waste must be stored temporarily. Waste packages can be left on the trucks for short periods of time until normal operations resume. However, if the delay in disposal operations lasts longer than a few days, interim storage will be required.

Waste containers in storage at a disposal site should be protected from water. One way is to provide a shelter, which could be as simple as placing tarpaulins over shipments for temporary protection or as permanent as creating a roof or other shelter over the storage area. Waste in storage should also be protected from surface runoff. The storage area should be graded to feed into the surface water management system. The waste containers can be placed on frames or platforms above the ground surface.

Site disposal operations should be suspended when water is visibly present in disposal units. Active disposal units should be allowed to drain into the sump or the inactive end of the unit and the collected water pumped out. When significant amounts of snow or ice accumulate in active units, they should be removed from the immediate disposal area before operations are resumed.

Procedures used to place waste packages in the trench will depend upon several factors including: (1) size and weight of the individual package, (2) radioactivity at the surface of the package, and (3) whether the disposal license specifically states that the waste packages should be "stacked" in the trench and prohibits "dumping" from the sides of the trench. Stacking waste packages in the trench makes it easier to fill void volume between packages, is less likely to puncture packages, and looks neater; however, it may increase personnel exposure due to the increased handling of waste packages. Dumping waste packages from the side of the trench is less time consuming and results in less personnel exposure, but requires more trench volume and is more likely to break open packaging material.

The off-loading crew has two options for transferring the waste from the truck to the trench; manual transfer where each package is lifted and placed in the trench by a man, and a mechanical method where heavy equipment (e.g., cranes or fork lifts) is used to unload the waste packages. Currently, large waste packages are mechanically transferred at commercial as well as government operated disposal sites. When packages arrive at the site in a condition to go directly into the trench (not packaged in reusable shielding containers), a fork lift or a crane is used to remove the packages from the truck and drive or lower them into the trench.

Waste placement should begin at the upslope end of the trench and progress towards the downslope end to minimize the chance of waste packages coming into contact with water should the trench drainage system fail or heavy rainfall occur while the trench is open and the waste is exposed. Using this procedure, along with stacking waste to the maximum height, means that a wall or front of waste packages will progress in the downslope direction until the trench is completely filled. If the trench is long, 150 to 300 m, waste can be covered continuously just behind the active waste placement zone. In the case of short trenches, where the length is approximately equal to the width, the trench is filled from the bottom up to the maximum allowable height and covering takes place when the entire trench is full.

Waste should be placed and covered in a manner which limits the gamma radiation at the surface of the cover to levels that are within a few percent of the natural background levels of the site. One option is to fill a disposal unit with waste to within one meter of the original grade, add compacted earth up to the original grade, and follow with a cap or cover about 1 m in thickness. This is suitable for most waste received at a typical near-surface disposal facility. Further shielding can be achieved by placing higher activity waste at the bottom of the disposal unit with lower activity waste above it or by using thicker trench caps. Techniques designed to protect against the inadvertent intruder, such as the use of caissons, concrete-walled disposal units, and grouting, will also aid in reducing surface radiation levels.

Regardless of how carefully the waste packages are placed in the trench, there will be a considerable amount of void space between packages. Void space, along with degradation of organic material, is the primary cause of cover subsidence in waste disposal trenches. For this reason the NRC has proposed that void spaces between waste packages be filled with earth or other material. A bulldozer is used to push the previously excavated and stockpiled soil over the top of the waste packages and a vibrating compactor or other specialized piece of equipment, such as a bulldozer or roller, is used to move soil into void spaces. After the void spaces are filled, additional fill material should be added and compacted to a height which will allow room for a clay cap and a final layer of topsoil.

A freely-draining, non-cohesive material, such as a clean sand or gravel, should be used to fill the spaces between waste containers as it can more readily fill void spaces and achieve a higher relative density than fine-grained material. These types of materials also promote rapid movement of water through the disposal unit and help minimize the length of time in which water would be available to leach the waste. In addition, if the backfill has a sufficient contrast in permeability to the material in the trench cap, capillary forces may promote unsaturated flow of interstitial water around the disposal unit instead of through it.

Alternatively, materials with extremely low permeability, such as grout or concrete, can be used instead of a free-draining backfill. Clay-soil mixtures may not be suitable for backfill because of the difficulty in filling void spaces and in achieving sufficient compaction to limit consolidation and permeability to acceptable levels.

Backfill should be added over and between the waste canisters as each layer of waste is placed. When an entire disposal unit is completed, a cover should be placed over the unit, shaped to facilitate drainage and stabilized. This would include grading and shaping the surface and, in humid areas, planting a shallow-rooted vegetative cover. Rip-rap or similar materials can be used on steep slopes to protect against wind and water erosion. In arid regions, it is difficult to maintain vegetative cover over the long-term and gravel or cobbles can be used to stabilize the disposal unit cover. Finally, the completed and covered disposal unit should be made part of the surface water management system for the overall site so that drainage off the cover is not allowed to form ponds in the disposal area, but is removed rapidly for discharge off-site.

The trenches should be closed in a way that leaves the site in as near a maintenance-free condition as possible. This will help ensure that after institutional control has ended, for example in 100 years, barriers (either natural or man made) will protect the inadvertent intruder from excessive exposure. The trench covers should be inspected regularly and revegetated if necessary.

The planned sequence of use for disposal units over the lifetime of the site should reflect the need to close and stabilize each unit as it is filled. The location of roads and disposal unit covers, use of heavy equipment, establishment of vegetative cover, and management of surface water should be planned so operations may be conducted at each disposal unit without damage to closed disposal units. Location and access to fill and borrow areas should also be planned to assure that they do not compromise the integrity of completed disposal units.

It is important to document waste locations in order to be able to retrieve waste packages, if required or requested. This can be accomplished by numbering each trench and dividing it into a three-dimensional grid consisting of length, width, and depth. As packages are buried, the off-loading crew records the grid location of each shipment relative to a selected datum, such as a permanent corner marker of the trench. This data is then entered into a permanent file itemizing trench contents.

Following the filling and closure of each trench, a permanent marker should be placed at the end of the trench to indicate the location of the trench, the dates the trench was used, the quantity of waste contained in the trench, the activity level and type of waste in the trench, and any other information required by federal and state regulations. The exact location of this monument and its relationship to the four corners of the trench should be determined and noted with permanent survey marker control points, referenced to United States Geological Survey or National Geodetic Survey survey control stations.

PROBLEMS ENCOUNTERED IN SHALLOW LAND BURIAL

Water Management

Several of the more serious technical problems in shallow land burial are related to water management. Water causes erosion which can reduce the structural integrity of the waste trenches and the trench caps and provide the vehicle for migration of radionuclides. Water has come into contact with waste in burial trenches at a number of facilities largely due to infiltration of precipitation through the trench backfill.[4,13-34]

Backfill used to cover the trenches has a greater permeability than the undisturbed portions of the formation. This results in an increased rate of infiltration of precipitation into the trench without a corresponding increase in the rate of movement away from the trench. Consequently, in humid regions, water levels in the trenches may rise and come into contact with the waste. Water continues to accumulate until the trench overflows at its low point or until the water level rises to a zone of high permeability, such as the interface between the backfill and the original ground surface. This effect is known as the "bathtub effect" and is most pronounced in formations of low native permeability.

At ORNL surface runoff and ground water flow from areas of higher elevation augment infiltration. Construction activities adjacent to the burial grounds have also altered hydraulic flow patterns. In addition, some early trenches were oriented with the long axis parallel to the fall line and with the bottom of the trench cut below the natural water table at the downslope end.[11]

At INEL the burial grounds are located in a natural topographic depression below the level of the channel of the Big Lost River. The burial ground area has been flooded in periods when rapid snowmelt combined wih high rates of precipitation.[32,33,35] Dikes and drainage ditches have been improved to reduce the problem.[33]

At a number of sites the waste material and backfill have compacted and sagged nonuniformly after waste was emplaced in trenches, creating breaches in the cover and zones where water can accumulate at the surface. This increases infiltration of water through the trench cover and can lead to exposure of the waste.

Trench cap erosion must be controlled to minimize contact between water and the buried waste. Practices for control of infiltration at a burial site may be at cross purposes with control of erosion and the appropriate balance will likely vary from site to site, reflecting differences in hydrology, climate, topography, soil mechanics, geology, etc. Therefore, water management programs must be carefully selected.

Periodic maintenance may be required to repair eroded trench caps. At locations such as ORNL, Maxey Flats, and West Valley where the climate is humid and the permeability of the backfill is greater than that of the undisturbed formation, periodic sealing may be required to limit infiltration of water. Both infiltration and erosion can be controlled by water management practices. Erosion control practices include a good vegetative cover, a permeable cap, and shallow slope; measures for infiltration control include limiting vegetation, an impervious cap, and a high slope to promote rapid drainage.

Unusual weather conditions must be considered, particularly in locales where they are dramatically different from the normal. In these areas management policy may need to be weighted toward a concern for abnormal weather events, such as flash floods or high winds.

Radionuclide Migration

In general, waste packages are designed to provide containment of waste during transport to the burial site and during emplacement, but not for long term containment after disposal. After wastes have been emplaced in trenches, the geologic formation is the primary barrier to radionuclide migration. Water is a major vehicle for radionuclide transport and the depth to the water table and the direction, flow paths, rate of movement, and dispersion of water coupled with the mechanisms and degree of interactions of specific radionuclides with the formation must be known to predict radionuclide migration. In humid regions, particularly in areas where the permeability of surface formations is low, the ground water lies at a short distance below the ground surface and the stream density is high.[36] In such cases the primary barrier to radionuclide migration may be the adsorptive property of the soil.[17] Radionuclides follow the general paths of movement of the transporting fluid but normally move at a slower rate due to interactions with the formation through which they are moving.

In the absence of moving ground water the rates of movement of nongaseous radionuclides by molecular or ionic diffusion are very small.

The major interaction between radionuclides and the formation is ion exchange. Other interactions, such as chemical precipitation and filtration, mineral replacement, surface exchange reactions, and physical adsorption may predominate but they are not always well understood.[15,18] In some cases even the physical characteristics and the direction of movement is not clearly defined.[20,23,26,37] Furthermore, the presence of fissures in the geologic unit may short-circuit opportunities for ion exchange and other retardation mechanisms for radionuclides mobilized during leaching by infiltrating water.

Normally, one assumes that the solids of the geologic strata and that fraction of the radionuclides adsorbed by them remain stationary over the time of concern. However, fine particulates can move through porous media and appreciable movement can occur if the formation contains channels.[38,39] For example, studies at ANL indicate that although the bulk of Pu and Am are very tenaciously retained by tuff, basalt, or limestone, a fraction of the Pu migrated at a much more rapid rate; it was conjectured that the rapidly migrating form was a polymer.[38] Over very long time periods, say several millenia, even the solid matrix can move significantly because of erosion or tectonic events.

Transport in the vapor phase may be important for some radioactive compounds with high vapor pressures, such as tritiated water, tritium gas, CO_2, noble gases, and short chain organic compounds. Diffusion coefficients for gaseous molecules in air are on the order of 0.1 to 0.5 cm^2/sec. Consequently, gaseous radionuclides may move appreciably due to molecular diffusion. Release of gaseous radionuclides through trench caps has been reported at West Valley.[24,28] Tritium is an important radionuclide in wastes, and studies are underway to determine its rate of release to both air and ground water.[40]

Another source of atmospheric transport exists at Maxey Flats and West Valley where water which accumulated in the burial trenches has been removed and evaporated.[20,21,22]

The concentrations of radionuclides migrating from shallow land-burial facilities are usually low and the rates of movement slow, making it difficult to obtain an accurate early depiction of radionuclide migration in the field.

Biological Intrusion and Translocation

Vegetation, particularly deep rooted plants, growing on the surface of covered burial trenches can accumulate radionuclides and provide a pathway for translocation of radionuclides from waste trenches to the uncontrolled environment.[4,16,30,41-46] Significant translocation of radionuclides by deep-rooted plants has occurred at Hanford[42] and specifically for plutonium at LASL.[43] As a result, efforts are generally taken to inhibit the growth of deep-rooted plants near the trenches.

Small animals may intrude into buried wastes and translocate radionuclides.[42] At LASL elevated levels of tritium have been found in honeybees.[34,47] The source of tritium is unknown, but it has been postulated that bees collect pollen from white clover which is abundant in a solid-waste burial area.

Space Utilization

Land burial of solid radioactive waste began in an era when the operations were expected to be temporary. The operations were conducted on extensive reservations so little concern was given to minimizing use of land resources. Current methods used for burial of low-level, solid radioactive wastes result in the use of large areas of land per unit of waste, and significant improvements in space utilization are possible. To date it has not been cost-effective to implement burial ground conservation measures as there are large reserves of land available for shallow land burial, particularly at most of the arid western sites.

Compaction could substantially reduce waste volume at a number of sites.[42,48,49] Incineration could also reduce waste volume but may lead to atmospheric releases of radionuclides. Other means for volume reduction include segregation and separate handling of non-radioactive wastes.

Standardized packaging and better placement of packages in trenches would decrease the void volume in trenches and improve space use.[9,48,49] Standardized packaging would also improve shielding and package handling for better safety to operators during waste receipt and emplacement at the burial ground, but might require additional handling at the waste source, especially if bulky items have to be dissembled to fit into standard packages.

At arid sites, such as Beatty, Richland, and Hanford, the depth of the trenches could be increased appreciably without coming undesirably close to the water table.[9,48] This could greatly increase the capacity of burial per unit land area. However, at the Hanford Reservation the near surface earth material has a shallow angle of repose which could limit the cost-effectiveness of deeper trenches at that site.[42] Increased width of trenches and closer attention to trench configuration could also provide improvements in land use.[9,48] A recent study has noted that improvements underway at INEL will at least double recent utilization.[48]

Management Practices

At sites where there has been no detectable migration of radionuclides, the problems that have occurred have resulted largely from lapses in good management practice. Codes of good practice would facilitate quality control for management practices at operating facilities. Such codes of practice, supplemented by performance standards and measures of performance, would be useful for groups wishing to make independent evaluations of the performance of burial ground operations.

REFERENCES

1. Duguid, J.O. and M. Reeves. July 1976. "A Comparison of Mass Transport Using Average and Transient Rainfall Boundary Conditions." Oak Ridge National Laboratory. Conference held in Princeton, NJ.

2. Browder, F.N. 1959. "Radioactive Waste Management at Oak Ridge National Laboratory," Oak Ridge National Laboratory, ORNL-2601.

3. Abee, H.H. 1955. "Problems in the Burial of Solid Wastes at Oak Ridge National Laboratory." A Seminar sponsored by the AEC and the Public Health Service, held at the Robert A. Taft Engineering Center, Cincinnati, Oh, December 6-9, 1955.

4. Webster, D.A. 1979. Land burial of solid radioactive waste at Oak Ridge National Laboratory, Tennessee: A case history. *In: Management of Low-Level Radioactive Wastes, Volume 2.* M.W. Carter, A.A. Moghissi, and B. Kahn, Eds. Pergamon Press, NY, 731-746.

5. Webster, D.A. 1976. "A Review of Hydrologic and Geologic Conditions Related to the Radioactive Solid-Waste Burial Grounds at Oak Ridge National Laboratory, Tennessee." U.S. Geological Survey. Open File Report 76-727.

6. U.S. Nuclear Regulatory Commission. September 1981. "Draft Environmental Impact Statement on 10 CFR Part 61 'Licensing Requirements for Land Disposal of Radioactive Waste.' " NUREG-0782.

7. Guilbeault, B.D. 1980. "The 1979 State-by-State Assessment of Low-Level Radioactive Waste Shipped to Commercial Burial Grounds." NUS Corporation. NUS-3440 (Rev. 1).

8. U.S. Department of Energy. 1983. "Low-Level Radioactive Waste Management Handbook Series, An Introduction, "National Low-Level Radioactive Waste Management Program. DOE/LLW-13Ta.

9. Mullarkey T.B., T.L. Jentz, J.M. Connelly, and M.P. Kane. October 1976. "A Survey and Evaluation of Handling and Disposing of Solid Low-Level Nuclear Fuel Cycle Wastes." Atomic Industrial Forum, Inc. National Environmental Studies Project. NUS Corporation.

10. *Report to the President by the Interagency Review Group on Nuclear Waste Management.* March 1979. TID-29442.

11. Jacobs, D.G., Epler, J.S., and Rose, R.R. March 1980. "Identification of Technical Problems Encountered in the Shallow Land Burial of Low-Level Radioactive Wastes." ORNL/SUB-80/13619/1.

12. U.S. Environmental Protection Agency. 1978. "Study of Engineering and Water Management Practices that will Minimize the Infiltration of Precipitation into Trenches Containing Radioactive Waste." SCS Engineers, Long Beach, CA. ORP LV-78-5.

13. National Academy of Sciences. 1976. "The Shallow Land Burial of Low-Level Radioactively Contaminated Solid Waste." Panel on Land Burial,

Committee on Radioactive Waste Management, National Research Council. TID-27341.

14. Duguid, J.O. July 1975. "Status Report on Radioactivity Movement from Burial Grounds in Melton and Bethel Valleys." ORNL-5017.

15. Dames and Moore. 1978. "Applicability of a Generic Monitoring Program for Radioactive Waste Burial Grounds at Oak Ridge National Laboratory and Idaho National Engineering Laboratory."

16. U.S. Energy Research and Development Administration. May 1976. "Alternatives for Managing Wastes from Reactors and Post-Fission Operations in the LWR Fuel Cycle." Volume 4. ERDA-76-43.

17. U.S. Energy Research and Development Administration. July 1976. "Development of Monitoring Programs for ERDA Owned Radioactive Low-Level Waste Burial Site." E(49-1)-3759.

18. Duguid, J.O. 1979. Hydrologic transport of radionuclides from low-level waste burial grounds. *In: Management of Low-Level Radioactive Waste, Volume 2.* M.W. Carter, A.A. Moghissi, and B. Kahn, Eds. Pergamon Press, NY, 1119-1137.

19. Committee on Government Operations. 1976. "Hearings before a Sub-Committee of the Committee on Government Operations—House of Representatives, Ninety-Fourth Congress, Second Session."

20. Meyer, G.L. 1976. "Preliminary Data on the Occurrence of Transuranium Nuclides in the Environment at the Radioactive Waste Burial Site Maxey Flats, KY." USEPA-520/5-76/020.

21. Montgomery, D.M., H.E. Kolde, and R.L. Blanchard. January 1977. "Radiological Measurements at the Maxey Flats Radioactive Waste Burial Site—1974-1975." EPA520/5-76/020.

22. Hardin, C.M. 1979. Operational experience of the Kentucky radioactive waste disposal site. *In: Management of Low-Level Radioactive Waste, Volume 2.* M.W. Carter, A.A. Moghissi, and B. Kahn, Eds. Pergamon Press, NY, 831-835.

23. U.S. Nuclear Regulatory Commission, and U.S. Energy Research and Development Administration. January 12, 1976. "Improvements Needed in the Land Disposal of Radioactive Wastes—A Problem of Centuries." Report to the Congress by the Comptroller General of the United States. RED-76-54.

24. Davis, J.F., R.J. Digman, R.H. Monheimer, and J.M. Matuszek, Jr. 1976. "Evaluation of Radionuclide Pathways at a Shallow, Low-Level Radioactive Waste Burial Site in Western New York." Presented at the National Meeting of the Geological Society of America, 10 November 1976, by J.W. Pferd.

25. U.S. Nuclear Regulatory Commission. March 1977. "NRC Task Force Report on Review of the Federal/State Program for Regulation of Commercial Low-Level Radioactive Waste Burial Grounds." NUREG-0217.

26. Kelleher, W.J. 1979. Water problems at the West Valley burial site. *In: Management of Low-Level Radioactive Waste, Volume 2.* M.W. Carter, A.A. Moghissi, and B. Kahn, Eds. Pergamon Press, NY, 843-851.

27. Prudic, D.E. and A.I. Randall. 1979. Groundwater hydrology and subsurface migration of radioisotopes at a solid radioactive waste disposal site, West Valley, New York. *In: Management of Low-Level Radioactive Waste, Volume 2.* M.W. Carter, A.A. Moghissi, and B. Hahn, Eds. Pergamon Press, NY, 853-882.

28. Matuszek, J.M., L. Husain, J.F. Davis, R.H. Fakundiny, R. Monheimer, J. Pferd, R. Digman, and A.H. Lu. Application of radionuclide pathway studies to short-and long-term management insights for shallow low-level radioactive waste burial facilities. *In: Management of Low-Level Radioactive Wastes, Volume 2.* M.W. Carter, A.A. Moghissi, and B. Hahn, Eds. Pergamon Press, NY, 901-916.

29. Dames and Moore. 1976. "Interim Status Report. First Stage. Development of Methods to Eliminate the Accumulation and Overflow of Water in the Trenches at the Radioactive Waste Disposal Site West Valley, New York." Prepared for the New York State Energy Research and Development Authority.

30. Horton, J.H. and J.C. Corey. 1976. "Storing Solid Radioactive Wastes at the Savannah River Plant." Savannah River Laboratory, E.I. DuPont de Nemours and Co., Aiken, SC. DP-1366.

31. U.S. Energy Research and Development Administration. October 1976. "Waste Management Operations, Savannah River Plant, Aiken, South Carolina." ERDA-1537.

32. Barraclough, J.T., J.B. Robertson, and V.J. Janzer. 1976. "Hydrology of the Solid Waste Burial Ground, As Related to the Potential Migration of Radionuclides, Idaho National Engineering Laboratory." USGS Open File Report 76-71.

33. Barraclough, J.T., J.B. Robertson, and V.J. Janzer. 1979. Geohydrologic study of a burial site for solid low-level radioactive wastes at the Idaho National Engineering Laboratory. *In: Management of Low-Level Radioactive Waste, Volume 2.* M.W. Carter, A.A. Moghissi, and B. Hahn, Eds. Pergamon Press, NY, 795-824.

34. U.S. Geological Survey. August 1975. "Evaluation of Monitoring of Radioactive Solid-Waste Burial Sites at Los Alamos, New Mexico." Prepared for the U.S. Energy Research and Development Administration. Open File Report 75-046.

35. U.S. Energy Research and Development Administration. 1977. "Waste Management Operations, Idaho National Engineering Laboratory, Idaho, Final Environmental Impact Statement." Washington, DC.

36. Richardson, R.M. 1963. Significance of climate in relation to the disposal of radioactive waste at shallow depth below ground. *In: Proceedings of*

the International Colloquium on Retention and Migration of Radioactive Ions in Soils, Saclay, France, 16-18 October 1962. Presses Universitaires de France, 205-211.

37.   Denham, (Rep.) M.B., M.D., et al. October 1977. "Report of the Special Advisory Committee on Nuclear Waste Disposal." Legislative Research Commission, Frankfort, KY.

38.   Fried, S., A.M. Friedman, J.J. Hines, R.W. Atcher, L.A. Quarterman, and A. Volesky. December 1976. "Annual Report for Fiscal Year 1976 on Project ANO115A: The Migration of Plutonium and Americium in the Lithosphere." ACS Symposium Series No. 35. *Actinides in the Environment.* Proceedings of a Symposium held in New York, NY, 19-46. ANL-76-127.

39.   Champlin, J.B.F. and G.G. Eichholz. February 1976. Fixation and remobilization of trace contaminants in simulated subsurface aquifers. *Health Physics*, 30(2): 215-219.

40.   E.I. DuPont de Nemours and Co. July-September 1974. "Savannah River Laboratory Quarterly Report." Waste Management. DPST-74-125-3.

41.   Adam, J.A. and V.L. Rogers. June 1978. "A Classification System for Radioactive Waste Disposal—What Waste Goes Where?" Ford, Bacon, and Davis, Utah, for the U.S. Nuclear Regulatory Commission. NUREG/-CR/0680.

42.   Geiger, J.F., D.J. Brown, and R.E. Isaacson. August 1977. "Assessment of Hanford Burial Grounds and Interim TRU Storage." RHO-CD-78.

43.   Environmental Group, H-8. 1977. "Development Activities on Shallow Land Disposal of Solid Radioactive Wastes, January-December 1976." Los Alamos Scientific Laboratory. LA-6856-PR.

44.   Cornman, W.R. 1979. Improvement in operating incident experience at the Savannah River burial ground. *In: Management of Low-Level Radioactive Waste, Volume 2.* M.W. Carter, A.A. Moghissi, and B. Hahn, Eds. Pergamon Press, NY. 787-794.

45.   E.I. DuPont de Nemours and Co. April-June 1978. "Savannah River Laboratory Quarterly Report." Waste Management. DPST-78-125-2.

46.   Ashley, C. and C.C. Zeigler. 1976. "Environmental Monitoring at the Savannah River Plant." Health Physics Department, Savannah River Laboratory, E.I. DuPont de Nemours and Co., Aiken, SC. DPSPU 77-302.

47.   Richmond, C.R. and E.M. Sullivan. May 1974. "Annual Report of the Biomedical and Environmental Research Program of the LASL Health Division, January through December 1973." LA-5633-PR, p. 165.

48.   Garrett, P.M. January 26, 1979. "An Evaluation of Low-Level Radioactive Waste Burial Ground Capacities at the Major DOE Reservations." ORNL/NFW-79/17.

49.   Burch, W.D., et al. September 1972. "Waste Management at ORNL: Present Practices—Immediate Needs-The Future." (The Final Report of the Committee on ORNL Waste Handling Practices.) ORNL CF 72 9 1.

Management of Radioactive Materials and Wastes: Issues and Progress. Edited by S. K. Majumdar and E. Willard Miller. © 1985, The Pennsylvania Academy of Science.

Chapter Seven

RADIOACTIVE WASTE MANAGE-MENT—A MANAGEABLE TASK

John B. Yasinsky[1], Ph.D. and Charles R. Bolmgren
Waste Technology Services Division
Advanced Power Systems Divisions
Westinghouse Electric Corporation
P.O. Box 286
Madison, PA 15663

Hazardous wastes are those wastes that exhibit one or more of the following properties: infectivity, ignitability, corrosivity, reactivity, toxicity or radioactivity. It is the last category of hazardous waste—namely radioactive or nuclear waste—that will be the focus of this paper.

Nuclear wastes are a product of the high technology society in which we live. They are the by-products of nuclear medicine, industrial processes, electrical power generation, a variety of research activities, and national defense material production. In addition to those wastes generated by the civilian nuclear power and defense programs in this country, radioactive wastes are derived from such varied sources as smoke detectors, discarded wristwatch dials, industrial testing equipment, and solutions used for medical diagnosis and research. As a matter of fact, of the approximately 700 facilities in Pennsylvania licensed to handle radioactive material and which therefore generate radioactive waste, only 6 are nuclear power reactors. Thus, effective radioactive waste management must adequately treat and safely handle and dispose of radioactive waste from all of these sources. Effective radioactive waste management is not just the responsibility of the Federal government, although the Federal government has the responsibility to establish overall waste management policy and regulations. Effective waste management must be the joint and cooperative responsibility of the generators, packagers, shippers, and disposers of radioactive waste. This includes industry, utilities, private and public institutions, and all levels of government.

Radioactive waste has two attributes that make its safe management simpler than other forms of hazardous waste. First, radioactive nuclides decay with time

---

[1]Formerly General Manager, Advanced Power Systems Divisions.

into stable, nonradioactive elements unlike many other waste materials that remain hazardous forever. Secondly, the volume of radioactive waste, compared to other forms of hazardous waste, is relatively small so that it is feasible to take exceptional measures to process and handle the radioactive waste in such a way as to provide very high assurance that the waste will remain isolated from the environment.

Atoms are radioactive because their nuclei are in an unstable energy state. Radioactive atoms will spontaneously change to a more stable state by the emission of energy in the form of subatomic particles (alphas, betas, or neutrons) or in the form of gamma radiation (similar to X-rays). Some nuclides or atoms decay to a nonradioactive state in a single step while others decay to a nonradioactive nuclide in a series of steps (decay chain). The rate at which radioactive decay takes place is usually specified in terms of halflife; that is, the length of time it takes for one half of a quantity of radioactive atoms to undergo a decay step. The halflives of radioactive atoms, called radionuclides, vary from a fraction of a second to thousands of years, but is a specific value for a particular radionuclide. For example, strontium-90 has a halflife of 27.7 years, americium-241 has a halflife of 458 years, and iodine-129 has a halflife of 17 million years.

The spent fuel from a nuclear power plant contains a wide variety of radionuclides each having its own halflife. The quantities of each of the radionuclides in the spent fuel as well as the decay chains and halflives are well known so it is possible to predict the amount of radioactivity present in the spent fuel with time. This prediction is shown in Figure 1. It can be seen that in the time period from one year to 100 years after the fuel is removed from the reactor, the amount of radioactivity in spent fuel (curve labeled Once-through Cycle Spent Fuel) decreases by about a factor of 50, and will decrease by about a factor of 1,000 after 1,000 years. For the waste left after reprocessing the spent fuel and removing most of the uranium and plutonium (curve labeled U and Pu Recycle High-Level Waste), the amount of radioactivity has decreased by almost a factor of 10,000 after 1,000 years and in fact its relative toxicity is about equivalent to that of uranium ore and is far less than that of such things as mercury, chromium, cadmium, and lead ores. As shown in Figure 1, the quantity of radioactivity is given in terms of curies per metric ton of heavy metal (MTHM) in the spent fuel (i.e. uranium and plutonium). One curie is the amount of radioactive material that results in 3.7×10^{10} radioactive decays per second and corresponds to the emissions from one gram of radium-226.

With regard to quantities of hazardous wastes produced, the U.S. Environmental Protection Agency has identified over 350 substances that are considered hazardous and estimates that about 60 million tons of these wastes are produced each year. Of this, less than 0.01 percent is waste from commercial nuclear power plants. To provide a perspective of the volume of radioactive waste produced, it has been estimated that the total volume of radioactive waste generated since the beginning of the national nuclear program would fill less than one

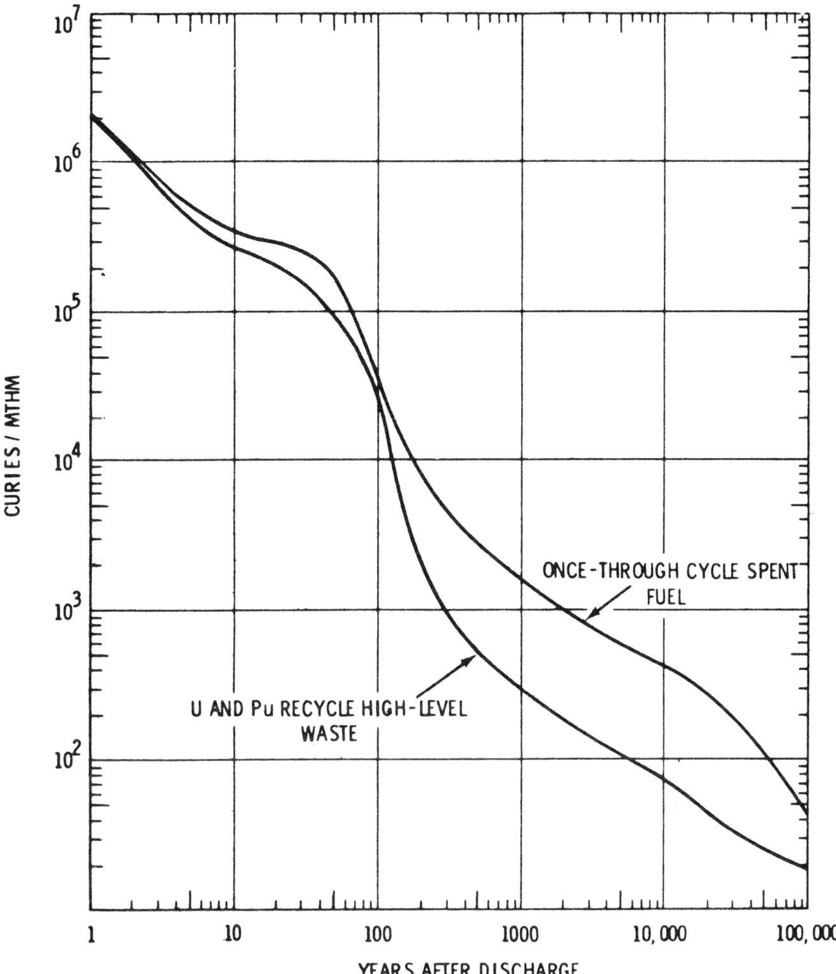

FIGURE 1. Decay of the radioactivity in spent fuel and high level waste with time.

typical football stadium. In contrast, it is estimated that the United States generates nonradioactive hazardous waste at a rate that would fill about three football stadiums *each day*.

To provide a further perspective with respect to volume, consider the spent fuel removed each year from a large nuclear power plant. This amounts to about 30 metric tons of spent fuel per year. If this spent fuel were reprocessed to recover the unused uranium and plutonium and if the remaining radioactive waste products were incorporated into the current reference waste form for commercial high level waste (CHLW), namely borosilicate glass, about 2.5 m³ of glass would result. This would be contained in 13 reference CHLW canisters, each 0.32 m in diameter by 3.0 m long (1.0 ft. in diameter by 10 ft. long) and containing 0.19 m³ of glass. Thus, a nuclear power plant generating electricity for about 400,000 customers for one year would produce a quantity of high level waste, processed into a form suitable for disposal, that occupies a volume of about 0.9 x 1.2 x 3.0 m (3 x 4 x 10 ft.).

Notwithstanding the fact that radioactive waste eventually becomes nonradioactive as discussed earlier, the public and the environment must be protected from the dispersal of radioactivity until such time as the radioactivity decreases to the level at which it is no longer of concern. To date, radioactive wastes have been adequately managed. Low level wastes have been disposed of at shallow land burial sites, high level defense wastes have been stored in tanks, and spent fuel assemblies from civilian power reactors have been stored in water pools at the power plants for many years. In spite of the fact that some of the shallow land burial sites have not proved to be totally successful from the standpoint of long-term, maintenance-free operation and that a few of the defense waste storage tanks developed leaks, remedial actions have prevented any impact on the public.

While safe storage of radioactive waste has been clearly demonstrated and is an important part of the overall waste management system, this demonstration does nothing to instill confidence in the public that safe, final disposal is possible. The fact of the matter is that it is almost the unanimous judgement of the technical and scientific community that the basic technology exists for the safe, long-term isolation of radioactive wastes and there are no insurmountable technological impediments to its implementation. Impetus for the implementation of this technology is provided by the Nuclear Waste Policy Act, passed by the Congress in 1982, and the subsequent formation of the Office of Civilian Radioactive Waste Management.

The following sections of this paper define the categories of radioactive waste and then describe for each category the sources of the wastes, the technologies for safe management, and the current status of the disposal programs for each waste category. The objective of this discussion is to establish the fact that radioactive waste management is indeed a manageable task.

DEFINITIONS OF RADIOACTIVE WASTE CATEGORIES

Currently, radioactive wastes are classified into three categories. These are high level waste, transuranic waste, and low level waste.

High level waste (HLW) is defined specifically as the liquid waste, or the solids into which such liquid wastes have been converted, resulting from the chemical reprocessing of irradiated (spent) reactor fuel.The reprocessing operation extracts about 99.5 percent of the unused uranium and plutonium that was left in the spent fuel, leaving a liquid waste stream containing about 0.5 percent of the uranium and plutonium and greater than 99.9 percent of the other radionuclides that were in the spent fuel. Consequently, the waste stream, which must be solidified for disposal, is highly radioactive and must be kept behind appropriate radiation shielding. If the option is exercised to dispose of spent fuel without reprocessing, then the spent fuel is also classified as HLW.

Transuranic (TRU) wastes are defined as radioactive wastes, other than high level waste, that contain greater than a very small quantity of transuranic elements that emit alpha particles and have a halflife of greater than five years. Transuranic elements are those that have an atomic number greater than that of uranium and include such elements as plutonium, americium, and curium. Many transuranic elements have very long halflives (e.g., plutonium-239 has a halflife of about 24,000 years). Currently, waste is classified as TRU if it contains in excess of 10 nanocuries of transuranic elements per gram of waste (one nanocurie equals one-billionth of a curie). Consideration is being given to increasing this limit to 100 nanocuries per gram. The total radioactivity content of TRU waste can vary over a considerable range. Therefore, TRU waste is further classified as contact-handled TRU, which means the radiation level is low enough to allow safe handling by personnel without shielding, and remote-handled TRU which has a sufficiently high radiation level to require shielding and/or remote handling.

Low level waste (LLW) is basically all other radioactive waste that is not either high level waste or transuranic waste. Because the LLW category is so broad, the radioactivity content can vary over a wide range so that LLW is further classified as contact-handled and remote-handled waste. In practice, LLW tends to contain mostly radionuclides that emit beta particles or gamma radiation and have relatively short halflives (compared to the long-lived transuranic elements).

HIGH LEVEL WASTE

As indicated earlier, high level waste (HLW) is the waste resulting from the reprocessing of irradiated reactor fuel. Because of the decision in 1977 by the Carter administration to prohibit the reprocessing of spent fuel from the nation's commercial nuclear power plants, only a small quantity of commercial high level waste (CHLW) exists in this country from the reprocessing of a small number of fuel assemblies prior to 1972. Reprocessing is currently being carried out in France and the United Kingdom and facilities are being planned in such countries as the Federal Republic of Germany and Japan. Also, the Reagan administration has reversed the ban on commercial reprocessing and is encouraging industry to resume reprocessing in the United States. However, as a result of this history, about 97 percent by volume of the HLW that current-

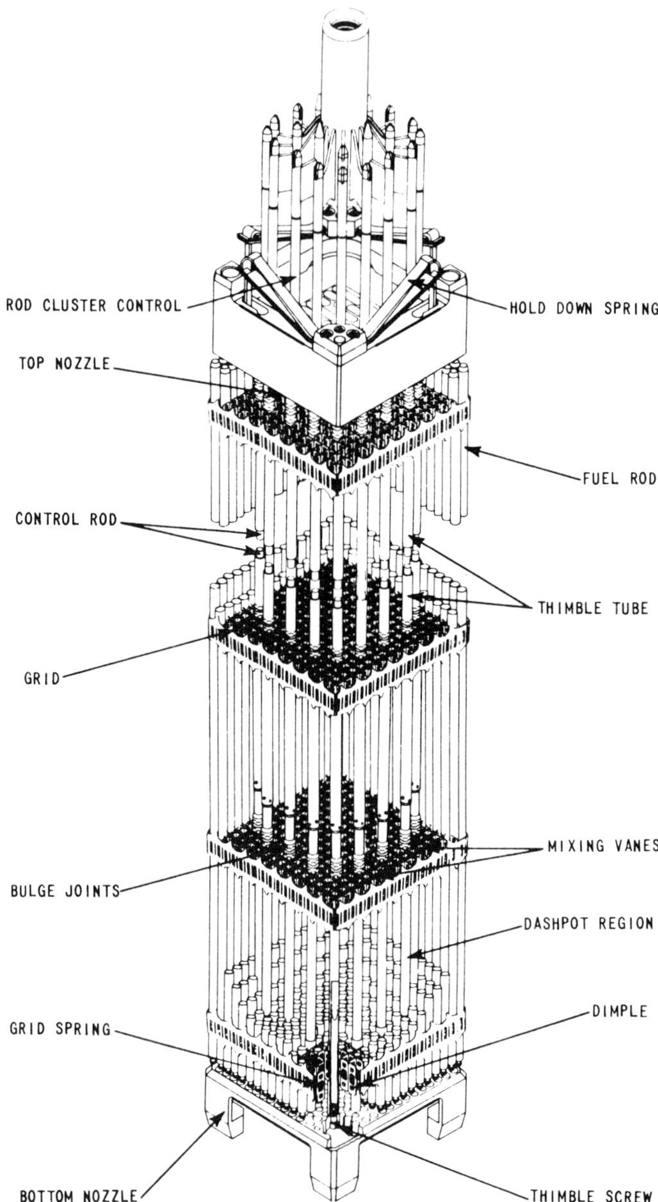

FIGURE 2. Typical Westinghouse PWR fuel assembly.

ly exists in this country was generated by Government defense-related activities. This defense high level waste (DHLW), which contains about one billion curies of radioactivity, is being safely stored in tanks at Government reservations until facilities are available to treat and finally dispose of these wastes. Currently, a major government program of equipment development and facility design is underway at the Savannah River Plant to incorporate the DHLW at that site into borosilicate glass suitable for disposal in a deep geologic repository. Although there are compositional differences between DHLW and CHLW, it is expected that the solidification and disposal methods for the two waste types will be essentially the same. Therefore the following discussion will deal only with CHLW.

The starting material for CHLW is the spent fuel from nuclear power plants which in the United States, with a very few exceptions, are light water reactors (LWR's) of two basic types; namely, pressurized water reactors (PWR's) and boiling water reactors (BWR's). Figure 2 shows a typical PWR fuel assembly about 190 of which make up the reactor core, depending upon the particular reactor design. The fuel assembly shown in Figure 2 is a 17 x 17 array consisting of 264 fuel rods plus thimble tubes and control rod tubes. The assembly is held together by the thimble tubes which connect the top and bottom nozzles, as well as by the grids located at several axial positions. Each fuel rod consists of a hollow tube, made of a zirconium alloy and called the fuel rod cladding, which is filled with cylindrical uranium dioxide fuel pellets. After loading the pellets into the tubes, the tubes are sealed at each end by welded end plugs. During reactor operation, the rod control cluster is moved up and down to regulate the power level in the core. To provide a perspective of size, a typical PWR fuel rod is 0.95 cm (0.37 in.) in diameter and 3.86 m (152 in.) long. The overall fuel assembly size is about 21.6 x 21.6 x 406 cm (8.5 x 8.5 x 160 in.). Each unirradiated fuel assembly contains about 450 kg of uranium which has been enriched to contain approximately 3 weight percent uranium-235 (natural uranium contains 0.7 percent uranium-235 and 99.3 percent uranium-238). BWR fuel assemblies are conceptually similar to the PWR assemblies except that they contain fewer but large diameter fuel rods and the assemblies are smaller in cross section (about 5.5 in. square). Each unirradiated BWR assembly contains about 190 kg of slightly enriched uranium.

During reactor operation, most of the energy produced by the core results from the fissioning or splitting of the uranium-235 nuclei by neutrons; that is, the uranium-235 atoms in the fuel are "burned up." As a result of splitting the nucleus of a uranium-235 atom, two or more new elements are formed called fission products. The uranium nuclei do not always split in the same way so that a wide variety of fission products are formed, most of which are radioactive. While it is not possible to predict how a single nucleus will split, the spectrum of fission products formed by a large number of fissions is well known. In addition to the fission products, new elements are produced as a result of neutron capture by the core materials. This leads to the formation of plutonium and other radioactive transuranic elements and other "activation products."

In order to sustain a chain reaction in the reactor, that is to keep the reactor operating, a certain minimum concentration of "fissionable atoms" must be present; that is, uranium-235, plutonium-239, and plutonium-241. Because the uranium-235 atoms are burned up faster than plutonium atoms are formed, fuel assemblies eventually have to be replaced with new ones (it is noted that in a breeder reactor plutonium atoms are formed faster than the uranium is burned up so there is a net gain in fuel quantity as the reactor is operated).

In general, a reactor is refueled once per year at which time about one-third of the fuel assemblies are replaced with new assemblies. The spent fuel assemblies removed from the reactor still contain about 95 percent of the uranium that originally was in the assemblies, but the weight percent of uranium-235 has been reduced to about one percent as a result of burnup. The spent fuel assemblies also contain, within the fuel rod cladding, the fission products, plutonium, other transuranic elements, and other activation products. Some of the plutonium atoms are fissionable (i.e., Pu-239 and Pu-241) so that they could take the place of uranium-235 atoms. It is readily apparent that if the uranium and plutonium could be recovered and some additional uranium-235 atoms added to make the mixture equivalent to the original slightly enriched uranium, then the mixture could be used to make new fuel assemblies; that is, the uranium and plutonium in the spent fuel could be "recycled." Therein lies the prime motivation for spent fuel reprocessing.

The first step in the reprocessing operation at nearly all operating and planned reprocessing facilities is to place a spent fuel assembly into a specially designed shear which is used to chop the fuel rods into lengths about 2.5 cm (1 in.) long. The fuel rod sections together with other fuel assembly hardware fall into a dissolver basket. The basket is then moved to a dissolver tank that contains a concentrated acid solution. The acid solution dissolves the uranium oxide fuel material from inside the cladding, but does not significantly attack the metal components. After drawing off the acid solution containing dissolved uranium, fission products, and activation products, the dissolver basket containing the cladding hulls and other metallic components is emptied into a container. The metallic components are then compacted in the container into a form suitable for disposal as solid low level waste or transuranic waste, depending on the concentration of transuranic elements.

The dissolver solution is then contacted with an organic solvent that does not mix with the acid solution. Because the uranium and plutonium are far more soluble in the organic solvent than in the acid solution, about 99.5 percent of the uranium and plutonium is taken up in the organic solvent while more than 99.9 percent of all the other constituents remains in the acid solution. This separation process is called solvent extraction. Since the organic solvent and acid solution are nonmiscible, they will physically separate into two distinct phases. The organic phase can be further partitioned to separate the uranium and plutonium if desired, and, after removal from the organic solvent, the uranium and plutonium are chemically treated into a useable form. The acid phase constitutes the high level waste stream that must be treated and disposed.

The Federal government, by law, is responsible for the safe disposal of commercial high level waste (CHLW). Within the Federal government, this responsibility has been assigned to the Department of Energy (DOE). Congress has assigned to the Nuclear Regulatory Commission (NRC) the responsibility for establishing regulations governing the disposal of CHLW, consistent with radiation protection standards promulgated by the Environment Protection Agency, and for licensing the disposal facility.

The method for disposing of the high level waste will be by burial deep under ground in a geologic repository. Technical Criteria for such geologic disposal were published by the NRC in the Federal Register in June 1983 for inclusion in Title 10, Part 60 of the Code of Federal Regulations (10CFR60) entitled, "Disposal of High-Level Radioactive Wastes in Geologic Repositories." The principal repository performance objectives contained in the Technical Criteria can be summarized as follows:

1. Provide reasonable assurance that releases of radioactive materials from the geologic repository following permanent closure conform to such generally applicable environmental radiation protection standards as may have been established by the Environmental Protection Agency.
2. Provide reasonable assurance that the radionuclides will be contained for a period up to 1,000 years after repository closure.
3. Provide reasonable assurance that, after the 1,000 year containment period, radionuclide releases to the geologic setting will not exceed one part in 10^5 per year of the radionuclide inventory calculated to be present at the end of the containment period.

The first performance objective places limits on releases to the environment from the overall repository system. These limits will be issued as part of new Part 191 of Title 40 of the Code of Federal Regulations (40CFR191). To date, this regulation has not been published by the EPA, but EPA has developed a number of preliminary drafts of the regulation. Using mathematical models of a repository developed by the EPA together with an assumed linear dose-effect relationship for health effects produced by exposure to radiation, EPA has estimated the health effects of a repository in which HLW is disposed of in accordance with these drafts standards. Assuming that the repository contains the waste from 100,000 metric tons of spent fuel (a quantity that would include all existing and future wastes from all currently operating commercial reactors), the EPA calculations indicate that no more than 1,000 premature deaths from cancer would occur in the first 10,000 years after disposal as a result of releases from the repository. This is an average of 0.1 premature death per

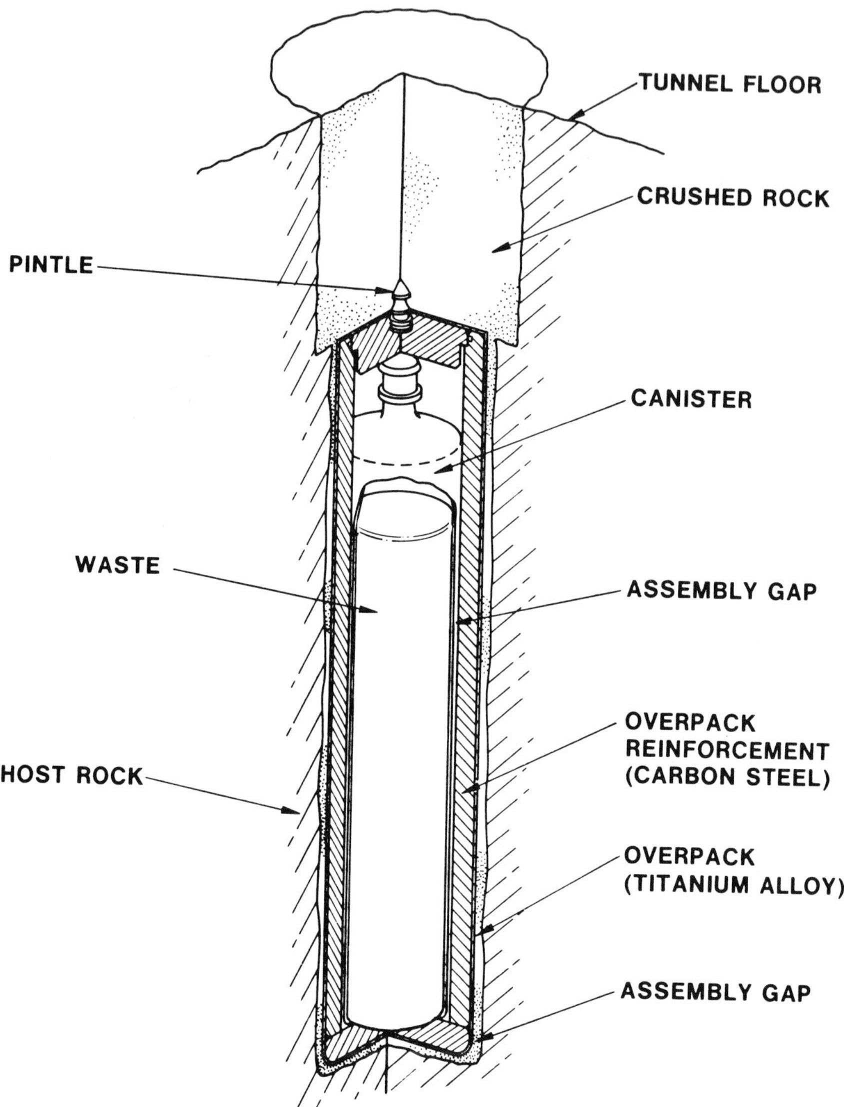

FIGURE 3. Conceptual design of a borehole-type waste package for the geologic disposal of high level waste.

year. By comparison, about 350,000 people currently die each year from all types of cancer. To provide additional perspective on the conservatism of the proposed repository standards, the EPA calculations predict about 4,000 premature cancer deaths per year due to exposure to natural background radiation. That is, releases from the repository would increase the number of predicted premature deaths by 0.0025 percent compared to those that would be predicted to occur from radiation to which everyone is normally exposed.

A number of mathematical models have been developed by DOE and its contractors to predict the performance of a geologic repository system. Even assuming very conservative characteristics with regard to the ability of the geologic medium to retard the migration of radionuclides from the repository to the environment, these models predict that the barrier provided by the geology is adequate to meet the draft EPA standard. Nevertheless, NRC has imposed the second and third performance objectives listed above; namely, 1,000 year containment and a slow release of radionuclides to the geologic medium thereafter. These two requirements imply a multibarrier repository system consisting of engineered barriers in addition to the natural geologic barrier so that control of releases to the environment is not dependent solely upon the geology. The containment function is assigned to the "waste package," while the slow release function is generally assigned to the "waste form." These two components of the repository system are described in the following paragraphs.

To meet the 1,000 year containment requirement, Westinghouse has developed, for the Office of Nuclear Waste Isolation, conceptual designs of engineered waste packages for use in a repository in a salt, basalt, or tuff medium. Two basically different design concepts have been developed, either of which, it is expected, can be adopted for any waste form and can be utilized in any of the currently proposed geologic media.

One of the designs (Figure 3) is called the "borehole concept" in which the waste package is placed in a hole bored in the floor or wall of the repository tunnel and utilizes the host rock for radiation shielding. In this design, a carbon steel overpack reinforcement provides structural support to preclude buckling due to hydrostatic or lithostatic pressure. The overpack reinforcement is surrounded by a thin, highly corrosion-resistant overpack, currently specified to be fabricated of a titanium alloy, that forms the long-lived containment barrier. A great deal of research and corrosion testing has been done on titanium alloys, especially in brine solutions. All testing to date indicates that only a few hundredths of a centimeter of the alloy will corrode in 1,000 years in typical repository groundwaters. Depending on site-specific repository conditions and the results of further corrosion testing of ferrous materials, it might be possible to eliminate the titanium alloy overpack and simply add an appropriate corrosion allowance to the wall thickness of the carbon steel component, thereby reducing cost. With this design, final assembly of the package must be done remotely in a shielded cell and the package must be transported and emplaced using a shielded cask. These types of operations have already been demonstrated.

The second design (Figure 4) is called the self-shielded package (SSP). In this design, a thickwalled cast iron or steel overpack forms the long-lived containment barrier and provides sufficient shielding to eliminate the need for remote operations. The wall thickness would be at least 38 cm (15 in.). For

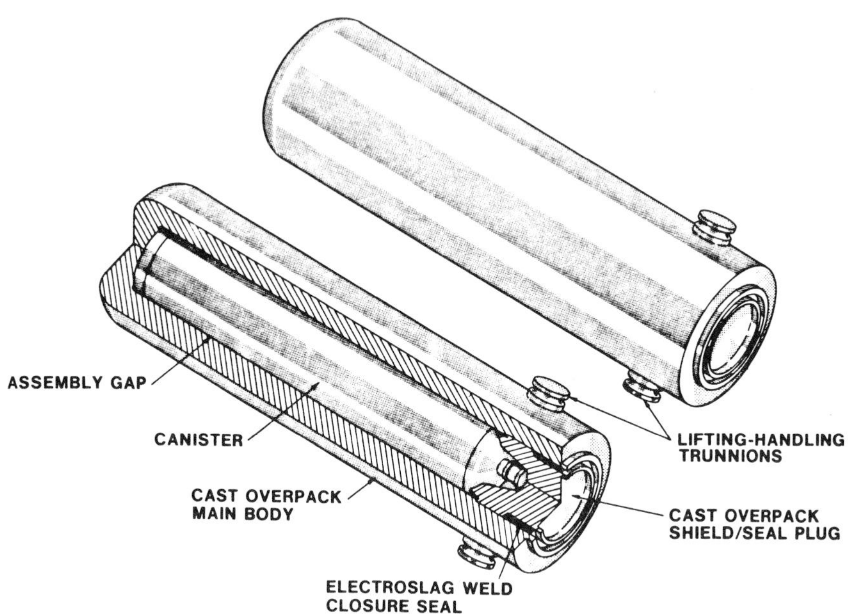

FIGURE 4. Conceptual design of a self-shielded package for the geologic disposal of high level waste.

emplacement, the SSP is simply placed on the floor of the repository tunnel. The SSP design trades off higher package component costs, compared to the borehole type package, against cost reductions that can be realized because of simplified repository facilities and lower mining cost.

Either of the two waste package types can be produced using existing fabrication methods, although some design-specific equipment will have to be designed and tested especially in the area of remote welding. Additional data on corrosion rates in specific repository groundwater is required to confirm design thicknesses of the containment barrier and to confirm that welding does not adversely affect the corrosion resistance of the materials. The final design choice will likely be based on economics and compatibility with repository system interface requirements since, based on existing data, either type of package will meet the 10CFR60 containment requirement.

Absorption by the waste form of some of the energy from the radioactive decay process results in generation of heat by the waste form (called decay heat). Since the decay heat generation rate is a function of the amount of radioactivity present, the heat generation rate will decrease with time as the radionuclides decay. The amount of waste per package, the design of the waste package, and the spacing of the waste packages in the repository must be coordinated to allow the decay heat to be dissipated into the host rock without causing excessive temperatures in the waste package and rock.

A number of waste form types, for the solidification of the high level waste stream from reprocessing, have been or are under development in several countries. However, the most advanced material is borosilicate glass which has been extensively studied for well over a decade and is the current reference waste form in the U.S. and foreign countries that are engaged in reprocessing or are planning to reprocess spent fuel. A production facility for the incorporation of HLW into borosilicate glass has been in operation in France for several years and a second facility is under construction and is scheduled to go into operation in 1985. The Defense Waste Processing Facility to be built at the Savannah River Plant in South Carolina for the incorporation of defense high level waste into borosilicate glass is scheduled to go into operation in 1990.

A few members of the scientific community have voiced objections to the use of borosilicate glass as a waste form because they claim that better waste forms could be developed and because experiments have shown that borosilicate glass degrades very quickly in hot brine (300°C). Neither of these arguments is valid. The waste form should be chosen on its ability to adequately perform its function, not on how well it could be made to perform. It also must be remembered that the waste form is an engineered barrier that is a backup to the geologic barrier which, by all indications, will perform adequately on its own. With respect to temperature, the use of the longlived waste package will prevent contact of groundwater with the glass for 1,000 years at which time the glass temperature will be only slightly above the ambient temperature of the rock. Furthermore, there are numerous repository and waste form design parameters that can be varied to reduce the glass temperature if necessary.

Considerable testing has been performed on borosilicate glass which indicates that it has good leach resistance in typical groundwaters at temperatures below 100°C. However, further confirmatory testing is required under simulated repository conditions including the presence of degradation products from the waste package components. In the unlikely event that this testing shows that borosilicate glass is not an adequate waste form, two alternatives are available. One is to use a higher integrity waste form. Several such forms are being investigated in the U.S. and some foreign countries. However, these are mostly at the laboratory stage of development and a number of years of process development and product testing would be required to bring the process to production scale. A second approach would be to add a "backfill" component to the waste package. The function of the backfill would be to provide a low permeability barrier around the waste form. This barrier would greatly reduce the quantity of groundwater reaching the waste form therefore reducing the degradation rate of the waste form. Furthermore, any water that reached the waste form would be retarded from moving back out to the geologic medium thereby reducing the migration rate of any radionuclides liberated from the waste form. One potential backfill material that has been extensively studied in the U.S. and Sweden is bentonite clay. In addition to its low hydraulic conductivity, it has good sorptive properties so that many of the radionuclides that might be released from the waste form would be trapped in the bentonite.

In accordance with the Nuclear Waste Policy Act, three potential sites for a repository have recently been nominated by the Department of Energy for detailed site characterization. This will include the sinking of an exploratory shaft and a program of under ground testing. One site is located in the large basalt deposits at the Hanford Site in the State of Washington, the second is in the tuff formations at the Nevada Test Site, and the third is in a bedded salt formation in the State of Texas. In addition, preliminary investigations are in progress to identify potential sites for a second repository in crystalline rock formations located in the central and eastern parts of the country. Thus, progress is being made toward qualification of one or more repository sites.

It is interesting to note that the ability of a geologic medium to provide long term isolation of radionuclides has been demonstrated. While mining uranium from a rich ore deposit in what is now West Africa, it was discovered by French scientists in 1972 that the ore deposit had sustained a chain reaction some two billion years ago. That is, the ore deposit had been a natural nuclear reactor. This event, called the Oklo phenomenon, was possible because two billion years ago the fraction of uranium-235 in natural uranium was significantly higher (uranium-235 has a shorter halflife than uranium-238). Since the discovery of this phenomenon, the fission products and activation products in the vicinity of the deposit have been exhaustively studied, and it has been determined that essentially all of the radionuclides produced by the chain reaction have remained in the immediate vicinity of the ore deposit even after two billion years. That occurred without the benefit of a carefully selected geologic medium and a high integrity wasteform.

In conclusion, the basic technology exists for the solidification of high level waste from spent fuel reprocessing and for its disposal in a geologic repository. We are in a period in which a focused program of confirmatory testing, site qualification, site-specific repository design, and specific equipment design and testing is required to expeditiously achieve the goal of a safe and cost effective disposal method.

SPENT FUEL

The spent fuel that is removed from the reactor during refueling is stored in a pool of water provided for this purpose at the power plant. The water provides radiation shielding and removes the decay heat generated by the spent fuel. In many plants, the size of the spent fuel storage pool was based on the assumption that the spent fuel would be stored for only a short time before being sent to a reprocessing plant and therefore the size of the pool was relatively small. The ban on reprocessing by the Carter administration eliminated this

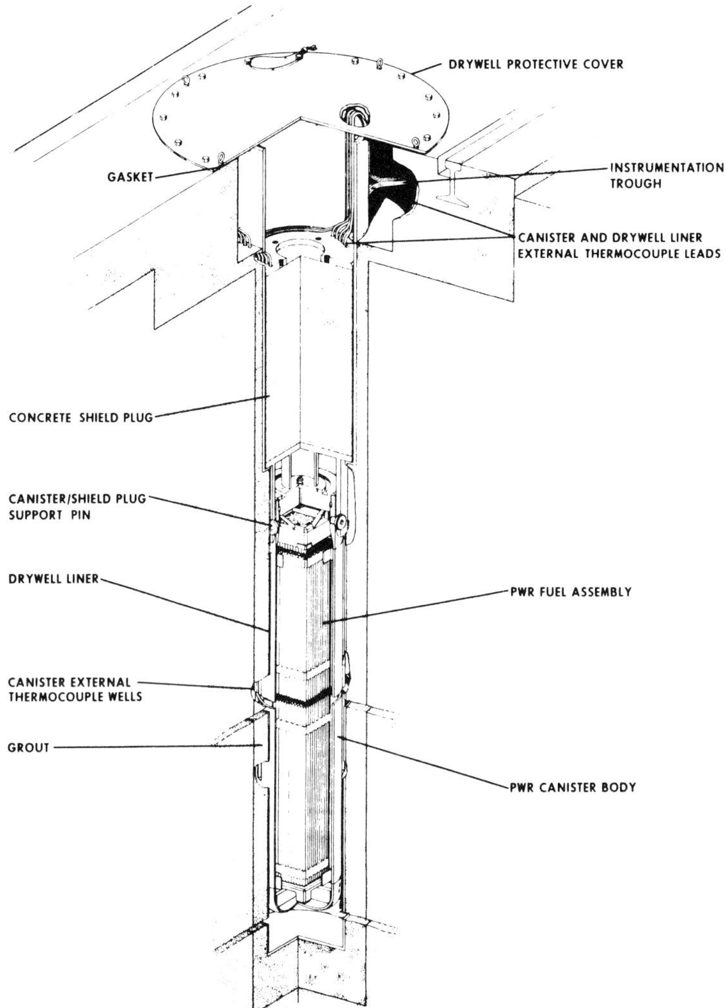

FIGURE 5. Cutaway view of the drywell being tested at the Nevada Test Site for the dry storage of spent fuel assemblies.

option for the utilities. In return, the Carter administration promised that Government-owned away-from-reactor (AFR) storage facilities would be provided to which utilities could send their excess spent fuel for a fee, but Congress has continually failed to provide funding for an AFR. Now, the Federal government has promised it will begin accepting spent fuel from utilities in 1998 when the first repository is scheduled to begin operation. In the meantime, with no

FIGURE 6. Concrete cask being tested at the Nevada Test Site for the dry storage of spent fuel assemblies.

AFR, no reprocessing capability in place, and no repository available, the utilities are faced with the dilemna of what to do with their spent fuel, since once the storage pool is filled up the power plant will be forced to shut down. Such a forced plant shutdown would, of course, be highly undesirable.

Several options for alleviating the at-reactor spent fuel storage problems have emerged, one of which is the technology for storing spent fuel that has decayed for a few years in the dry state. Dry storage has been studied extensively in the U.S. and other countries because it is more economical and more flexible than building additional water pool capacity. Three basic designs for dry storage have been developed: these are called drywells, casks, and air cooled vaults. A drywell is a lined hole in the ground which contains a canisterized spent fuel assembly just below the ground surface. Radiation shielding is provided by a shield plug and the surrounding soil. The decay heat from the fuel assembly is dissipated through the soil to the atmosphere. Drywells of the design shown in Figure 5 and containing PWR spent fuel assemblies have been under test

FIGURE 7. Ductile iron cask for the dry storage and shipment of spent fuel assemblies.

at the Nevada Test Site since early 1979 and have demonstrated very satisfactory operation.

Storage casks can be made of a variety of materials such as concrete, cast ductile iron, forged steel, or stainless steel encapsulated lead. The cask provides radiation shielding and conducts the decay heat to the surrounding air. The concrete storage cask shown in Figure 6 has contained spent fuel since 1978. Ductile iron casks of the type shown in Figure 7 have been developed and extensively tested in the Federal Republic of Germany for both storage and shipping.

An air cooled vault is a large concrete structure which is designed to direct a flow of air, either by natural or forced circulation, over encapsulated spent fuel assemblies to remove the decay heat. The concrete structure provides radiation shielding. A large facility of this type has been in operation in the United Kingdom since 1979.

Current planning is directed toward the once through fuel cycle (direct disposal of spent fuel without reprocessing) even though this would mean throwing away the valuable energy resource residing in the uranium and plutonium left in the spent fuel. If deep geologic disposal of spent fuel is undertaken, the same regulations would apply as for high level waste (i.e., 10CFR60) and the same technology would be used as described earlier. There are as yet insufficient data available to establish that the spent fuel pellets would be capable of meeting the 10CFR60 slow release rate criterion. If the fuel pellets themselves should prove to be an unsatisfactory waste form, the spent fuel could be encased in a corrosion resistant stabilizer such as glass or lead and/or a backfill material such as bentonite could be added to the waste package.

In conclusion, the technology for the safe long-term storage of spent fuel, whether dry or in water pools, is well established. The choice of the particular mode of storage is largely a matter of economics. The status of the technology for the direct burial of spent fuel, if that option should be taken, is essentially the same as for high level waste.

TRANSURANIC WASTE

Because so little commercial reprocessing of spent fuel has been performed in this country, only small quantities of commercially generated transuranic (TRU) waste exist. What does exist has come primarily from research projects and consists of such things as TRU contaminated equipment, air filters, rags, and protective clothing, shoe covers, and gloves. Commercial TRU waste will

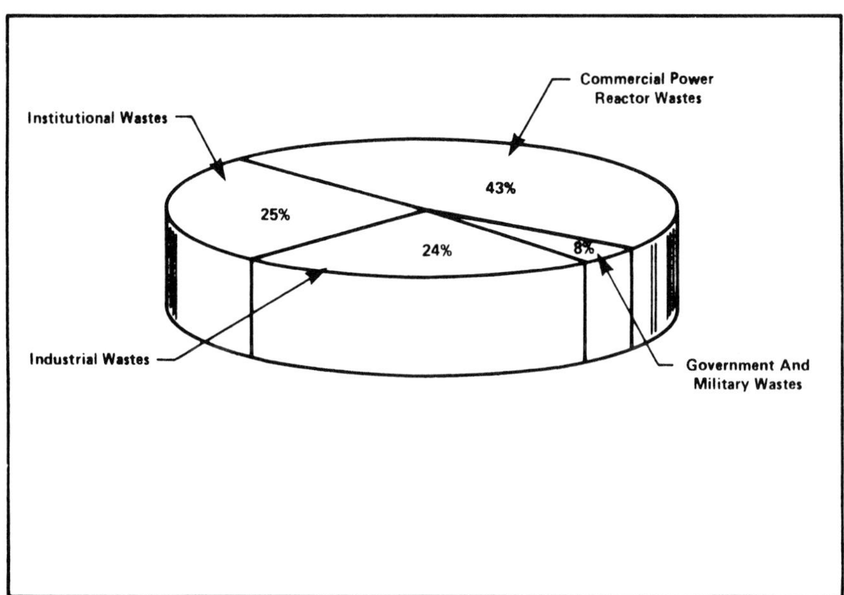

FIGURE 8. Percentage of low level waste produced by various types or organizations.

undoubtedly be disposed of in a geologic repository to provide long term isolation, but specific regulations governing its disposal have not been formulated as yet by the NRC. Because much of the TRU waste is in the form of combustibles that are only slightly contaminated, methods of reducing the volume of the waste, such as incineration and acid decomposition, have been developed to facilitate disposal.

A considerable quantity of TRU waste has been generated by defense-related activities. These wastes are currently being stored on Government reservations. A prototype geologic repository for the disposal of defense TRU waste is under development in a salt formation in southeastern New Mexico. This facility is called the Waste Isolation Pilot Plant (WIPP). TRU wastes will initially be emplaced in a retrievable manner in mined tunnels approximately 2,100 feet under ground. After a period of confirmatory testing and repository evaluation, a decision will be made relative to turning the facility into a permanent repository. To date, two shafts have been sunk to the repository horizon and considerable under ground excavation has been completed. In addition, surface facilities are being constructed and waste processing and handling equipment is being developed to support the test program.

LOW LEVEL WASTE

Low level waste is generated in a wide range of forms such as shoe covers and gloves; paper, rags, and other trash; ion exchange resins used to purify water used in nuclear power plants; radioactive solutions used for medical treatment and diagnosis; air and water filters; irradiated metal; and large pieces of contaminated equipment. Also, low level wastes originate from a variety of sources such as defense programs, commercial power plants, and institutions including hospitals and universities. Figure 8 depicts the percentage of low level waste produced in the U.S. by each of these sources. It is seen that slightly less than half of the low level waste results from commercial nuclear power generation.

Low level waste (LLW) constitutes over 80 percent of the volume but only 2 percent of the total radioactivity contained in all nuclear waste. Furthermore, it contains short halflife radionuclides compared to transuranic waste. Accordingly, LLW can be adequately disposed of by shallow land burial. This has typically involved digging a trench 6 to 12 m (20 to 40 feet) deep. After the waste is placed in the trench, the top portion of the trench is backfilled to form a "trench cap." The primary purpose of the trench cap is to minimize surface water infiltration into the trench since the only mechanism by which radionuclides could migrate from the trench is by transport by water flow. Therefore, a low permeability soil such as clay is often used for the trench cap and the cap is elevated above the surrounding land surface to promote runoff of the surface water. For final decommissioning of the site, it is likely that more sophisticated trench caps will be required, with materials being added to prevent penetration of the water barrier by plant roots and burrowing animals and insects. Testing of such trench cap systems has been underway for several years.

At some LLW sites in the past, little attention was given to the methods of waste emplacement and backfilling around the waste and to the form of the waste. The packages of waste were placed in such a way as to leave voids among the packages. These voids allowed the cover soil to settle. Also, much of the waste consisted of uncompacted trash which was compacted by decay of the waste and the weight of the overburden. These effects allowed subsidence of the trench cover which in turn allowed surface water to enter the trenches. Although this water infiltration has not resulted in any hazard to the public, it has resulted in the need for costly maintenance programs including pumping of the water from the trenches, treatment and disposal of the pumped water, and backfilling and repair of the trench caps. In some cases, it has led to the closing of the disposal site. The lessons learned from this experience have been addressed in the new NRC regulation, 10CFR61, covering licensing requirements for land disposal of LLW. In addition, more advanced disposal technologies have been developed, including the highly engineered integrated systems approach recently introduced by Westinghouse called the SUREPAK (subsurface recoverable package) system.

Approximately 3 million cubic feet of low level waste are generated annually in the United States. There are currently only three sites operating in this country for the disposal of these low level wastes. One of these sites is located at Barnwell, S.C. This site has been in operation for a number of years, but it is currently restricted by the State of South Carolina to receiving no more than 1.2 million cubic feet of waste per year. A second site is located near Richland, WA. Slightly over half of the low level waste generated in the United States, much of it from the eastern part of the country, is being shipped to this site. The third site, which currently disposes of only a small fraction of the waste, is located at Beatty, NV. The State of Nevada has taken action to close the Beatty site, but this action is being contested by the site operator.

Recognition of the crucial need to develop additional disposal sites to assure adequate national low level waste disposal capacity led to a number of significant actions in 1980. One of these actions was the issuance of a report in mid-1980 by the National Governor's Association Task Force on Low Level Waste. This report, endorsed by the National Governor's Association, proposed the formation of state compacts (Figure 9) to dispose of low level waste on a regional basis by the development of one or more disposal sites within each state compact region. A second significant action occurred in December 1980 when the Congress passed the Low Level Waste Policy Act, Public Law 96-573. This Act basically assigns responsibility for low level waste disposal to the states and encourages, but does not require, the states to do this on a regional basis by the

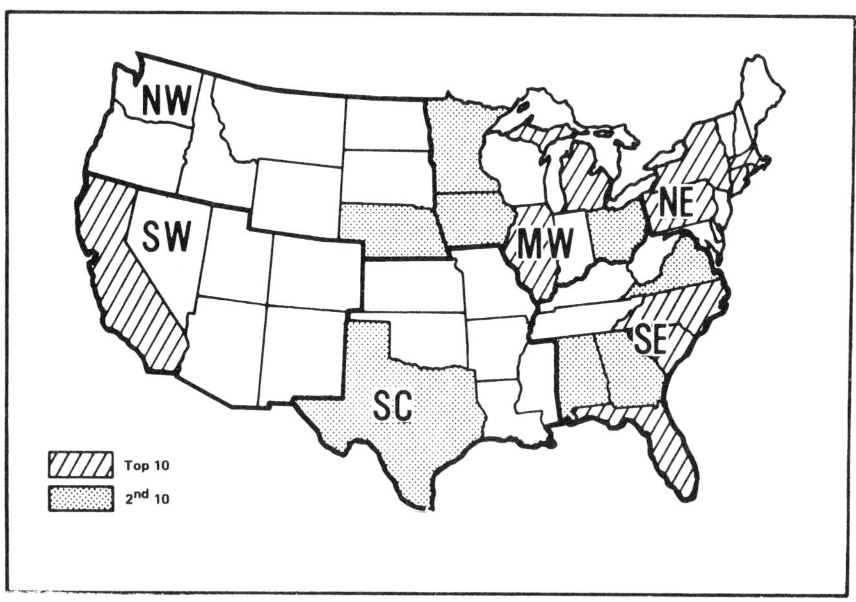

FIGURE 9. Regional low level waste compacts proposed by the National Governor's Association 1980 Task Force Report.

formation of state compacts. However, the Congress retained the authority to approve any proposed state compact. The Act also provides that after January 1, 1986, the member states within a compact can prohibit waste shipments to a regional disposal site by non-member states.

Since passage of the Act in 1980, the states have been attempting to deal with the problem of compact formation and negotiation of mutually acceptable terms. However, the intended initiative provided by the Act has not resulted in the development of additional waste disposal sites. Five compacts have been proposed to the Congress, but Congress has taken no action to enact these proposals. This stems largely from the fact that approval of compacts having an operating disposal site could potentially leave states outside of these compacts with no place to ship their wastes after January 1, 1986.

Notwithstanding the above, some progress is being made. For example, Texas, having decided not to enter a compact, is actively working to identify a suitable disposal site. California has entered into a compact with Arizona and is in the process of designating a contractor to develop and operate a disposal site in California. Pennsylvania, one of the largest generators of low level waste, has taken the initiative to form a compact with West Virginia and Maryland and is actively investigating potential disposal sites within the Commonwealth.

Because of the long lead time required to site, design, construct, and license new disposal facilities—on the order of 5 to 7 years—it is highly desirable to minimize the volume of low level waste for disposal so as to make the best use of the limited burial sites available today. Several technologies are available to achieve this goal, including incineration, compaction, and solidification. The treatment method used is dependent on the nature of the waste to be treated.

Organic wastes such as paper, rags, clothing, and filters can be volume reduced by incineration. A typical incinerator can provide a volume reduction of 80 to 1 for dry, combustible, uncompacted waste. Furthermore, equipment is available to incorporate the ash into a concrete matrix to form a high quality waste form. Compactors that operate on the principle of an ordinary household "trash masher" are an alternative to incineration. Compactors are simpler than incinerators, but offer lower volume reduction factors.

LLW that is in liquid form must be solidified prior to shipment into a form that eliminates all free water. Evaporators are often used to reduce the volume of water before solidification. Equipment is available for incorporating the liquids or evaporator sludge into a concrete matrix for shipping and disposal.

In summary, low level waste treatment and disposal technology is readily available and well understood. But to handle the volumes of waste that will continue to be produced in the future, the states and the Congress must act responsibly to expedite the development of new disposal facilities.

SUMMARY

In summary, inventories of nuclear waste exist today, and will grow even if no nuclear power plants are built in the future—a situation which hopefully will not occur because of the importance of this energy source to the economic viability of this country. Thus, it is not only mandatory, but clearly in the national interest to move forward without delay on a focused national program for the management of all radioactoce wastes.

The principal aim of the nation's long term nuclear waste program is the permanent disposal of high level wastes. The volume of these wastes is minimal and there are no insurmountable technical obstacles to preclude safe disposal of these wastes in geologic repositories. Reference methods for solidification, packaging, and emplacement of the waste have been developed. If confirmatory testing should show that any of these methods are unacceptable, backup techniques and materials are available. The same is true for transuranic wastes.

The radiation hazards of low level wastes do not represent a long term problem and safe disposal by shallow land burial can be accomplished by the use of existing technology and good engineering practice. However, additional burial sites will be needed in the near term to accommodate the increased quantities of waste that will be generated over the next ten years.

As we look to the future, a major challenge for those associated with the energy business is to see that adequatge energy is available to support a healthy economy. In an industrialized society such as ours, an adequate energy supply at a reasonable price is imperative for economic growth. More and more of our energy must come from electricity made from domestic fuels... coal and uranium. This electricity-made-in-America can be substituted for costly imported oil, and over several decades can lead this nation towards real energy security. If this is to happen, nuclear power must play a key role. However, one of the most important barriers to increased use of nuclear power in the future is the perceived problem of nuclear waste disposal. Several states, including California, Oregon and Connecticut, have passed laws that forbid the building of additional nuclear power plants until an acceptable solution for high level waste disposal is demonstrated. The fact of the matter is that the basic technology is available to dispose safely of all waste produced in connection with the generation of electricity from nuclear energy. The establishment of a firm national policy on radioactive waste by the U.S. Congress will enable this technology to be fully developed into practical disposal systems that will finally demonstrate that radioactive waste management is indeed a manageable task.

BIBLIOGRAPHY

U.S. Department of Energy, "Final Environmental Impact Statement on the Management of Commercially Generated Radioactive Waste," DOE/EIS-0046F, October 1980.

Intgeragency Review Group, "Report to the President by the Interagency Review Group on Nuclear Waste Management," TID-29442, March 1979.

U.S. Nuclear Regulatory Commission, "10CFR60, Disposal of High-Level Radioactive Wastes in Geologic Repositories, Technical Criteria," Federal Register, Vol. 48, No. 120, page 28194, June 21, 1983.

U.S. Department of Energy, "General Guidelines for the Recommendation of Sites for the Nuclear Waste Repositories (10CFR960)," Federal Register, Vol. 49, No. 236, page 47714, December 6, 1984.

Slate S.C., Ross, W.A., Partain, W.L., "Reference Commercial High-Level Waste Glass and Canister Defnition," PNL-3838, September 1981.

Nuclear Waste Policy Act of 1982, Public Law 97-425.

U.S. Department of Energy, "Mission Plan for the Civilian Radioactive Waste Management Program," DOE/RW-0005 Draft, April 1984.

U.S. Department of Energy, "NWTS Program Criteria for Mined Geologic Disposal of Nuclear Waste; Program Objectives, Functional Requirements, and System Performance Criteria [NWTS-33(1)]," OWI/NB-1, May 1982.

U.S. Department of Energy, "NWTS Program Criteria for Mined Geologic Disposal of Nuclear Waste; Repository Performance and Development

Criteria [NWTS-33(3)]," OWI/NB-6, July 1982.

Coman, G.A., "A Natural Fission Reactor," Scientific American, July 1976.

Walton, R.D., Coman, G.A., "The Relevance of Nuclide Migration at Oklo to the Problem of Geologic Storage of Radioactive Waste," IAEA-SM-204/1, June 1975.

U.S. Nuclear Regulatory Commission, "Data Base for Radioactive Waste Management, Review of Low-Level Radioactive Waste Disposal History," NUREG/CR-1759, Volume 1, November 1981.

U.S. Nuclear Regulatory Commission, "Licensing Requirements for Land Disposal of Radioactive Waste (10CFR61)," Federal Register, Vol. 47, No. 248, page 57446, December 27, 1982.

PART 2
Storage, Transportation and Disposal

The general public is frequently not aware of the wide use of radioactive materials in diagnosing and treating health problems. A diagnostic specialty, nuclear medicine imaging, is now a well-developed field. By using a carrier molecule with a radiotracer without disturbing its biologic affinity, a radioactive substance can be directed to a specific organ for diagnostic purposes.

The use of radionuclides within the body presents a special problem so that general health is not harmed. The key to the medical use by doctors of external sources of radiation is length of time to exposure, distance from exposure, and the development of adequate shielding. In all medical use there are radioactive wastes that must be disposed. Three general methods are now practiced. If the waste material has a reasonably short half-life of a few weeks, the usual procedure is to store the wastes until the radioactivity has disappeared. If there are small quantities of radioactive waste, it may be diluted and dispersed in the environment. The disposal of long-lived radioactivity is to concentrate the waste and then bury the material.

The radioactive wastes from nuclear plants have increased more rapidly than early estimates indicated. As a consequence, the problems of treating the waste, transporting it to a disposal site, and developing satisfactory disposal sites have received major attention by the industry. Of all nuclear power industry problems, none is more sensitive to the general public than the management of wastes generated within the power plants. The management problems include the initial processing of the waste material to separate radionuclides from all plant effluents. The second step involves the temporary storage, with further reduction if needed, and preparation for transportation to the disposal site. The next step is the shipment of the wastes in suitable forms and licensed packages, and the final procedure is disposal at a licensed burial facility.

Because the transportation of radioactive wastes has potential dangers, a set of regulations has evolved over the past three decades. The fundamental goal here is to ensure that radiation exposures during transportation remain as low as reasonably achievable.

It is now recognized that the operation of a low-level waste facility is an ever increasing problem nationally. The operation and closure in 1977 of the Maxey Flats, Kentucky disposal facility illustrates some of these problems. The problems included the leaking of the waste packages and the subsidence of the trenches into which storm and surface water entered with resulting migration of the waste contaminents. From these failures, however, lessons have been learned in the development of more adequate disposal sites.

Management of Radioactive Materials and Wastes: Issues and Progress. Edited by S. K. Majumdar and E. Willard Miller. © 1985, The Pennsylvania Academy of Science.

Chapter Eight

DISPOSAL PROBLEMS OF LOW-LEVEL RADIOACTIVE WASTES AND IMPLICATIONS FOR HOSPITALS

David R. Brill, M.D.
Chief, Nuclear Medicine
Geisinger Medical Center
Danville, PA 17822

The general public often reacts with surprise at the wide usage of radioactive isotopes in medical diagnosis and treatment. Such a response is perhaps to be expected, given the perception that radiation is somehow harmful or evil. In addition, activities of this sort do not attract the attention or adulation of the news media. They are, nonetheless, a standard part of modern medical care. In Pennsylvania during 1981, over four million diagnostic procedures which employed radionuclides were performed. An estimated 1600 persons are employed and conservatively 135 million dollars expended annually to provide these services. Nearly all of the commonwealth's 241 acute care general hospitals are involved to some extent. There are three clinical disciplines which depend on radioisotopes; nuclear medicine imaging, radioimmunoassay, and radiation therapy.

Nuclear medicine imaging, a diagnostic specialty, involves the direct administration of a small amount of a radioisotope to a patient. Many organs can concentrate certain chemicals or drugs. If one labels such a carrier molecule with a radiotracer without disturbing its biologic affinity, one can direct a radioactivity to specific organ. By "viewing" the organ with a sodium iodide detector which has been specifically adapted to reproduce spatial information (Figure 1), a picture (often called a scan or scintigram) of that organ is obtained. The pattern of radioactivity may be altered by disease. A good example of this principle is the bone scan, a procedure used for detection of bone cancer. This study, commonly performed with 20 milliCuries of Technetium-99m labeled methylene diphosphonate (a carrier which adsorbs to the organic matrix of living bone) is capable of determining bony malignancy with a sensitivity unrivaled by any other imaging technique. While diseases other than cancer can also pro-

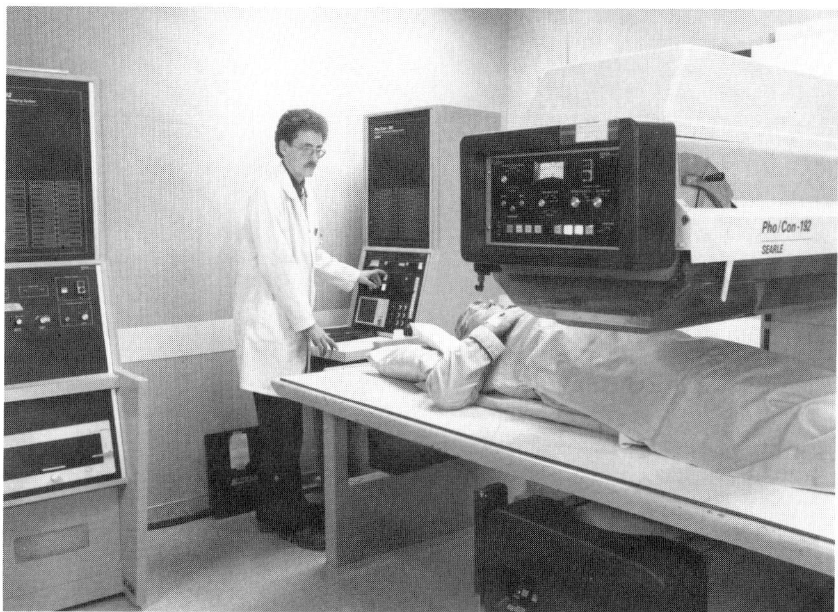

FIGURE 1. Tomographic Scintiscanner (Siemens Pho-con 192). This instrument, shown performing a total body bone scan, yields 12 simultaneous images of the skeleton, each at a different depth plane parallel to the long axis of the patient.

duce increased isotope deposition, the bone scan (Figure 2) has become the predominate screening tool for detection of skeletal cancer in susceptible individuals in whom symptoms are nondescript or nonexistent, and whose X-rays are frequently normal.

Application of this principle is far more widespread than just the bones. Many carriers are available which may be labeled Technetium-99m, Indium-111, Iodine-123, and other short-lived radionuclides. The emission spectra of these radioactive species are such that it is possible to use milliCurie quantities with very modest radiation exposure to the patient. As a general rule, in fact, nuclear medicine imaging procedures impart less radiation to the patient then comparable X-ray studies. Another generic advantage is the ability to detect physiologic, rather than just anatomic alterations. Since physiologic alterations often precede anatomic changes, nuclear medicine imaging may commonly detect early disease before abnormalities are present on standard X-ray studies, ultrasound or even CT scans. The variety of carriers make almost every organ system in the body accessible with these techniques.

There are, of course, generic weaknesses to such studies. Although equipment has improved dramatically in the last decade and the diagnostic yield has increased immensely with the use of computers, nuclear medicine imaging can-

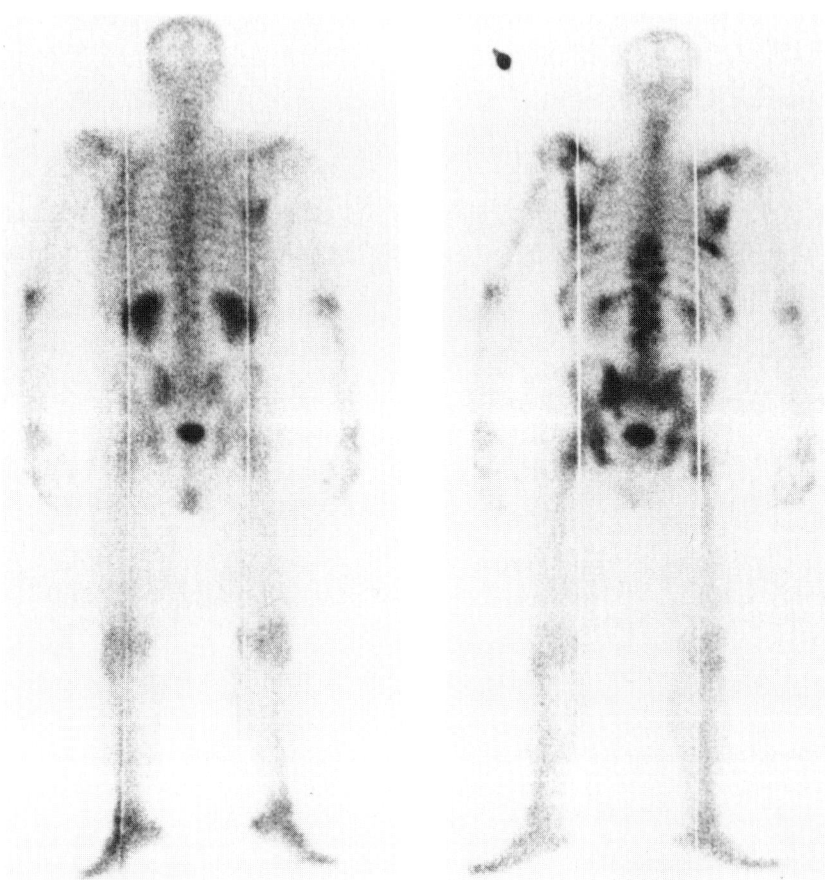

FIGURE 2. Typical bones scans. The image performed on 4-28-78 shows the posterior aspect of the patient and is normal. That performed on 12-5-78 on the same patient shows multiple areas of abnormal concentration, especially in the spine, ribs and pelvis. Concurrent X-rays were normal, but the patient subsequently proved to have bone cancer in the distribution revealed by the bone scan.

not demonstrate anatomic detail nearly as well as conventional X-rays. Moreover, differentiation of one disease state from others may be difficult. In practice, this has limited the use of nuclear medicine to a screening tool or as a compliment to other procedures for more specific diagnoses.

Computer analysis, to which nuclear medicine imaging is well adapted, has dramatically increased information yield and application. It has become extremely important in the assessment of cardiac disease (Figure 3), leading to improved utilization of other more invasive procedures such as coronary

FIGURE 3. Computer keyboard with two images of the beating heart being displayed on the screen to the right.

arteriography. Digitization has also improved management of diseases of many other organs, such as the kidney, most especially in transplant patients.

While other new modalities such as nuclear magnetic resonance imaging (which uses no radiation) will undoubtedly have some impact on nuclear medicine imaging, the future appears bright. Labeling of metabolites with ultra short-lived positron emitters is opening a new chapter in research, diagnosis and treatment of many diseases of the brain.[1] Imaging with radiolabeled tumor specific monoclonal antibodies appears to hold enormous promise for the diagnosis of visceral cancers.[2]

In 1981, 650,046 nuclear medicine imaging procedures were reported to the Pennsylvania Bureau of Health by 190 hospital departments.[3] Nationwide, over nine million of these studies are done annually.

Radioimmunoassay (RIA) is a laboratory based technique (Figure 4) for quantitation of minute amounts of hormones, drugs and other chemicals in the blood of patients. The principle, developed in 1960 by Berson and Yalow,[4] involves the extreme specificity with which antibodies react with their antigens. The antibodies are isolated, radiolabeled, and allowed to combine with their specific antigens which are present in a sample of the patient's serum. The antigen-antibody complex will precipitate out. By comparing the radioactivity in the precipitate with values obtained on a standard curve, one can determine how

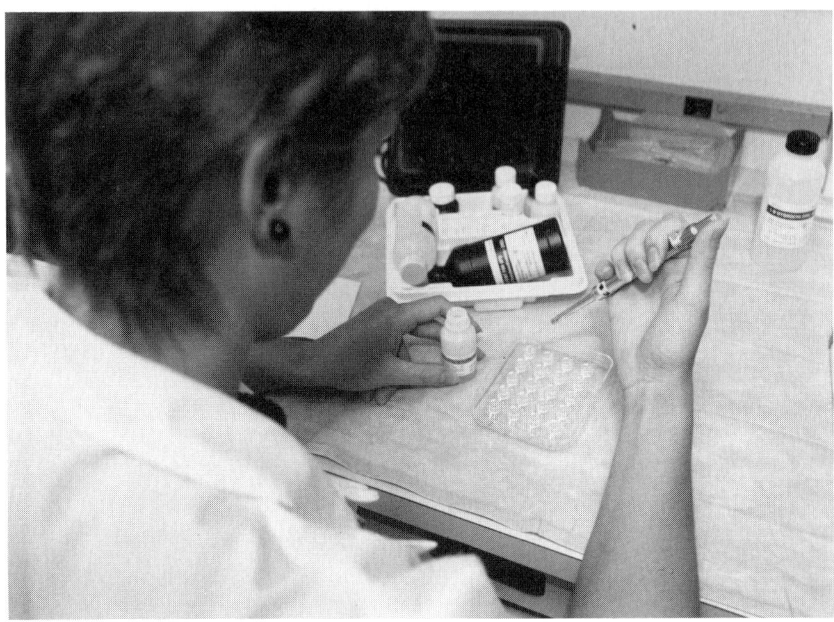

FIGURE 4. Radioimmunoassay technologist preparing samples for an insulin assay.

much of that particular substance was present with a sensitivity of parts per million and often per billion. The most commonly used radionuclide is Iodine-125, although Tritium and Carbon-14 are occasionally used. Typically, activities are in the nanoCurie range and the patient himself receives no radiation. The number of substances for which radioimmunoassays have been developed is staggering. The Pennsylvania Bureau of Laboratories listed 63 different RIA's in 1981.[5] This list does not include studies being developed or experimental work.

Radioimmunoassay is very accurate generically, but is being partially replaced by other techniques such as EMIT (enzyme mediated immunoassay technique), a nonradioisotopic method of comparable sensitivity. As of this writing, EMIT is only partially developed. There are many agents which can be studied with RIA for which comparable tests do not now exist with EMIT. Some substances, such as large proteins like human growth hormone, may be impossible to assess with EMIT. While EMIT and other nonradiometric assays will undoubtedly make intrusions into the area now occupied by RIA; RIA will continue to be a vital part of medicine for years to come.

In 1981, 3,381,086 RIA procedures were reported to the Pennsylvania Bureau of Laboratories.[3] Two hundred forty licensed labs operate in Pennsylvania.[5] Over 40 million RIA's are done annually in the US.

Therapy of cancer with ionizing radiation is a concept familiar to many people and indeed, was one of the earliest applications of isotopic technology. Currently, three modalities of isotopic therapy exist—teletherapy, brachytherapy, and therapy with unsealed sources.

Teletherapy consists of exposing tumor tissue to ionizing radiation from a distance. In many institutions, the source of this radiation is a source of Cobalt-60 of about 5 kiloCuries (Figure 5). Brachytherapy involves direct application of a sealed radioisotopic source to the surface of or actually into the body of a tumor. Some of the first radiation therapy was undertaken in this fashion using Radium-226 and its daughter radionuclide Radon-222. While still in use, radium is disappearing in favor of shorter lived agents such as Cesium-137, Iridium-192, and Iodine-125 which allow for more accurate tailoring of the radiation field to the contours of the tumor with fewer side effects arising from radiation of normal tissue.

FIGURE 5. Head and table of a Cobalt-60 therapy machine. The source (about 5400 Ci of Cobalt-60) is contained in the head.

In some cases, a tumor may actually metabolize enough of a radioisotope to deliver a therapeutic exposure. Iodine 131 can be accumulated by certain cellular types of thyroid malignancies and is generally accepted as a valuable adjunct to surgery and medical management. Even more commonly, this same radionuclide is used in place of surgery or medical management of benign hyperthyroidism. The radionuclide is given orally and is colloquially known as the "atomic cocktail."

In many large cancer centers, teletherapy is performed with linear accelerators capable of delivering megavoltage radiation, electron beams and other densely ionizing particles. These machines are very expensive and are inappropriate for smaller installations. Despite the advantages of larger cancer centers, the economic and logistic hardships imposed upon cancer victims make therapy in their own locale an attractive option.

Chemotherapy and more sophisticated cancer surgery have had and will have an impact on radiation therapy for cancer. Recent experience has suggested that all of these modalities have their place and that optimum cancer therapy may involve a combination of all of these methods.

A new approach to therapy with unsealed sources currently being investigated at the University of Washington, involves the use of around 100 milliCurie doses of Iodine-131 labeled tumor specific monoclonal antibodies.[6] It is too soon as of this writing to know whether this technique will bear fruit, but some preliminary success has been reported.

In 1981, 77,271 patients were reported to the Pennsylvania Bureau of Health as having undergone radiation therapy with isotopic sources.

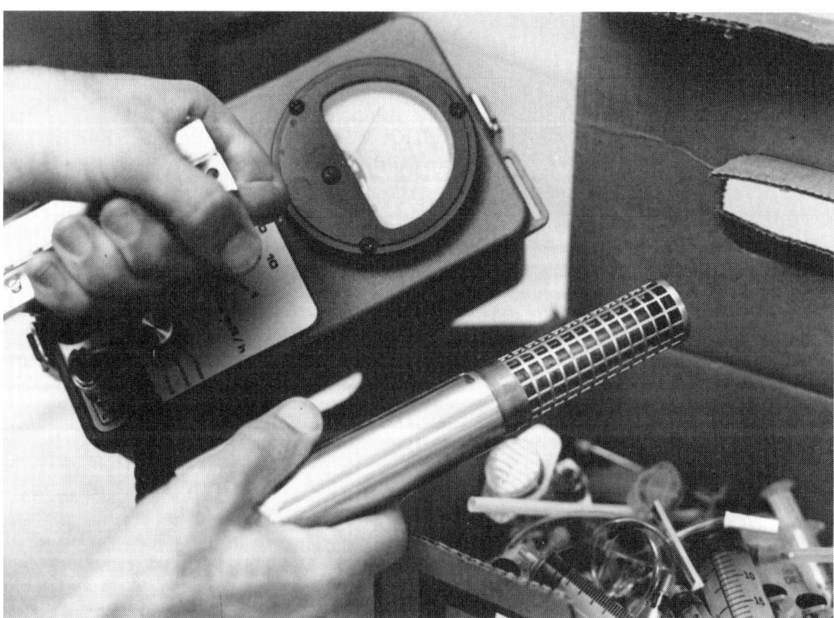

FIGURE 6. Assaying for residual radioactivity in box containing typical hospital generated low-level radioactive wastes. The box contains mostly syringes and needles and is about one-half a cubic foot in volume. This represents about a week's worth of waste.

There are other health care implications of radioisotope use which do not directly involve patients. Basic science and medical research depends heavily upon isotopic methods for elucidation of biochemical pathways, autoradiography, radioimmunoassay, and other information. Pharmaceutical houses must use similar methods in deriving the mechanisms of action of almost all new drugs, information required by the FDA before approval for human use. Finally, some prosthetic body parts can only be sterilized with radiation—heat sterilization can damage plastics and the gas used in gas sterilization may impregnate some synthetics.

Relatively few hospitals use commercial facilities to dispose of their low-level radwastes. Since the isotopes are general short half-lived, a combination of techniques such as segregation by half-life, volume reduction by limiting contamination of large articles, compaction, and incineration of extremely small amounts of RIA wastes allows for onsite decay to background. At the Geisinger Medical Center, the volume of wastes stored temporarily is about one-half cubic foot per week, consisting mostly of syringes, needles, Band-Aids, alcohol sponges, and small dressings (Figure 6). Several radiopharmaceutical manufacturers recycle their Molybdenum-Technetium generators, thus freeing the hospital of storing that bulkier and more radioactive item.

Still, health care activities are very dependent upon having disposal sites available for low-level radwastes, because the suppliers of the radioisotopes operate cyclotrons and reactors, activities which produce large volumes of contaminated wastes. One manufacturer bears a cost of $60 per cubic foot to dispose of his wastes,[7] money that must be made up in price to his customers and their patients. Radwaste disposal problems will very quickly manifest as increased costs and diminished availability of related services to patients.

Medical and pharmaceutical research is in double jeopardy. Not only are they dependent on the radiopharmaceutical firms the isotopes they use, but they generate large volumes of wastes which must be shipped off-site at considerable expense. An estimated 283 cubic meters of low-level radwaste is shipped by university and medical school research facilities per year.[8] This contains milliCurie amounts of Tritium and Carbon-14.

Thus, the health profession has a considerable stake in seeing the issue of disposal of low-level radioactive wastes resolved. To this end, in May and June of 1983, the Pennsylvania Association of Clinical Pathologists, Pennsylvania Radiological Society, Pennsylvania College of Nuclear Medicine, and the Pennsylvania Medical Society adopted positions calling for the "rapid development of a safe and economical means" of disposing of such material.

The medical profession does not advocate speed at the expense of care and safety, but recognizes the value of public education in the ultimate acceptance of a disposal facility wherever one is located. Such a process will take time and needs to be begun soon. Economy is an issue because of new reimbursement mechanisms developed by the federal government and other financial pressures new to the health care industry which will diminish monetary resources for accomplishing health delivery. Indeed, the relative economy of nuclear medicine imaging studies vis a vis those of the CAT scanning and NMR may increase dependency upon isotopic methods. Additional disposal costs may cripple this resource. There is some debate as to whether the site should serve only Pennsylvania or should offer disposal to geographically contiguous states. The health care industry does not advocate any particular approach. Some form of interstate cooperation is probably necessary since no large radiopharmaceutical house is based in this state. All medicinal radionuclides are made in states like Massachusetts, New York, New Jersey, Illinois, Minnesota, Missouri, and California. If any of these states refuse to deal with their own wastes, Pennsylvanians as well as many others will suffer.

The key word for all concerned, including hospitals and health care professionals, is safety. Above all else, adequate planning, regulation, development, and operation is essential to insure that no one is inadvertently exposed to unnecessary radiation in either the short term or in the future as long as such a site exists.

BIBLIOGRAPHY

1. Alavi, A., Reivich, M., Greenberg, J., et al: Mapping of functional activity of the brain with ^{18}F-Fluoro-deoxyglucose. Semin. Nucl. Med. Jan. 1981; 11:24-31.
2. Larson, S.M., Brown, J.P., Wright, P.W., et al: Imaging of melanoma with I-131 labeled monoclonal antibodies. J. Nucl. Med. Feb. 1983; 24:123-129.
3. Brill, D.R.: Nuclear waste disposal facts revealed. Pennsylvania Medicine Nov. 1983; 86:53-54.
4. Berson, S.A., Yalow, R.S.: Immunoassay of endogenous plasma insulin in man. J. Clin. Invest. 1960; 39:1157.
5. Pennsylvania Bureau of Laboratories. Personal Communication Nov. 1982.
6. Larson, S.M.: Monoclonal antibodies. Lecture given before the American College of Nuclear Physicians. March 11, 1983.
7. New England Nuclear Corporation-Personal Communication.
8. Pennsylvania Radwaste Working Group—Low-Level Radioactive Waste Disposal. Sept. 1983, page 39.

Management of Radioactive Materials and Wastes: Issues and Progress. Edited by S. K. Majumdar and E. Willard Miller. © 1985, The Pennsylvania Academy of Science.

Chapter Nine

MANAGEMENT, TREATMENT AND TRANSPORTATION OF NUCLEAR PLANT WASTE: SOLID, LIQUID AND GASEOUS

L.H. Barrett[1] and P.J. Grant[2]

[1]Deputy Director and [2]Chief Technical Support
Three Mile Island Program Office
U. S. Nuclear Regulatory Commission
P. O. Box 311
Middletown, PA 17057

I. INTRODUCTION

Operating nuclear power plants generate various types of solid, liquid, and gaseous waste containing radioactive materials. The quantity and form of these waste varies greatly between plants and is generally a function of the type of plant (e.g., PWR and BWR's), and the management philosophy and processing concepts employed. This chapter will focus on the various types of waste generated within the power plants, the principle processing technologies deployed for each, the major waste management/volume reduction techniques utilized, and the packaging of the final waste forms for transporting and disposal.

Of all the problems associated with the nuclear power industry, probably none is more sensitive to public opinion, and its solution so controversial as that of management of the radioactive wastes generated within nuclear power plants. Management of these wastes is complicated not only because of their diverse physical and chemical characteristics but also because of the environmental limits and the controls of containment required for these radioactive materials. The waste management solution is generally conceded to require acceptable performance in four equally important areas: (1) processing of waste stream (liquid and gaseous) to separate radionuclides from all plant effluents that are to be released to the environment; (2) treatment of these separated radionuclides to place them in suitable forms (solids and liquids) for temporary storage, further volume reduction if needed, and preparation for disposition; (3) shipment of these conditioned wastes in suitable forms and licensed package for transport to storage and disposal sites; and (4) final disposition at a licensed burial facility. The purpose of this chapter is to provide the reader with background and current state-of-the-art in the waste management areas (1) and (2) described above with emphasis on waste types, processing strategies, waste segregation, volume reduction, temporary on-site storage and suitable forms for transporting these radioactive wastes.

II. SOLID WASTE MANAGEMENT

A. WASTE CLASSIFICATION/DEFINITION.

Solid waste at nuclear power plants is classified[1] "wet" or "dry" depending upon the amount of residual water associated with them. Wet waste consists mainly of concentrated evaporator bottoms, filter sludges, reverse osmosis concentrates, spent ion-exchange resins, etc., all of which derive from liquid waste treatment and purification from the various fluid streams in the nuclear plant. (See discussion on Liquid Waste Management). Wet waste have specific limits for transporting packages and burial requirements on residual water.[2] The majority of dry waste consists of contaminated clothing, trash, air filters, rags, etc., which can be compacted/volume reduced and shipped as Low Specific Activity (LSA) waste for burial. Some dry waste, such as cartridge filters, can become highly contaminated and require additional packaging/shielding conditions. Additionally, plants may generate *highly* contaminated equipment and components (viz., control rods, incore internals, instrumentation, and decontamination equipment) which require special handling and decontamination. Since these are special cases they will not be included in this section.

B. WASTE SYSTEM DESIGNS

A significant number of nuclear power plants have shifted toward the concept of "maximum recycle" or near "zero release" for radioactive liquids. These modes of operation have resulted in an increased volume of solid waste. Historically, many of the earlier plants shipped solid waste in the form of dewatered sludges and resins or evaporators concentrates immobilized by sorbent material. Because of burial requirements these plants are required currently to demonstrate that dewatered solid waste contains no more than 0.5 percent of the waste volume as free liquid or (< 1.0 gallon),[3] as a minimum.

The boundary between liquid and solid radwaste systems is usually defined that the solid waste systems start at the receiving and collection tanks which contain slurries and concentrates from the demineralizers, filters, evaporators and reverse osmosis equipment. Treatment of these wastes can be further subdivided into; (1) waste collection, (2) pretreatment and volume reduction, (3) solidification/immobilization and (4) packaging and storage. Figure 1 shows a flow diagram for the management of solid waste at a typical light water reactor (LWR) plant.[4] The interface between the solids pretreatment and the solidification subsystem is a critical area in radwaste treatment because the amount of residual water associated with the treated solids is a major factor in integrating the systems and in determining what solidification method is most economical. Likewise, the suitability of the pretreatment subsystem would impact on the prior choice of a particular solidification process.

System design consideration for all radwaste management systems should address some general areas including the maintainability; the placement and location of handling equipment in low-radiation areas to ensure ALARA and minimize exposures to operating personnel;[5] the compatibility of equipment with the chemical and physical properties of water and solidification material; and the reliability of system performance on demonstrating acceptable solidification product over a wide range of waste chemical/radiochemical concentrations.

C. WASTE COLLECTION

Sufficient tankage for waste collection is essential to insure un-interrupted power plant operations. Typical waste tankage designs for LWR include provisions for at least 45-60 days of radioactive decay for the reactor coolant purification resins or filter sludges prior to solidification. Usually, other wet waste such as evaporator/RO concentrates, lower activity radwaste filter sludges only require 20-30 days to insure sufficient decay. In addition to radioactive decay, the waste storage tanks also provide surge capacity to accommodate periods of abnormally high waste generation or outages in the solid waste processing system. Usually an adequately designed system will include sufficient tankage to receive all waste stream inputs for anticipated events (viz., solidification system maintenance, system upsets/surges and extended outages).

D. PRETREATMENT/VOLUME REDUCTION.

Waste pretreatment is basically a volume reduction process serving to minimize the quantity of waste to be solidified and shipped off site. The wet wastes generated at power plants usually contain a significant quantity of water. Removal of this water can provide substantial volume reduction and associated savings with smaller quantities of solid waste which require storage, solidification and disposal. Mechanical processes for water removal from spent resins and filter sludges can be accomplished by methods such as: decantation, centrifuge or filtration (see Figure 1). Centrifuging systems which are the most efficient systems, can reduce water content below 50 w/o. While decanting and filtration usually can only reduce the residual water in resins and filter sludge solids to 70-80% by weight.

Other types of dewatering/solidification devises utilized by the industry include heating and chemical processes. These devices can produce solids from wet waste that have a final water content ranging from zero (0 w/o to 50 w/o). The fluid-bed dryer, the fluid-bed incinerator/calciner, the thin-film evaporator, and the crystallizer are typical examples of these devices. References[6] through[9] provide a detailed description of these volume reduction devices including operating parameters and performance characteristics.

The volume reduction equipment and the associated residual water content has to be integrated and married with the solidification equipment and procedures selected. For example, waste containing 25 w/o water (i.e., 75 w/o solids) is about optimum for the cement solidification process. However, if the water content of the pretreated slurry is 50 w/o or less some water or liquid waste would have to be reintroduced to maintain the product form required in meeting the criteria established in the process control program and burial requirements.

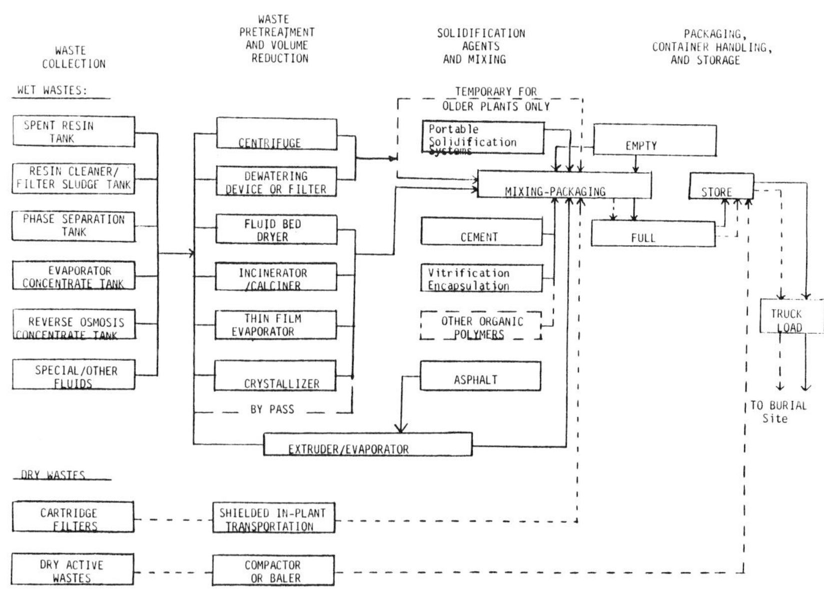

FIGURE 1. Flow Diagram for Radwaste Management at Light Water Reactor Plants.

Over the past five years volume reduction has become an increasingly important factor in nuclear plant waste management. This is primarily attributed to the restriction and limits imposed at the three burial sites in 1980,[10] as well as actual waste volumes being much larger than originally expected[11,12] while disposal costs have continued to increase dramatically. The prime advantage of a large volume reduction is the substantial savings in reduced transportation and handling costs between the plant and burial site. An industry wide implementation of VR could also significantly increase the available disposal capacity at the burial sites. Another associated cost savings with volume reduction is a lower plant construction and operating costs due to the smaller on-site space requirements for storing and staging waste.

E. SOLIDIFICATION/IMMOBILIZATION

The major factors which influence the selection of an adequate solidification/immobilization process include *burial requirements, safety, economics* and *space constraints within the plant.*

As previously discussed, a large portion of the waste produced at nuclear power plants is in a wet solids form and requires further processing to achieve an acceptable solid; stabilized monolithic form for burial. In order to assure that the solidification process will consistently produce a product which is acceptable for disposal/general requirement have been established by the NRC.[13] Additionally these requirements and conditions for process control programs are provided in NRC Standard Review Plan 11.4, "Solid Waste Management Systems," (NUREG-0800) and its accompanying branch technical position ETSB 11-3, "Design Guidance for Solid Waste Management Systems Installed in Light-Water-Cooled Nuclear Power Reactor Plants" (revised in July 1981)[14].

The *Process Control Program* establishes boundary conditions in the form of processing parameters for the solidification system such that operation within these limits will provide reasonable assurance that solidification is complete and all stability requirements are satisfied. The boundary conditions for each solidification system will be determined by tests with chemical wastes that could be found in the liquid wastes at the nuclear power plant. These boundary conditions will be established as measurable physical parameters important to the solidification process such as chemical content of the liquid/slurry waste being solidified (e.g., pH, soluble/insoluble concentrations, oil content, etc.,), chemical quality of solidification agents (e.g., type cement, binder, pH, etc.,) and liquid waste to solidification agent radios. Once the boundary conditions are fixed the operator will be expected to stay within these limits since they will be part of the solidification system operating procedures. In general, process control programs for solidified class A[15] waste products need demonstrate that the product is a free standing monolith with no more than 0.5 percent of the waste volume as free liquid. All class B and C[15] waste products must be stabilized (i.e., solidified or equivalent). One equivalent or alternative to solidification being the use of a high integrity container (See requirements for HIC in reference.[16])

In the late 1970's the most common radwaste solidification agents used in the United States have been cement and Urea-formaldehyde (UF) resins. However, significant problems have developed with the UF process including excess water production and inability to meet the NRC and burial site requirement for stabilized waste. Because of these problems UP systems have been removed from the market. In the 1980's different types of organic polymers have entered the domestic radwaste service market as well as asphalt (bitumen), which has been extensively used in Europe. The chemical and physical properties of each of the major solidification processes and the methods used are described in the following section.

F.   POWER PLANT SOLIDIFICATION PROCESSES

1. Incorporation in Cement

Cement solidification generally mix the concentrated waste form with portland type I or type II cement in a ratio of approximately 30 gallons of liquid waste with 55 gallons of cement to form 55 gallons of solid.[17] The mixing may take place in a drum by rolling, mixing blade or by a mixing screw, or in an in-line line mixer or vessel before injection into the drum or container. Various additives may be used in the cement mix; calcium chloride, sodium silicate, vermiculite, bentonite, grundite, lime or slag. These additives are used as volume fillers and absorbants and to reduce problems in setting up of the cement with certain chemical compositions of wastes. The American Society for Testing and Materials ASTM has defined the restrictions on the chemical composition of portland cements as imposed in all national standards specifications.[18] Incorporation in cement, one of the oldest solidification techniques, has been used at commercial power plants for over 20 years.

A process schematic diagram of a typical cement system, is shown in Figure 2, and various alternatives are schematically represented in Figure 3. The process is extremely simple. Cement powder is pneumatically conveyed from delivery into a storage hopper. A vibrator and screw conveyor feed the cement to a screw mixer while waste liquid is simultaneously metered from a holding tank into the mixer. The mixed cement is discharged into a drum or larger shipping liner where it solidifies. Generally, the operation is a batch process.

Because of the alkaline nature of portland cement, basic wastes are readily solidified, but acidic wastes (such as boric acid) and highly con-

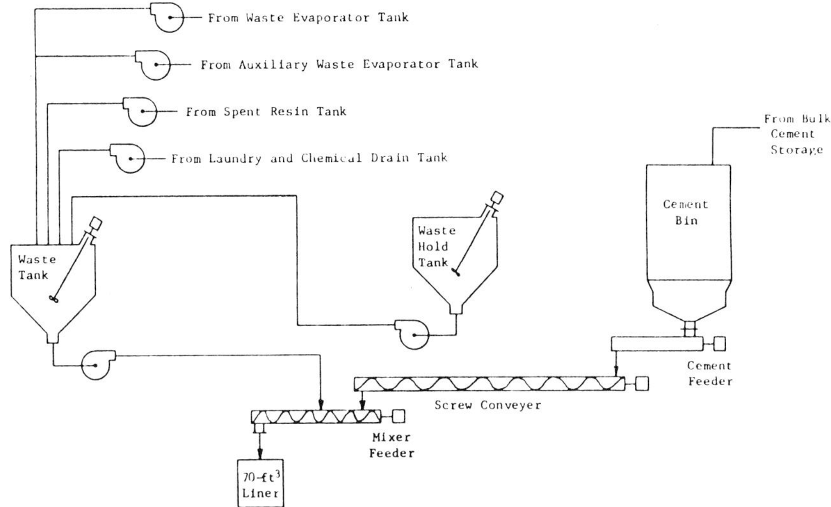

FIGURE 2. TVA Cement System.

centrated sodium borate solutions may cause problems producing an acceptable final product.[19,20] Sodium silicate has been used successfully as an additive to solidify these waste liquids. Acidic wastes should be pH-adjusted prior to mixing with cement to minimize these problems. Residual water sometimes forms on top of a solidified drum and must be usually absorbed by adding vermiculite.

Operating characteristics are tabulated in Table 1 for comparison with other solidification systems.

2. Incorporation in Bitumen (Asphalt Extrusion)

Bitumenization has been used in Europe for more than a decade at nuclear power reactors, and currently exist at two U.S. plants.[21] The process involves mixing in a screw extruder streams of concentrated liquid waste and hot liquid bitumen. Steam generated from the evaporation of the liquid is vented at various locations on the extruder and is condensed. The solids formed as the liquid evaporates within the extruder are intensively mixed into the liquid bitumen by the screw's shearing action. A liquid stream of bitumen is discarded from the extruder into a drum or liner. The Werner & Pfleiderer Corporation, a manufacturer of extruders for thermoplastic molding operations, in cooperation with the Nuclear Research Center at Karlsruhe, developed this process and offers it in the U.S. Up to 250 gallons of liquid with 12 wt % solids can be evaporated and incorporated into a 55-gallon drum of solidified bitumen.

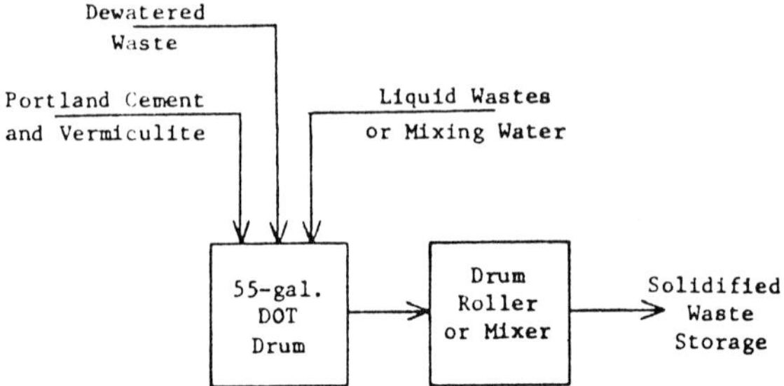

PORTLAND CEMENT DRUM MIXING SOLIDIFICATION SYSTEM

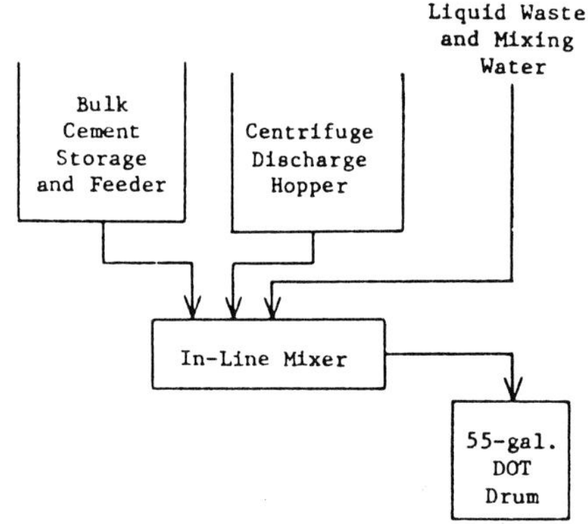

PORTLAND CEMENT IN-LINE MIXER
SOLIDIFICATION SYSTEM

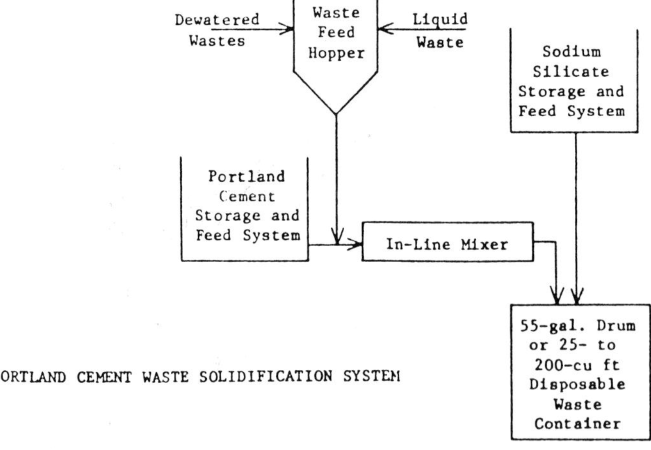

PORTLAND CEMENT WASTE SOLIDIFICATION SYSTEM

FIGURE 3. Alternative Cementation System.

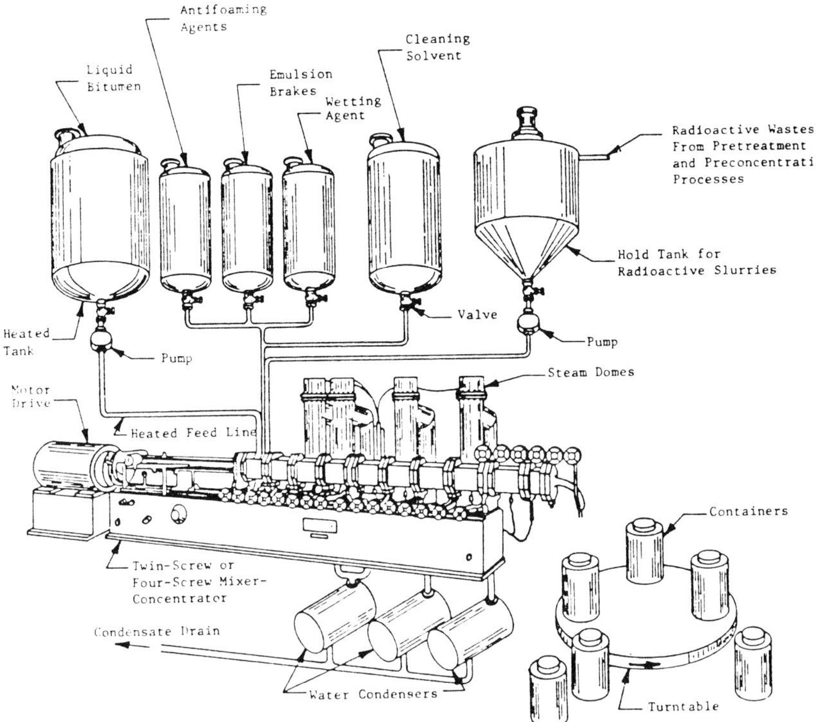

FIGURE 4. Extruder Bitumenization Schematic.

The extruder bitumenization process is shown schematically in Figure 4. Liquid wastes from a storage tank are pumped and metered into the extruder together with liquid bitumen at a temperature of about 140 C. Prior treatment of the liquid wastes may include pH adjustment and co-precipitation of some radionuclides. Cesium-137 is precipitated with nickel-ferrocyanide, strontium-90 is treated with barium sulfate, and cobalt-60 is precipitated as cobalt sulfide. The extruder is heated to 200 C steam to evaporate the water in the waste liquid. The twin or four screws within the extruder shear, heat, and mix the liquid wastes (as it evaporates to dryness) with the bitumen matrix. The screw action is self-cleaning,, which avoids caking and buildup. Various temperature regions are maintained along the length of the screw with the final product exiting at about 175 C into a container where it is allowed to cool and solidify.

3. Aerojet Volume Reduction System/Solidification System (VR-20)

The Aerojet VR-20 process is a continuous process which can dry to solid 20 gal/h of liquid waste in a fluidized bed dryer operated at 800

TABLE 1

Operating System for Drying and Solidifying Concentrated Wastes

Operating Data	Cement, cement with additives	Bitumen extrusion	Aerojet VR-20	Drying-polyester	Extruder polyethylene	Wiped film, stirred evaporator bitumen
Packaging efficiency, max wt % solids	3 to 4	50	80	40	50	40
Density, sp. g of solid	2.3	1.3	2.3 cement 1.1 aeropep	1.4	1.1	1.3
Leachability[101-103] (cesium), g/cm^2	10^{-4} to 2 x 10^5	10^{-5} to 10^{-6}		10^{-5} to 10^{-6}	10^{-5} to 10^{-6}	10^{-5} to 10^{-6}
Radiation stability, rads	$>10^9$ 10^8 aeropep	10^8	$>10^9$ cement	10^8	10^8	10^8
Tolerance for						
Boric acid wastes	Poor, good	Good	Good	Good	Good	Good
Basic wastes	Good	Good	Good	Good	Good	Good
Regenerate wastes	Good	Good	Good	Good	Good	Good
Laundry, decontamination wastes	Problems	Good	Good	Good	Good	Good
Free-standing Solid	Yes	Yes				
Residual water	Sometimes, none	None	None	None	None	None

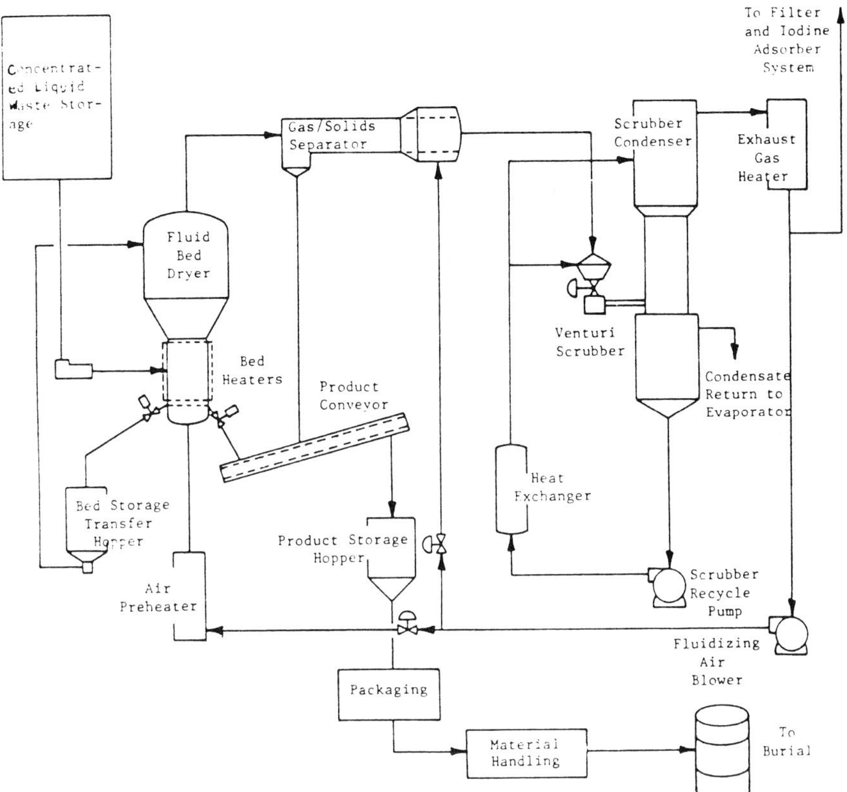

FIGURE 5. Aerojet Volume Reduction System.

to 1000 F. The solids are granular, from 0.3 to 0.5 mm in size. The solids may be embedded in various media (cement or aeropep) after drying. Aeropep is a thermoplastic polymer similar to low-molecular-weight polyethylene. An incorporated solid content of up to 80 wt % in cement can be produced. This is equivalent to incorporating about 360 gallons of liquids with 12 vt % solids in a single 55-gallon drum. The fully developed process is based on developmental work in waste calcining performed at the National Reactor Testing Station in Idaho.

The VR-20 process schematic is shown in Figure 5. Concentrated/slurried liquid wastes are sprayed into an electrically heated air stream. Water vapor is driven off in the fluid-bed dryer, leaving granules of anhydrous waste calcine, which are discharged into the product conveyor. Solid carryovers from the fluid bed are separated from the vapor stream and are also fed to the product conveyor. The granular solids are collected in a hopper and fed to an embedding or packaging station. The air and water

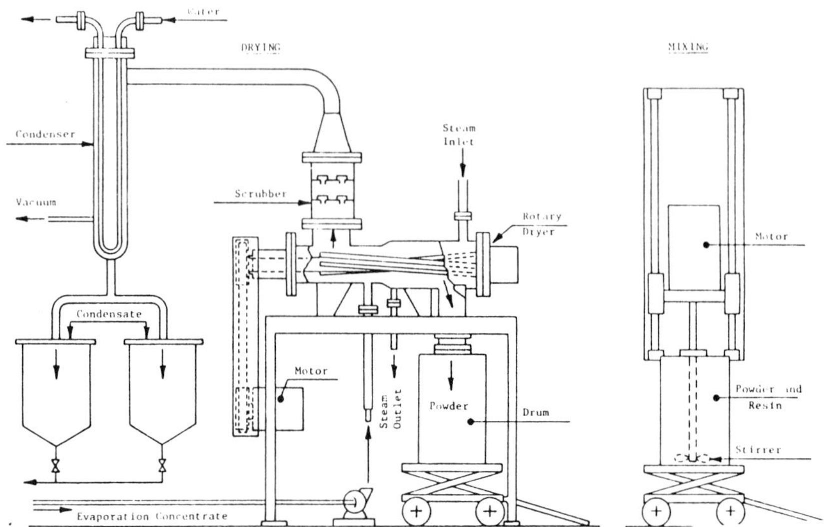

FIGURE 6. Thermosetting Plastic Solidification Drying and Embedding Station.

vapor stream enters a venturi scrubber followed by a scrubber condenser. Condensate water is recycled to the plant's radwaste evaporator. The air is continuously recycled through the condenser scrubber for decontamination, while a portion of the stream is reheated and fed back to the fluid bed by a blower. Offgas air is run through absolute filters and charcoal iodine absorbers before being vented through the plant stack.

4. Drying Thermosetting Resin Encapsulation

This process was developed by the French Commissariat a l'Energie Atomique (CEA), and industrial development is being handled by the Groupement pour les Activities' Atomiques (GAAA). Liquid wastes are evaporated to dryness in a rotary vacuum dryer. The dried powder is then mixed in a drum with a polyester resin, a solution of a maleophtalate of propylene glycol in styrene. Solidification is accomplished by adding a catalyst to initiate polymerization. Up to about 200 gallons of 12 wt.% liquid wastes can be incorporated into a 55-gallon drum of solidified polyester.

The drying-thermosetting resin process is shown schematically in Figure 6. Concentrated wastes are pumped and metered into a steam-heated rotary vacuum dryer. Blades within the shell scrape and advance the dried solids to an air lock with capacity for one drum load of powder. The air lock periodically discharges loads of powder to the drums. The water vapor evaporated in the dryer is cleaned in a scrubber, condensed, and can be returned to the plant evaporator. Offgas drawn through the

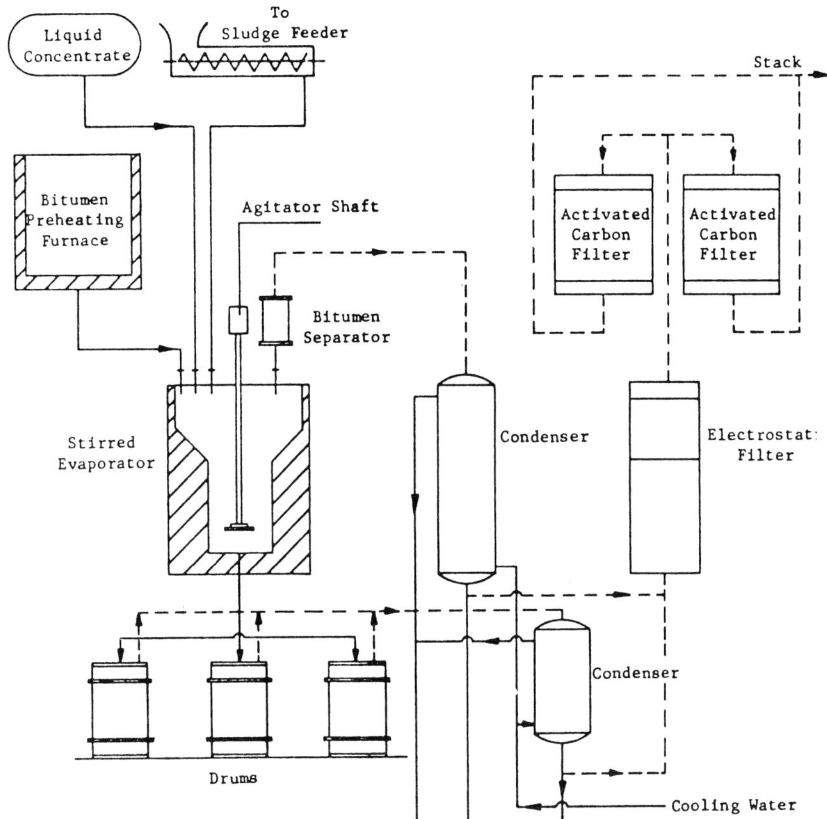

FIGURE 7. Stirred Vaporator Bitumenization.

vacuum supply must be filtered and passed through charcoal before being released. The drum of powder is moved to a mixer, where it is filled with a polyester and mixed in the drum using a rotary and up-and-down motion. When a homogeneous mixture is obtained, a catalyst (usually methyl-ethyl keytone peroxide) is added and the mixer is raised. Solidification ensues. The polymerization process is exorthermic. Table 1 provides comparison lists the operating characteristics and product quality, of the drying-thermosetting resin encapsulation process.

5. Stirred Evaporator Bitumenization

In this type of bitumenization process, waste liquids or solid slurring are stirred in a heated vessel with liquid bitumen. The liquid evaporates, and the remaining solids are coated with bitumen. About 40 wt % solids can be incorporated. This is equivalent to incorporating 180 gallons of liquids with 12 wt % solids in a 55-gallon drum. Stirred evaporator

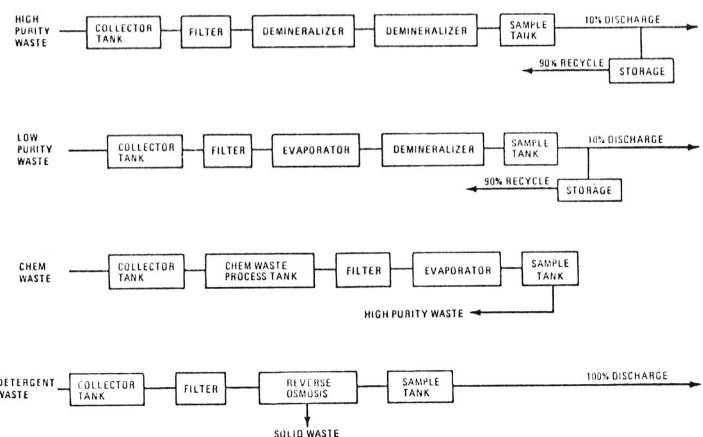

FIGURE 8. Radwaste Treatment Systems for BWR.

bitumenization has been developed at Belgo-Nucleaire in Mol. and Harwell (AERE). A process schematic diagram is shown in Figure 7. Evaporator hot inner surface, giving good heat transfer. This mixes and evaporates the water from the bitumen. Vapor and gases are condensed, filtered, and released. The encapsulated bitumen mixture discharges into drums on a carousel.

Residence times within the evaporator are short. However, the rotating blade assembly tends to be mechanically delicate and the close wall clearances are hard to maintain. Mixing is better than that in the stirred evaporator but not as good as an extruder. Operation characteristics are tabulated in Table 1.

III. LIQUID WASTE MANAGEMENT

The development of facilities and equipment to collect and process liquid radwaste has given the nuclear industry the capability to hold releases of radioactive material in liquid effluents within applicable regulatory limits and the guidelines established in Appendix I of 10 CFR 50. These limits are most readily met by reducing the volume of liquids discharged or by processing (decontaminating) these liquids to a high degree before discharging them to the environment. A liquid radwaste processing system should be designed to allow the maximum reuse of waste water consistent with the overall plant water balance. There are a number of NRC Regulatory Guides, Standard Review Plans and Effluent Treatment System Branch Position[22-32] which collectively provide the basis for requirements which must be considered in the design, operation and management of liquid waste processing system.

Liquid radwastes are generally classified and processed according to their physical and chemical properties. As with solid waste, the management of liquid processing is a function of the type of plant, the waste forms encountered and additionally, the capability for recycling both water and chemicals. The selection of processing techniques usually contain a combination of a number of physical and chemical separation processes. The most commonly used processes are evaporation, filtration and ion-exchange. Boiling Water Reactors (BWR's) and Pressurized Water Reactors (PWR's) typically generate different waste forms and therefore have differing waste processing streams. The following provides a summary of typical BWR and PWR liquid waste processing systems.

A. *BWR LIQUID RADWASTE PROCESSING SYSTEMS*

Boiling Water Reactors typically have four (4) classes of liquid waste sources which they process and recycle if possible.

Figure 8 provides a description of a liquid waste treatment system representative of that used in many current BWR plants. The following is a summary of each treatment stream and the processing strategy.

1. High Purity Waste

   Major sources of high purity BWR waste include equipment drains from the dry well; the reactor, turbine, auxiliary and fuel pool buildings; ultrasonic resin cleaner; resin and filter backwash and transfer water; phase separater decant liquid; waste evaporator condensate. This waste is generally low conductivity (< 50 unbo/cm at 25°C) but potentially contains suspended solids and dissolved oils. Because of the particulate solids and oils this waste is usually processed by filtration and ion exchange. If the quality of the processed water is determined acceptable the majority of this waste water is recycled (> 90%).

2. Low-Purity Waste

   Major sources of low-purity waste include floor drains from the various BWR buildings which contain radioactive fluids including the dry well, reactor, turbine, radwaste and fuel-pool buildings. Additional it includes uncollected valve and pump seal leaks and waste from dewatering of slurry waste. This waste is generally higher conductivity; if the conductivity is sufficiently low (< 100 unbo/cm at 25°C) this waste is normally processed by filtration and ion-exchange but if the conductivity is higher, evaporation is used. Whether or not a filter precedes the evaporator depends upon the susceptibility of fouling the evaporator and the properties of the feed stream. When required, condensate from the evaporator can be treated further by the high purity waste ion-exchange system or by a separate ion-exchange polishing unit.

3. Chemical Wastes

The major source of chemical waste in a BWR is the product of the regenerate solution (Na_2SO_4 - Sodium Sulfate) from the deep bed in ion-exchange beds. Other significant sources include laboratory drains and non-detergent chemical decontamination wastes. Because these waste contain a high concentration of dissolved and undissolved solids, chemical waste are usually processed by evaporation in most cases. Polishing of the condensate by ion exchange may be needed to meet recycle or discharge requirements.

4. Detergent Waste

The major sources of BWR detergent waste are on-site laundries, personnel showers, and detergent type decontamination waste. These wastes usually contain relatively small amounts of radioactivity but significant amounts of suspended solids and organic chemicals which can adversely effect the operation of conventional evaporator (priming and foaming) and ion exchanger fouling. It is for this reason that processing is usually by filtration and reverse osmosis systems.

B. PWR LIQUID RADWASTE PROCESSING SYSTEMS

Pressurizing Water Reactors have segregated waste streams similar to BWR's, however, because of the reactor coolant system boron shim control and the steam generator designs some additional waste stream are generated and processed differently. Figure 9 provides a description of a typical waste treatment system representative of that used in many current generation PWR plants. The five (5) separate processing streams are as follows:

1. Clean Waste

The major source of clean waste includes reactor coolant bleed and feed waste generated during the boration and deboration operations for reactor power shim control. This waste stream is relatively clean (<50 unbo/cm) and contains high concentrations of boric acid which can be readily concentrated and recycled by use of evaporation. Two ion-exchangers (cation and mix-bed) upstream of the evaporator are used to insure that the recycled boric acid is relatively pure of ionic impurities. Ion exchange is also provided for the evaporator condensate treatment to reduce effluent activity and to provide further purification for water recycle applications. Typically, PWR's which maximize reuse of boric acid waste water consistent with the overall plant water balance and chemical purity can recycle as much as 90% of both the waste water and boric acid.

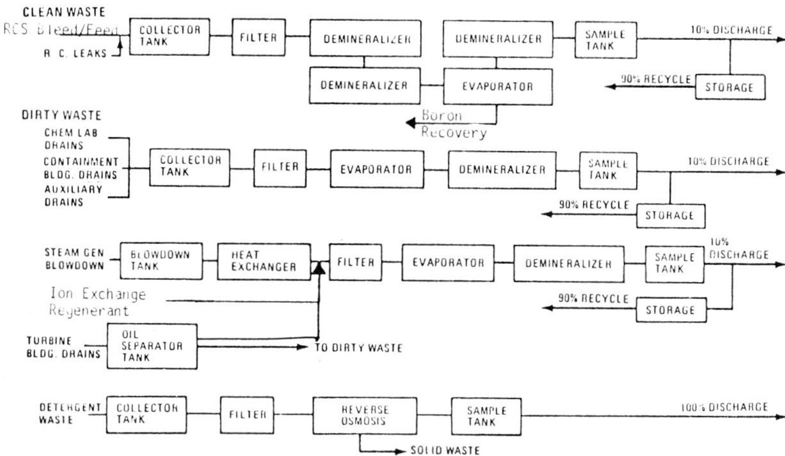

FIGURE 9. Radwaste Treatment Systems for PWR.

2. Dirty Waste

The major source of dirty waste are Containment, Auxiliary and Fuel Handling Building floor drains, outdoor controlled area wastes, sampling sink waste, deaerated systems and equipment drains, primary system ion-exchange and filter wastes. This waste is usually processed by filtration and evaporation followed by condensate polishing by ion-exchange. Evaporation is used since it can efficiently separate water from both soluble and unsoluble impurities. Filtration is used to protect the heat transfer surfaces and prevent fouling of the evaporators. If water quality is met, water is typically recycled.

3. Secondary/Steam Generators Waste

The major source of this waste is steam generation blowdown and secondary system drains. This waste is usually of low conductivity and low radionuclide concentrations if a primary to secondary steam generator tube leak exists. The steam generator blowdown is commonly cooled through a heat exchanger and processed by filtration and ion-exchange. Some PWR plants have alternate waste treatment schemes for treating S.G. blowdown through an evaporation process or reverse osmosis. Turbine building drains, if contaminated due to a S.G. tube leak, should have provisions for processing through the dirty waste system. Additionally, some PWR's will utilize the secondary/SG blowdown treatment system for processing ion exchange regenerant waste.

4. Chemical and Ion Exchanger Regenerant Waste

The primary sources of this waste are primary/secondary ion-exchange regenerant solutions, chemical cleaning waste, non detergent type decontamination waste, laboratory drains, etc. This waste usually contains high concentrations of soluble inorganic substance and relatively low concentrations of radioactivity. PWR plants typically neutralize this prior to considering processing or discharge. Maximum recycle PWR plants will process this waste through a filter, evaporator (or RO) and polish with an ion exchanger, if required, for recycle purposes.

5. Detergent Waste

The major source of PWR detergent wastes are similar to BWR's, and they include on-site laundries, personnel showers and detergent type decontamination wastes. Because these wastes contain significant quantities of suspended solids and organic chemicals they generally are processed via particulate filters and reverse osmosis systems. Some plants also install charcoal filters and polish with an ion-exchange system prior to discharge. Other plants prefer not to establish laundry facilities (but utilize off-site laundry service) which substantially reduces the amount of detergent waste generated on-site.

C. WASTE GENERATION AND PROCESSING DESIGNS PARAMETERS

The quantities of waste generated in the various PWR and BWR waste streams identified above are usually a function of the size and design of the plant and are influenced by upset conditions caused by equipment failures, extended shutdowns for maintenance, and abnormal occurences. The American National Standards N197/ANS-55.3 and N199/ANS-55.2 has generated an itemized list of probable waste stream inputs which includes expected average daily input sources and the design basis occurences for BWR and PWR plants respectively.[33,34]

A more detailed description of a current generation PWR liquid radwaste system is shown in Figure 10. This represents the Catawba PWR plant which is scheduled to go into operation in late 1984. The plant will contain a boron (resin) thermal regenerative system which will supplement the clean waste boron recycle evaporator described above. The design parameter and expected input sources for this plant are shown in Tables 2 and 3.

A similar description of current generation BWR liquid radwaste system is shown in Figure 11. This represents the LaSalle County Station units Nos. 1 and 2 which went into operation in June 1982. The design parameters and expected input sources for this plant are shown in Table 4.

The resulting cleanup of liquid waste streams contaminated with chemical and radiochemical activity is obtained by the selection of one or more combinations of the physical and chemical separation processes. The composi-

TABLE 2

*Design parameters of principal components considered in the evaluation of liquid and gaseous radioactive waste treatment systems**

Component	Number	Capacity, each
LIQUID SYSTEMS		
Chemical and volume control system		
Volume control tank	1 per reactor	400 ft³
Mixed-bed demineralizers	2 per reactor	120 gpm
Cation demineralizer	1 per reactor	75 gpm
Thermal regeneration demineralizers	5 per reactor	120 gpm
Boron recycle system		
Recycle evaporator feed demineralizers	2 shared	120 gpm
Recycle holdup tanks	2 shared	112,000 gal
Recycle evaporator package	1 shared	15 gpm
Recycle evaporator condensate demineralizer	1 shared	35 gpm
Liquid waste processing system		
Reactor coolant drain tank	1 per reactor	350 gal
Waste drain tank	1 shared	5,000 gal
Waste evaporator feed tank	1 shared	5,000 gal
Floor drain tank	1 shared	10,000 gal
Recycle monitor tanks	2 shared	5,000 gal
Waste monitor tanks	2 shared	5,000 gal
Chemical drain tank	1 shared	600 gal
Laundry and hot shower drain tank	1 shared	10,000 gal
Waste evaporator	1 shared	15 gpm
Waste evaporator condensate demineralizers	1 shared	35 gpm
Steam generator drain tanks	2 shared	50,000 gal
Ventilation unit condensate drain tanks	1 per reactor	5,000 gal
GASEOUS SYSTEMS		
Waste gas system		
Waste gas system compressors (design pressure, 150 psig)	2 shared	40 scfm
Gas decay tanks (design pressure, 150 psig)	8 shared	600 ft³

*Design classification and seismic design criteria per Regulatory Guide 1.143.

tion of the waste stream and the quality of the effluent required will dictate which separation processes are chosen. Presently the most frequently used processes for treating liquid radwaste are evaporation, filtration, ion exchange and reverse osmosis. Evaporation is the process by which volatile and non-volatile (dissolved and undissolved) components of a solution or slurry are separated via boiling away and removing the volatile component. The principal elements involved in evaporation and a review of the evaporator designs and operating experience are provided in references 35 through 36. A similar compilation for filtration (Ref. 37-38), Ion-Exchange (Ref. 39-40) and reverse osmosis (Ref. 41-42) are provided.

TABLE 3

PWR LIQUID WASTES

EXPECTED DAILY AVERAGE INPUT FLOW RATE (in Gal/day)

SOURCE	Type of treatment of blowdown recycled to secondary system (U-tube steam generator plants) or type of treatment of condensate (once-through steam generator plants)				Fraction of Primary Coolant Activity (PCA)
	Deep-bed cond. demineralizers with ultrasonic resin cleaner	Deep-bed cond. demineralizers without ultrasonic resin cleaner	Filter-demineralizer	Plant with blowdown treatment. Product not recycled to condenser or secondary coolant system	
1. REACTOR CONTAINMENT					
a. Primary coolant pump seal leakage	20	20	20	20	0.1
b. Primary coolant leakage, miscellaneous sources	10	10	10	10	1.67
c. Primary coolant equipment	500	500	500	500	0.001
2. PRIMARY COOLANT SYSTEMS (OUTSIDE OF CONTAINMENT)					
a. Primary coolant system equipment drains	80	80	80	80	1.0
b. Spent fuel pit liner drains	700	700	700	700	0.001
c. Primary coolant sampling system drains	200	200	200	200	0.05
d. Auxiliary building floor drains	200	200	200	200	0.1

3. SECONDARY COOLANT SYSTEMS					
a. Secondary coolant sampling system drains	1400	1400	1400	1400	10^{-4}
b. Condensate demineralizer rinse and transfer solutions	3000	12000	—	—	10^{-8}
c. Condensate demineralizer regenerate solutions	850	3400	—	—	Calculated in GALE Code
d. Ultrasonic resin cleaner solutions	15000	—	—	—	10^{-6}
e. Condensate filter—demineralizer backwash	—	—	8100	—	2×10^{-6}
f. Steam generator blowdown	—	—	—	Plant dependent[a]	Plant dependent
g. Turbine building floor drains	7200	7200	7200	7200	Calculated in GALE Code
4. DETERGENT AND DECONTAMI—NATION SYSTEMS					
a. On—site laundry facility	300	300	300	300	**
b. Hot showers	Negligible	Negligible	Negligible	Negligible	—
c. Hand wash sink drains	200	200	200	200	**
d. Equipment and area decontamination	40	40	40	40	**
TOTALS	29,700	26,300	19,000	10,000	

**GALE Code uses release data calculate releases from this source.
[a]Input parameter

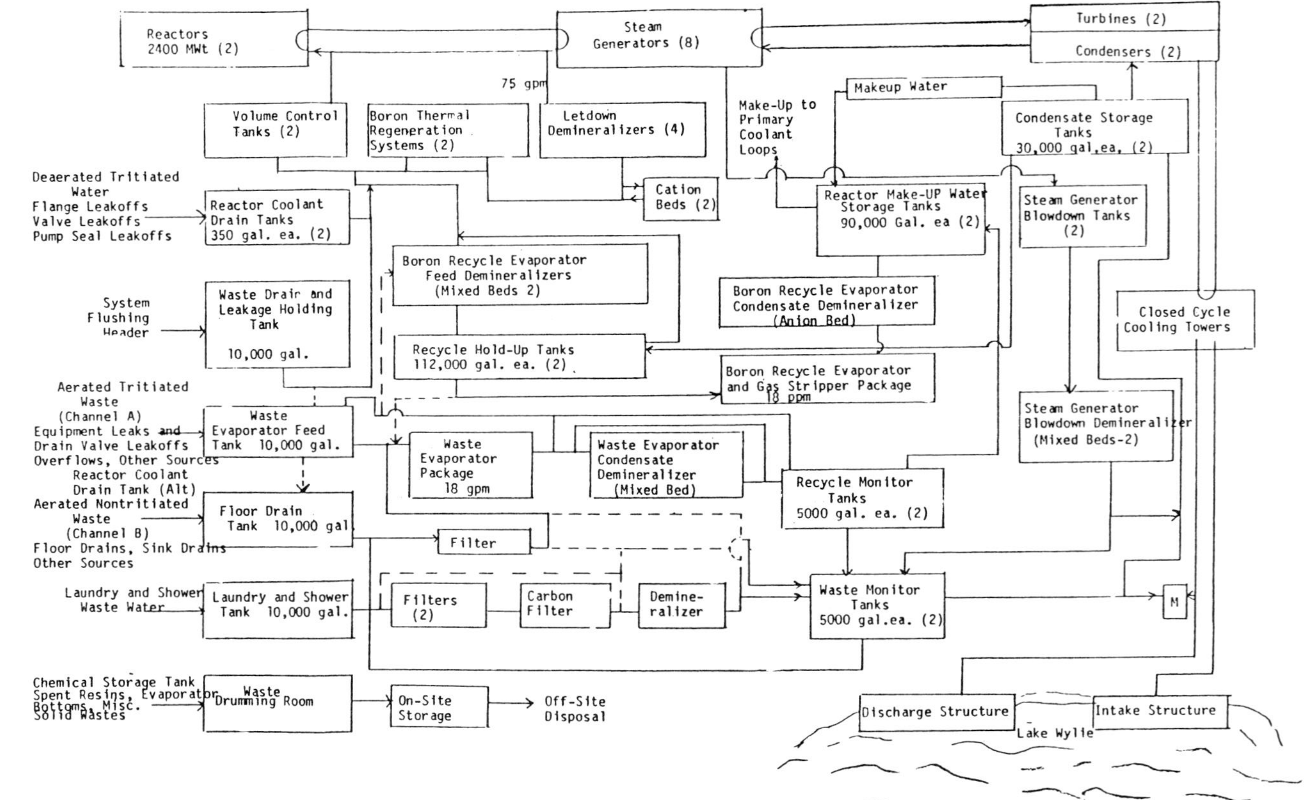

FIGURE 10. Liquid Radwaste System, Catawba.

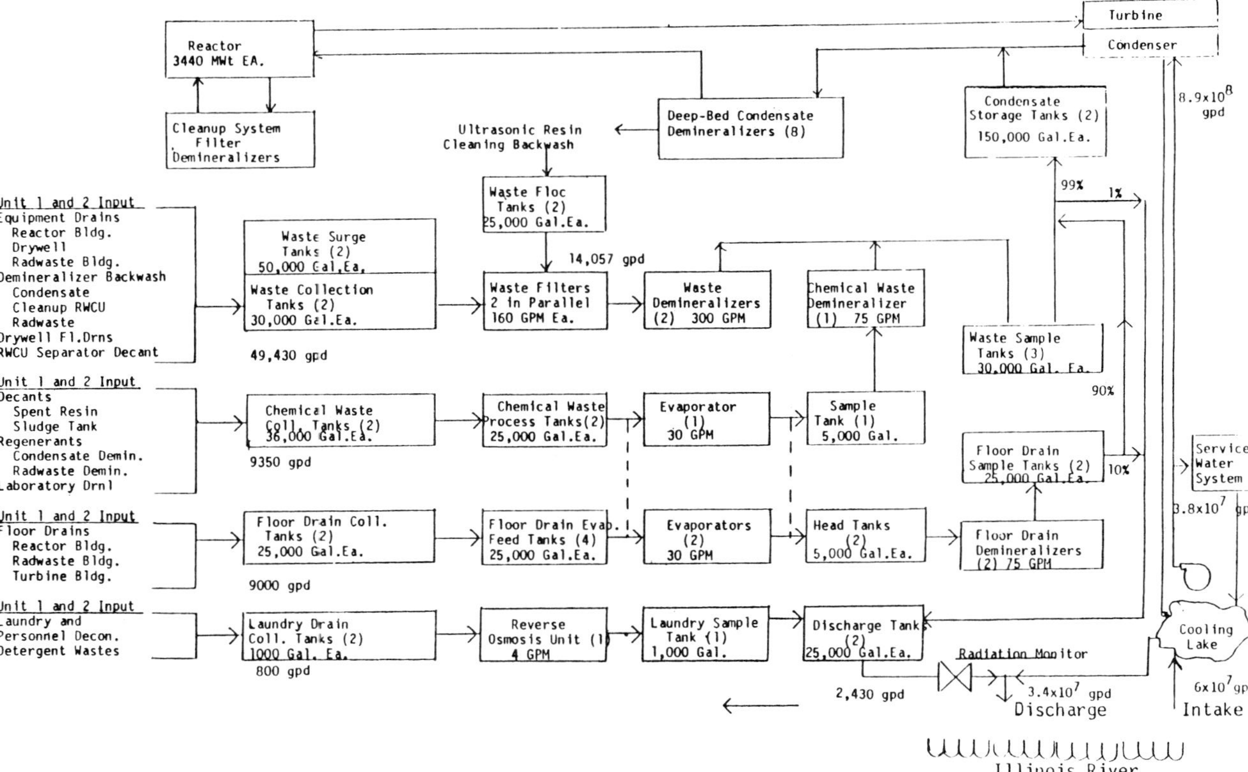

FIGURE 11. Liquid Waste System, LaSalle County Station Units 1 and 2.

IV. GAS WASTE MANAGEMENT

Nuclear power reactors generate radioactive fission gases (predominately Xenon and Krypton), activation gases (predominately Argon-41 and Nitrogen-16) and combustible radiolytic gases (predominately hydrogen and oxygen). Each of these gaseous waste have different properties and associated problems, and therefore they require differing waste treatment strategies. Because of their volatility and solubility in liquids these gases are communicated throughout various plant systems and require special controls and safety features. Specially designed off-gas treatment systems are required to limit radioactive release to the environment as well as to provide safety and limit radiation exposures to plant operators.

Typically LWR plants do not concentrate, transport and dispose of gaseous waste as was discussed in the previous sections for solid and liquid waste management. Gaseous radwastes are normally collected, stored for radioactive decay and released to the environment. Since LWR waste gases do not constitute a long-term, hazardous radioactive waste source this section will only provide a brief summary of how these gases are processed and managed at nuclear power plants.

A. BWR OFF-GAS SYSTEM

Radiolysis of the cooling water in the reactor core, generates gaseous hydrogen and oxygen. These gases, along with fission product noble gases released from the fuel are carried by steam to the condenser. These gases plus any activation gas and air inleakage are removed from the main condenser by the steam jet air ejectors. The air ejector off-gas is typically treated in a three stage process. First, hydrogen and oxygen which represents about 80 percent of the exhaust must be removed from the process stream for both safety and to reduce the size of the required process equipment. This is accomplished by passing these gases through a catalytic recombiner, followed by removal of the dilution steam. Secondly, any residual water vapor is then removed by a disicant dryer or freeze-out heat exchanger. In the third stage the radioactive gases are removed from the gas stream by separators and/or collected for radioactive decay. At this point, the fission product gases remaining can be either released to the environment or placed in cylinders for long-term storage. The state-of-the-art in BWR off-gas treatment systems range from Compressed Gas Storage, Charcoal Delay, Cryogenic Distillation and Cryogenic Charcoal.

Adequate shielding must be incorporated to protect plant personnel from radiation exposure during system maintenance. Shielding requirements depend on the quantity of fission product gases released and their time out of the reactor. In a typical BWR off-gas system N_{-16} dominates the source term for shielding design for the recombiner condenser.

TABLE 4

BWR LIQUID WASTES

Source	Expected Daily Average Input Flow Rate (in gal/day)		Fraction of the Primary Coolant Activity (PCA)
	Deep Bed Plant with Ultrasonic Resin Cleaner	Deep Bed Plant without Ultrasonic Resin Cleaner or A Filter/Demineralizer Plant	
Equipment Drains			
Drywell	3,400	3,400	1.00
Containment, auxiliary building, and fuel pool	3,700	3,700	0.1
Radwaste building	1,100	1,100	0.1
Turbine building	3,000	3,000	0.001
Ultrasonic resin cleaner	15,000	-	0.05
Resin rinse*	2,500	5,000	0.002
Floor Drains			
Drywell	700	700	0.001
Containment, auxiliary building, and fuel handling	2,000	2,000	0.001
Radwaste building	1,000	1,000	0.001
Turbine building	2,000	2,000	0.001
Other Sources			
Cleanup phase separator decant	640	640	0.002
Laundry drains	1,000	1,000	-
Lab drains	500	500	0.02
Regenerants*	1,700	3,400	**
Condensate demineralizer backwash	-	8,100	2×10^{-6}
Chemical lab waste	100	100	0.02

*Deep-bed condensate demineralizers only.
**Calculated by BWR-GALE Code.
Filter/demineralizer (Powdex) condensate demineralizers only.

Because of the Nitrogen-16, 7.6 MeV high energy gamma, approximately five feet of concrete shielding is typically required for the recombiner components. For components located further downstream, the source term depends on the time since leaving the reactor, the holdup time in the component, and the buildup of decay daughter products. Most off-gas equipment is inaccessible during plant operation because of high contact dose rates which can be up to 200 rem/hr for the recombiner vault and the charcoal adsorber vessels.

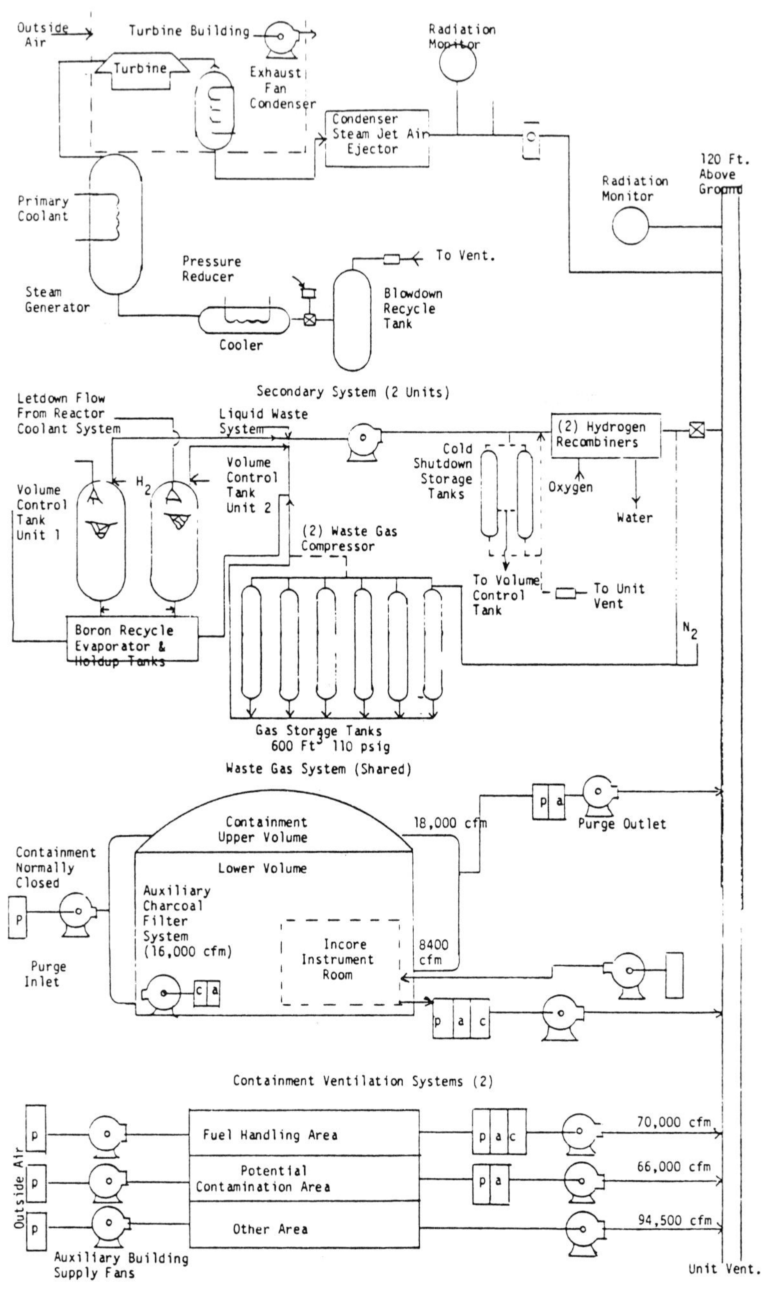

FIGURE 12. Gaseous Waste System, Catawba.

B. PWR OFF-GAS TREATMENT

Fission product and activation product gases generated in the core of a pressurized water reactor could be released to the containment building by primary coolant leakage, to the auxiliary building through the letdown system or to the turbine building through tube leaks in the steam generator. Waste gas control is a lesser problem in PWR's than BWR's because of the primary system boundary and suppression of radiological gas generation by hydrogen over-pressure. The release of fission produce gases to the environment from leak pathways is best controlled by reducing the inventory of gases in the reactor coolant systems. This is accomplished to a large extent by better quality fuel cladding and by passing the reactor coolant letdown flow through a gas stripper. Additionally, to prevent gaseous releases when the reactor vessel and RCS are opened for maintenance, gases are also removed via the gas stripper following plant shutdown but before opening the reactor system. PWR waste gases systems, much like BWR's, contain provisions for combustible gas control (viz., recombiner and/or nitrogen purge) and utilize components such as waste gas decay tanks, Ambient and Cryogenic Charcoal, Nitrogen recycle compressors, etc. Figure 12 shows the waste gas system for the Catawba plant.

Although off-gas systems are not regarded as safety-class systems, some additional design requirements over those for non-safety-related systems are imposed. These systems are designed in accordance with Section VIII of the ASME Boiler and Pressure Vessel Code[43] and with the American National Standards Institute (ANSI) standard B31.1.[44] In addition, other requirements include mandatory pressure testing and seismic analysis of the equipment supports. These requirements, alone, with restrictions on the use of certain materials are discussed in NRC Branch Technical Position 11.1.

V. CONCLUDING DISCUSSION

Radioactive Waste Management in nuclear power plants is complicated because of the diverse physical and chemical characteristics of the waste forms and also because of the environmental limits and controls required to properly handle, process, transport and dispose of these wastes. The ideal radwaste system, as an auxiliary system to the reactor, should be so conceived, so engineered, so selected and constructed, so operated, and so maintained that there is no unexpected occurences within it which would force the reactor to shutdown.

Historically, many operating reactors have found it necessary to retrofit, modify and augment their radwaste system shortly after they went into commercial operation. Additionally, many that are presently under construction are already obsolete and must be upgraded. This is partly due to regulatory developments as well as deficiencies with plant designs and solidification technology. These modifications have been costly both from a time element and economics. Plant designers and operators should focus their engineering and operational considerations towards such objectives as: constructability, availability, maintainability, economy, operability, safety, and licensability. Significant improvements are needed in all of these areas for better power plant radwaste management.

REFERENCES

1. U.S. Atomic Energy Commission, Final Environmental Statement Concerning Proposed Rule Making Action, Volume 1, WASH 1258 (July 1973).

2. U.S. Nuclear Regulatory Commission, "Design Guidance for Solid Radioactive Waste Management Systems Installed in Light-Water-Cooled Nuclear Reactor Power Plants," NUREG-75/087 Standard Review Plan, USNRC, Sect. 11.4, Effluent Treatment System Branch Technical Position, BTP ETSB 11-3, (1980).

3. USNRC letter to all Commission Licensees from L.G. Higinbotham, NMSS, "Final Classification and Waste Form Technical Position Papers, dated May 11, 1983.

4. Kibby, A.H. and Godbee, H.W., A critical Review of Solid Radioactive Waste Practices at Nuclear Power Plants, ORNL-4924, March 1974.

5. U.S. Nuclear Regulatory Commission, Information Relevant to Maintaining Occupational Radiation Exposures As Low As Reasonably Achievable (Nuclear Power Reactors), Regulatory Guide 8.8 and 8.10, USNRC, Washington, D.C.

6. International Atomic Energy Agency, "The Volume Reduction of Low-Activity Solid Waste," Technical Report Services No. 106, STI/DOC/10/106, IAEA, Vienna (1979).

7. R.H. Burns, "Solidification of Low-and-Intermediate-Level Waste," Atomic Energy Rev., Vol. 9, No. 3 (1971).

8. International Atomic Energy Agency, Management of Low and Intermediate Level Radioactive Wastes, STI/PUB/264, IAEA, Vienna.

9. Holcomb, W.F. and Goldberg, S.M., Available Methods for Solidification of Low-Level Radioactive Waste in the United States, U.S.E.P.A., ORD/TAD-76-4, December 1976.

10. USNRC Information Notification Bulletin 80-24 Low-Level Radioactive Waste Burial Criteria May 30, 1980, USNRC, Washington, D.C.

11. Blomeke, J.O., and Kee, C.W. "Projections of Wastes to be Generated," Proceedings of the International Symposium on the Management of Wastes, Denver, Colorado, July 11-16, 1976, CONF-76-0701, pp. 96-117.

12. Blomeke, J.O., and Nichols, J.P., Projections of Radioactive Waste to be Generated by the U.S. Nuclear Power Plant Industry, ORNL-TM-3965.

13. USNRC, Code of Federal Regulations, 10 CFR 61, LAND Disposal of Low-Level Radioactive Waste, Washington, D.C.,1982.

14. USNRC, NUREG-0800, Solid Waste Management System, and Design Guidance for Solid Waste Management System Installed in Light Water Cooled Reactor Plants (revised July 1981).

15. USNRC Code of Federal Regulations, 10 CFR 20.311 and 10 CFR 61, Classification of Radioactive Waste for LAND Disposal, Washington, D.C., December 1982.

16. State of South Carolina Radioactive Material to Chem-Nuclear System Incorporated, License No. 097, Amendment 28, dated November 13, 1980.

17. Boque, R.H., "The Chemistry of Portland Cement," 2nd Edition, Reinhold, New York, 1955.

18. ASTM C:150-72, "Standard Specification for Portland Cement," Annual Book of ASTM standards: 1972k Part 9, American Society for Testing and Materials, Philadelphia, PA.

19. Colombo, P., and Neilson, Jr., R.M., "Properties of Radioactive Wastes and Containers, Quarterly Progress Report, April-June 1976, BNL-NUREG-50571 October 1976.

20. Colombo, P., and Neilson, Jr., R.M., "Critical Review of the Properties of Solidified Radioactive Waste Packages, BNL-NUREG-50591, December 1976.

21. Palisades, Consumer Power, Jackson, Michigan; Midland 1 and 2, Consumer Power, Midland, Michigan.

22. 10 CFR 20, "Standard for Protection Against Radiation."

23. 10 CFR 50, "Licensing of Production and Utilization Facilities."

24. 10 CFR 50, Appendix A, "General Design Criteria for Nuclear Power Plants."

25. 10 CFR 50, Appendix I, "Numerical Guides for Design Objectives and Limiting Conditions for Operation to Meet the Criterion ALARA for Radioactive Material in Light Water-Cooled Nuclear Power Plant Effluents."

26. Regulatory Guide 1.21 "Measuring, Evaluating and Reporting Radioactivity in Solid Waste and Releases of Radioactive Materials in Liquid and Gaseous Effluents from LWR's."

27. Regulatory Guide 1.112 "Calculation of Releases of Radioactive Materials in Liquid and Gaseous Effluents from LWR's."

28. NRC Standard Review Plan, Section 11.2, "Liquid Waste Management Systems," NUREG 75/287.

29. NRC Standard Review PLAN, Section 11.5"Process and Effluent Radiological Monitoring and Sampling Systems."

30. NRC Branch Technical Position ETSB 11-1, "Design Guidance for Radioactive Waste Management Systems Installed in LWR Plants."

31. USNRC, "Calculations of Radioactive Materials in Gaseous and Liquid Effluents from Boiling Water Reactors (BWR-GALE Code)', NUREG-0016.

32. USNRC, "Calculation of Radioactive Materials in Gaseous and Liquid Effluents from Pressurized Water Reactors (PWR-GALE Code), NUREG-0017.

33. American National Standards N197/ANS-55.3, "Design Inputs to Liquid Waste Processing System of a BWR, 1976.

34. American National Standard N-199/ANS-55.2, "Design Inputs to the Liquid Waste Processing System of a PWR," 1976.

35. Godbee, H.W., "Use of Evaporation for the Treatment of Liquids in the Nuclear Industry," ORNL-4790.

36. Bidwell, G.P., "Technical Tradeoffs in Designing Evaporators for Nuclear Power Stations," Paper #74-WA/NE-7, ASME Winter Meeting, New York, NY. 1974,

37. Brooks, N.S. and Chow, E.R., "Cartridge Filtration in Pressurized Water Reactors," AICHE/Filtration Society Annual Meeting, Houston, TX 1976.

38. Wedig, C.P., and Coplan, B.V., "Selection of Filters for Radioactive Liquids, IEEE-ASME Joint Power Conference Paper, Buffalo, NY, Sept. 1976.

39. Lin, K.H., Use of Ion Exchange for the Treatment of Liquids in Nuclear Power Plants, ORNL-1792.

40. Kunin, R., Ion Exchange Resins," 2nd Edition, Robert E. Krieger Publishing Co., Huntington, NY, 1972.

41. Kamp, E.C., "Design Factors in Reverse Osmosis," Chemical Engineering, 80, April 1980.

42. Kniazewycz, B.G., "The Treatment of Liquid Radwaste by Reverse Osmosis Conception to Operation," Paper No. 75-PWR-27, ASME-IEEE Joint Power Conference, Portland, OR, Sept. 1975.

43. 1974 ASME Boiler and Pressure Vessel Code,—Section VIII, America Society of Mechanical Engineers, New York.

44. Power Piping, American National Standard Code for Pressure Piping, ANSI-B31.1, The American Society of Mechanical Engineers, New York.

Management of Radioactive Materials and Wastes: Issues and Progress. Edited by S. K. Majumdar and E. Willard Miller. © 1985, The Pennsylvania Academy of Science.

Chapter Ten

OPERATION OF A LOW-LEVEL WASTE DISPOSAL FACILITY AND HOW TO PREVENT PROBLEMS IN FUTURE FACILITIES

Ralph DiSibio, Ed. D.
Hittman Nuclear & Development Corporation
9151 Rumsey Road
Columbia, Maryland 21045

Operation of a low-level waste facility is an ever increasing problem nationally, and specifically one that could grow to crisis proportion in Pennsylvania. There has been, nevertheless, a variety of changes over the years in the management of low level radioactive waste, particularly with regard to disposal facilities that can avert a crisis condition. A number of companies have been organized thru possible a broad range of services to the nuclear industry, including those that emphasize solidification of waste materials, engineering services, waste management, and transportation to disposal sites across the United States.

Nationally a checkered record exists with respect to proper siting and disposal practices over the last two decades. By 1971, six disposal sites were in operation nationally. They were dispersed regionally throughout the country. Should they have remained open, they would likely have been capable of handling the increasing amount of low level waste generated from the nuclear industry, the medical community, and a variety of industrial sources. Unfortunately, by 1979, three of the facilities were closed for a combination of political and environmental reasons.

This paper addresses one particular site and the problems which evolved at that site from an environmental perspective. It is important that it is clearly understood that, although these problems are resolvable, the lessons learned here are critical for the prevention of problems at future facilities.

The focus of this paper is on the Maxey Flats, Kentucky disposal facility which was closed in 1977. It must be understood that the regulations for siting, management, burial techniques, waste classification, and the overall management of disposal sites were limited when this facility was in operation. Many of the problems have now been precluded as a result of much more stringent regulations as identified in NRC regulations 10 CFR 61. It is incorrect to infer that while this facility was in operation, the operators were in any way responsible for the resulting conditions.

An aerial view of the Maxey Flats facility shows it is located on a knoll approximately 60 miles east of Lexington, Kentucky. The facility consisted of several buildings, standard earth moving and material handling equipment, and a series of trenches approximately 30 feet in depth, and varying in length and width depending on the operational purpose. One building on the site was an evaporator to process liquids which have accumulated in the trenches. A steady stream of white smoke belched from this evaporator and contained tritium which had to be strictly controlled.

At the time the plant was built, the state-of-the-art for material handling consisted of dumping material either by crane of by hand into the trenches. Because of the humid conditions, plastic was used to cover the material on a temporary basis until the trench was completely filled. The trenches were unlined and dug to the depth of a sandstone lens, surrounded by a fractured strata. The trenches were backfilled with waste materials, and a cap was put in place, followed by revegetation. Many lessons have been learned in terms of this method of operation. In hindsight it can be recognized that these procedures ultimately led to the problems presently faced. As a consequence of these practices, the following conditions evolved.

The materials in the trenches were in many forms, including cardboard containers, wooden boxes, and metal packages of several configurations, ranging from 55-gallon drums to railroad tank cars full of liquids. As these packages lost their integrity, and in some cases disintegrated, voids occurred in the covered trenches. Earth and other materials filled those voids, eventually causing subsidence in the trenches into which storm water and surface water entered. The sandstone lens acted to hold this water, creating, what is called, a "bathtub effect." In short, the trenches began to fill with water which became contaminated with the contamination migrating from trench to trench.

In summary then, these are the conditions which led to closure and the ongoing problems faced by the Commonwealth of Kentucky. Specifically, the problem areas included:

Quantity of Water in the Trenches

Contaminated Water Infiltration

Limited Water Processing Rate

Time Required to Remove Contaminated Water

Accumulation of Concentrate and Sludges as a Result of
    Treating the Liquids

Continued Subsidence of the Area

Additional Subsidence after Dewatering

Continued Potential for Lateral Migration

Lack of Overall Funding for Unanticipated Costs

Subsidence is one of the key problems that must be avoided in the future. Large sink holes actually opened on the surface and large amounts of water entered the trenches. There is no question that water is the major enemy to be avoided in future site operations. Large portions of the trenches began to subside until, in one case, an entire trench section began to settle. In addition, the waste became close to the surface as the earth began to settle. In places exposed waste even appeared on the surface.

When the Commonwealth of Kentucky contracted the Hittman Nuclear and Development Corporation to remedy the immediate problems, the following were considered necessary steps. The contaminated liquids had to be removed from the trenches. The liquids has to be processed that had accumulated for several years. The plant operating 24 hours a day had pumped and processed more than 1 million gallons of liquids per year. In addition, two overflow ponds had been created on site and had to be decommissioned. So that this condition would not persist the site had to be graded more effectively, the trench covers improved and all subsidence areas repaired. Interestingly, when the trenches were dewatered there was a tendency for additional subsidence because of new voids created. Typical construction techniques were used to grade the site, construct runoff drains. Weirs and sewer lines were also installed. One of the most important health physics problems was how to pump the residual liquids from the trenches. This was accomplished by driving 50 sump pumps directly into the trenches and pumping the liquid out.

Ultimately, all of the regrading was completed, but the next challenge was how to keep, even incidental, water from entering the trenches during runoff. To proceed, more than 30 acres of plastic were laid by hand over the entire site so that it resembled a baseball field during a rain delay. The pumping and evaporation operation were then temporarily closed down. The release of the tritium into the air which had been occurring for several years at maximum EPA allowable limits was now stopped.

The next problem was the permanent solution to not only the Maxey Flats problem, but development of a protocol that would ultimately apply to all of the disposal sites presently in operation and previously closed. Such a decommissioning plan must include the implementation of a long range drainage plan and a variety of other techniques such as the installation of barrier walls to avoid lateral migration which may bring the final chapter to a close in the Maxey Flats operation.

Experimentation is now in progress with an *in-situ* grouting technique in an attempt to fill all the voids in the trenches. This technique includes pumping a grout under high pressure into each of the trenches.

The story of Maxey Flats continues, but it is strongly felt that a resolution is imminent, at least in terms of an ultimate plan for closure. The Maxey Flats experience is extremely valuable because future sites must avoid the problems of management, siting and waste packaging procedures that were allowed in the past.

There has been a tremendous number of studies, reports and new regulations which have evolved over the last 5 years to provide a satisfactory solution to the management of low level waste in this nation. There are now many safe alternatives for future land disposal of radioactive waste, based on past experience. They include improved shallow land burial, intermediate depth disposal, the use of sanitary landfills for innocuous materials, engineered disposal facilities, geologic disposal, and combined engineered disposal and burial, which appears as one of the best future method. To assure safe disposal of low-level wastes, past burial experience as well as the inadequacy of past decommissioning plans, the flexibility of those decommissioning methods, and the time required to decommission must have consideration.

In shallow land burial the following must be considered:

1. It is critical that in site selection, a water management plan, waste containment, and an effective plan for decommissioning be of the highest priorities.
2. Intermediate depth disposal offers even more security with respect to waste containment.
3. An above-ground facility has inherent within it untried maintenance problems which may be resolved by an earth cover and must also contain sophisticated monitoring equipment.
4. Geologic disposal is often considered in foreign countries, as well as in the United States. These methods have inherent problems in material handling procedures and often do not resolve the problem of water infiltration.
5. A combined engineered disposal and burial site is frequently considered the most cost effective and safe method for future disposal in the United States. Such a system would have built in intrusion barriers, barrier walls and water management systems, as well as a simple waste handling scheme, and would be adaptable to most siting environments. Most importantly, the higher lived low level waste materials could be emplaced in such a fashion as to be retrievable, which makes this design much more palatable from a public and political perspective. Such a concept would have the appearance of an underground mausoleum, and again, take into consideration all of the shortcomings of past procedures.

New waste handling techniques devised by the Westinghouse organization and others will allow for not only the ease of handling, but better insure against void creation and subsidence. Each day, waste disposal organizations experiment with new techniques which will be incorporated into new engineered facilities, including such procedures as the caisson technique which allows for sophisticated monitoring devices and perhaps more critically, retrievability.

Laymen as well as experts must come to the realization that past problems are indeed history; the effective management and operation of future facilities must use present-day technology. It is hoped that this paper has revealed some of the procedures that have been applied in the past and what the future holds with regard to the protection of the environment, and equally important, for taking immediate action to prevent any disruption in medical and industrial services and in the provision of energy to our citizens.

Pennsylvania has the opportunity to control its disposal of waste materials. A clear understanding of the issues have evolved, and private industry is developing the techniques to resolve what could be a pending crisis.

PERTINENT LITERATURES

U.S. Department of Energy, *Understanding Low-Level Radioactive Waste,* LLW MP-2, November, 1980.

Picazo, E., J. Clancy, M. Cody, *Specifications for Evaporator Concentrate Solidification at the Maxey Flats Low-Level Radioactive Waste Disposal Site,* Task 1.1 Report. Dames & Moore, April 2, 1981.

Wild, R.G., J. Clancy, M. Ziskin, *Water and Sludge Management Alternatives for the Maxey Flats Low-Level Waste Disposal Site*, Task 2 Report, Dames & Moore, March, 1980.

Clancy, J. et al., *Evaluation of Alternatives for the Future of Facilities at the Western New York Nuclear Service Center*, Dames & Moore, 1978.

U.S. EPA, *Background Document for Hazardous Waste Disposal Regulations: Availability and Cost of Third-Party Liability Insurance for Permitted Hazardous Waste Disposal Sites*, February 20, 1980.

Radioactive Waste Management, "Editorial Comment," 1 (3): i-ii (1981).

Abrams, Nancy E., and Joel R. Primack, "Helping the Public Decide: The Case of Radioactive Waste Management, *"Environment*, 22:14-20, 39-40, April, 1980.

Carnes, S.A., "Confronting Complexity and Uncertainty: Implementation of Hazardous Waste Management Policy," in*Environmental Policy,* Dean Mann, ed. (Lexington Books: 1981).

Kasperson, R., "The Darkside of the Radioactive Waste Problem," in T. O'Riodan and R. Turner (eds.), *Progress in Resource Management and Environmental Planning,* Volume 2, Chichester/New York, John Wiley & Sons.

National Governor's Association (NGA), Siting Hazardous Waste Facilities, Final Report of the NGA Committee, March, 1981.

Rochlin, Gene I., *The Role of Participatory Impact Assessment in Radioactive Waste Management Program Activities* (Berkeley: Institute of Governmental Studies, University of California, Berkeley, Report IGS/RW-002 (Draft) August, 1980.

State Planning Council on Radioactive Waste Management (SPC) *Interim Report to President*, February 24, 1981.

Management of Radioactive Materials and Wastes: Issues and Progress. Edited by S. K. Majumdar and E. Willard Miller. © 1985, The Pennsylvania Academy of Science.

Chapter Eleven

TRANSPORTATION, DISPOSAL AND CONDITIONING OF LOW-LEVEL RADIOACTIVE WASTE

Charles W. Mallory, B.S. in Mech. Engin.
Senior Technical Advisor
Hittman Nuclear and Development Corporation
9151 Rumsey Road
Columbia, MD 21045

In the planning of the early nuclear power plants, the quantities and activity levels of the low-level radioactive waste that would be generated were generally underestimated. In addition, the problems associated with the land disposal of radioactive waste were not fully appreciated. Over the past twenty-five years, improved methods and equipment for the conditioning, transportation and disposal have evolved. Technology is now in place to assure the continued safe, effective and economical handling of low-level radioactive waste from nuclear power plants in the future.

LOW-LEVEL WASTE SOURCES

Nuclear power plants produce both high level waste and low level waste. The high level wastes are the fission products produced within the fuel elements which are either reprocessed or disposed of as high level waste. The radioactivity in low level wastes comes from leaking fuel elements, activation of corrosion products as they pass through the reactor core or by corrosion of materials that have become radioactive by neutron activation. Processing systems are used to remove and concentrate the radioactivity. Processes used include filtration, ion exchange, absorption, evaporation, etc. The principal waste forms prior to conditioning for disposal are:

      Expended ion exchange resin
      Filter sludges
      Concentrated ion exchange regenerant
      Concentrated liquid waste
      Compactible dry active waste, and
      Non-compactible waste

Although classified as "low-level" waste, the concentrated waste from nuclear power plants may have specific activity levels greater than 1,000 microcuries per milliliter and external gamma radiation in excess of 1000 R per hour.

WASTE TRANSPORTATION

The equipment used to transport the waste from the power plants to the disposal sites represents a vital interface in the overall low-level waste management system. The transportation equipment must be compatible with both the power plant and the disposal site and meet the applicable regulations for the transport of radioactive materials. In the U.S., the transportation of radioactive waste is covered by Department of Transportation, DOT, and Nuclear Regulatory Commission, NRC, regulations. The DOT regulations apply to the lower activity materials classified as "Type A" materials. The NRC regulations apply to the materials having higher levels of radioactivity and classified as Type B materials. Both sets of regulations are based on standards and guidelines established by the International Atomic Energy Agency, IAEA.

Transportation Limits

The present regulations are based on using Transport Groups to define the quantities of radioactive materials that can be transported as Type A, Type B and low specific activity, (L.S.A.) radioactive materials. Most of the radioisotopes in nuclear power wastes are in Transport Group II, III and IV. Table 1 shows the radionuclides normally found in power plant wastes by transport group and lists the Type A quantity and the maximum specific activity for L.S.A. materials.

TABLE 1

Transport Group, Type A and Type B Quantity and L.S.A. Concentrations

Transport Group	Radionuclides	Type A Quantity (C_2)	Type B Quantity (C_1)	L.S.A. Concentrations mCi per gm.
II	Sr-90	0.05	20	0.005
III	Ce-144, Cs-134, Cs-136 Co-60, I-131, I-133 Ag-110^m, Sr-89	3	200	0.3
IV	Cr-51, Co-58, Fe-59, La-140, Tc-99, Sn-125, Zn-65	20	200	0.3

When the amount of activity exceeds the Type B quantity specified, the shipment must be handled as a "Large Quantity" shipment. The Large Quantity classification does not apply to low specific activity radioactive materials.

Regulatory Changes

A number of regulatory changes are pending which would affect the transportation of radioactive materials. In 1973, the IAEA adopted a change in the standards which classified each radionuclide on the basis of toxicity. The number of curies of each material which can be handled as Type A material is specified. In addition, the concentrations of radioactive material that can be handled as Low Specific Activity material or L.S.A. are based on the Type A quantity for the individual isotopes. The U.S. has not adopted the IAEA classification system but is expected to do so in 1983 or 1984.

The most significant regulatory change in the offing is a proposed change in the criteria for the classification of Low Specific Activity materials. Heretofore, L.S.A. classifications were based on ingestion and inhalation of the material. The allowable concentrations were limited to assure that no individual would receive excessive exposure even if the material were accidently released to the environment. The 1984 IAEA Standards include a provision that would limit the external radiation from unshielded and unconfined L.S.A. materials to one R per hour at three meters. For gamma emitting isotopes this will severely limit the concentrations of radioactive material that can be shipped as L.S.A. Since the waste from nuclear power plants contains primarily gamma emitters, the new regulations can restrict operations. Materials which cannot be classified as L.S.A. materials because of external radiation will have to be shipped as Type B shipping using casks capable of withstanding a relative severe impact test and prolonged exposure to flame.

Type A and Type B Casks

Radioactive waste shipping casks are classified by regulations as Type A or Type B packages. In addition, certain Type A packages are certified to handle greater than Type A quantities of L.S.A. materials. Type A shipping casks must be capable of withstanding the normal conditions of transport, whereas, Type B casks must withstand hypothetical accident conditions. Table 1 shows the

TABLE 2

Design Conditions

Case Type	Type A	Type B
Water	Spray for 30 min.	Immersion > 3 feet for > 8 hours
Free drop	One foot drop (> 30,000 lbs.)	30 foot drop
Penetration	Bar 1.25 dia. & 13 lbs. from 40 inches	Cask drop from 40 inches onto 6 in. dia. bar
Compression	Five time weight	—
Thermal	—	Exposure to 1.475°F for 30 minutes

pertinent conditions which must be met by each type of cask.

Figure 1 shows a Type A cask and Figure 2 shows a Type B shipping cask. Figure 3 shows a cask designed to handle greater than Type A quantities of L.S.A. material. Figure 4 is a box cask designed to handle 55-gallon drums.

FIGURE 1. Type A Radwaste Shipping Cask.

FIGURE 2. Type B Radwaste Shipping Cask.

FIGURE 3. Radioactive Shipping Cask—Low Specific Activity Material.

FIGURE 4. Box Cask for 55 Gallon Drums.

Inner Container/Cask Cavities

Fifty-five gallon steel drums were used at most of the early power plants to contain radioactive waste. Radioactive waste shipping casks were designed to handle varying numbers of drums. These casks were generally of two types, i.e., cylindrical or box. With the cylindrical casks, drums are placed on pallets and lifted into the cask. Two or three pallets of drums can be placed in a cask. Specially fabricated steel containers were introduced at an early date to replace steel drums. The containers, called liners, provide more effective use of the internal volume of shipping casks and thereby reduce the unit costs of shipping radioactive waste. As a result, standard sizes for radioactive waste shipping casks have evolved based on the casks being capable of handling either 55-gallon drums or liners. Table 3 shows the shipping cask sizes which have evolved.

The higher utilization of shipping cavity volume by using liners can be illustrated with the 14-drum cask. Fourteen drums would have a maximum internal volume of 14 x 7.5 = 105 cubic feet. A liner for this cask would have an internal volume of 162 to 192 cubic feet or 54 to 83 percent more.

TABLE 3

"Standard" Cylindrical Radioactive Waste Shipping Casks

Cask Cavity			Liners		
Diameter (in)	Height (in)	Number of Drums	Diameter (in)	Height (in)	Volume (CF)
26-27	38	1	24	33-36	8.5-9.5
26-27	73-75	2	24	69-72	18-19
54-56	38	3	52	33-36	40-44
54-56	62-64	4	52	58-61	71-75
54-56	73-75	6	52	69-72	84-88
60-62	73-75	8	58	69-72	105-110
74-77	75-80	14	72-75	69-72	162-192
74-83	109-110	21	72-78	105-108	247-300

High Integrity Containers

A new type of inner container was introduced starting in 1981. These containers were developed to allow ion exchange resins and filter media to be shipped and buried in a dewatered form. The containers are generally constructed of high density cross linked polyethelene to provide a burial life of 300 years in a burial environment. The containers are also designed to withstand the full weight of 35 feet of overburden having a density of 120 pounds per cubic foot. They are capable of withstanding the impact of a twelve foot drop and to meet the other requirements for Type A containers. The containers must also be capable of withstanding the radiation exposure from the contained waste. Figure 5 shows a drum size and a large size high integrity container.

FIGURE 5. High Integrity Containers.

Shielding Requirements

Practically all low-level radioactive waste shipments are made by 18-wheel trucks with trailers. Low-level radioactive waste shipping casks must be shielded to meet the requirements shown in Table 4.

The amount of shielding that can be used in a shipping cask will be limited by the allowable weight of the truck, tractor, cask and contents. At the present time, practicaly all states allow trucks with gross vehicle weights up to 80,000 pounds to be operated without permits. Routine overweight permits can be obtained however, this may not be effective because of restrictions on movement of permitted vehicle. The weight of the truck tractor is generally about 28,000 pounds and the weight of the trailer used with cask is about 12,000 pounds. Accordingly, the weight of the cask and the contents must be limited to 50,000 pounds to avoid overweight permitted shipments.

The amount of shielding required to reduce the radiation levels to the limits for transport shown in Table 4 will depend upon the specific activity of the radio-nuclides in the waste, the energy and fractional abundance of their emissions, the packaging of the waste and the amount of self-shielding of the waste. Figure 6 shows the amount of lead required to reduce the radiation levels at two meters from a seven foot diameter cask to 10 mR per hour.

Since total weight must remain nearly constant, cask design involves trading off capacity to provide shielding for high level shipments. Figure 7 shows the

TABLE 4

Shielding Requirements for Low-Level Radioactive Waste Shipping Casks
(Exclusive-Use Shipment)

External surface of cask	200 mR per hr.
Two meters from side of trailer	10 mR per hr.*
Amp normally occupied position	2 mR per hr.

*To meet 10 mR per hour at two meters surface radiation levels must be generally limited to:
Small diamter casks — 80 mR per hour
Large diameter casks - 60 mR per hour
Box cask and large boxes - 40 mR per hour

relationship between capacity and surface radiation levels of the inner containers for a number of Type A and Type B casks presently in operation.

Transportation Costs

There are a number of cost elements which must be considered in transport of low-level radioactive waste. These include:

Radwaste shipping cask usage charge
Disposable liners and drums
Transportation to (and from) disposal site
Special permits and fees
Transportation contractor service fees

The least element of cost will generally be the usage charge for the radwaste shipping cask. The items of cost in this element include:

Present replacement cost for the cask
Useful life
Allowance of technical or regulatory obsolence
Interest rate and required capital recovery
Cask utilization factor
Licensing and certification costs

The cask usage charges are generally combined with the transportation contractor service fees and billed as a percentage of the transportation costs. Fees of 50 to 70 percent of the transportation costs for cask usage and service contractor fees are quite common.

The next largest element of cost will generally be the transportation to and from the burial site. Figure 8 shows the current transportation costs for one-way and two-way shipments between nuclear power plants and disposal sites.

The largest element of transportation cost will be the disposable liners or drums. The drums themselves are relatively inexpensive, but cost more overall because of ineffective use of the cavities in radwaste shipping casks. Custom-built steel containers and high integrity containers can cost $3,000 to $10,000 depending upon the conditioning equipment incorporated into the containers.

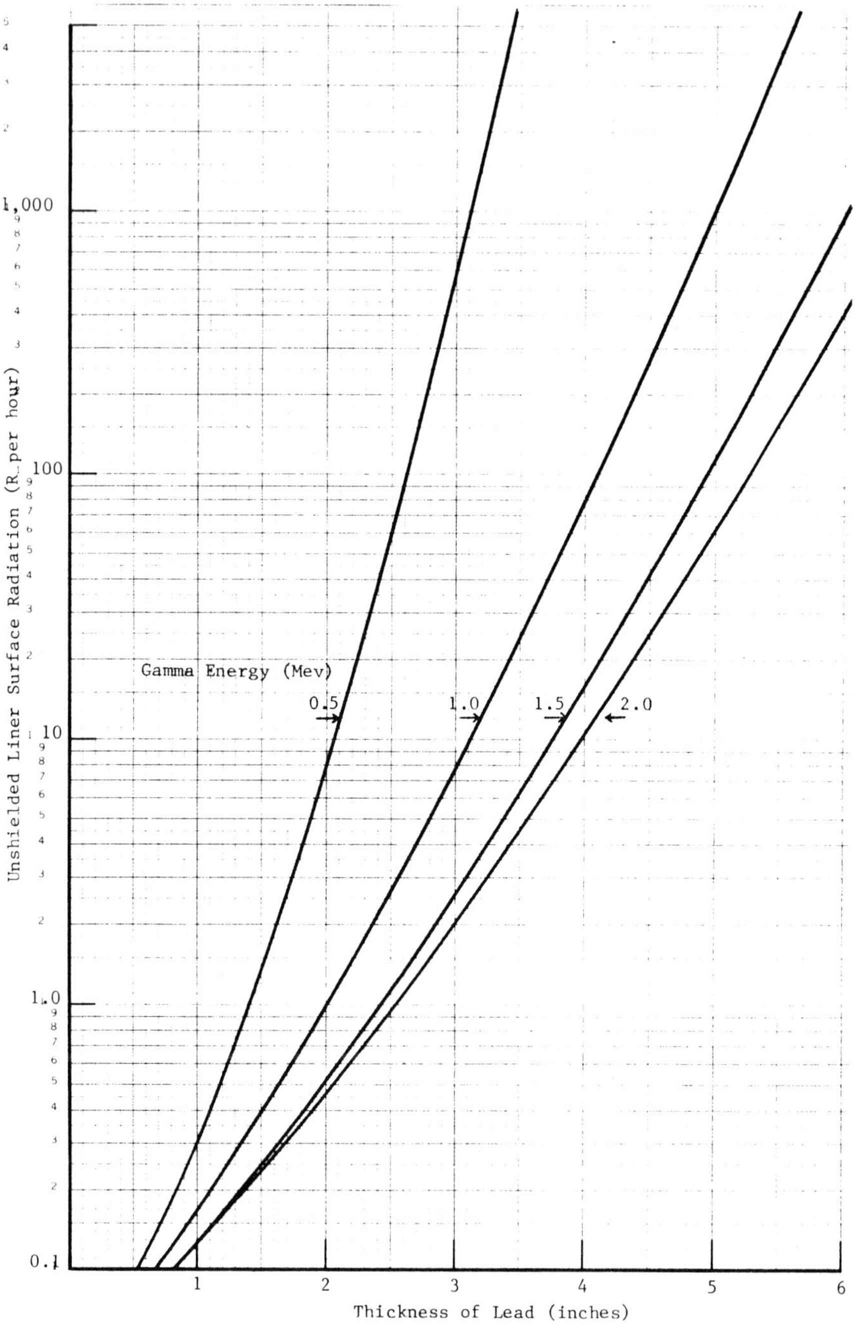

FIGURE 6. Thickness of Lead Required to Reduce Radiation Levels to 10 mR per hour at 2 Meters.

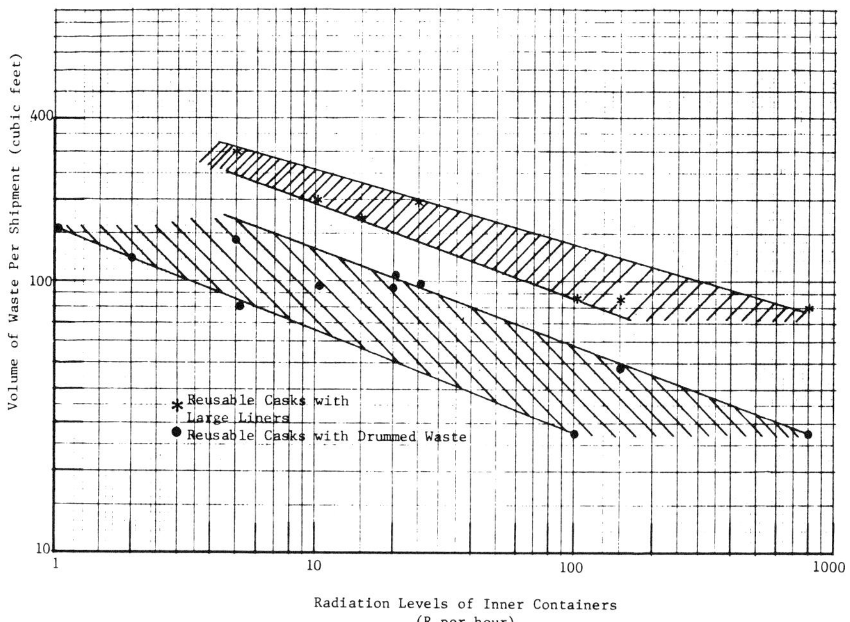

FIGURE 7. Shielding & Capacities of Present Radioactive Waste Shipping Casks.

Typical Transportation Costs

The cost of typical shipment of radioactive waste in shielded cask are shown in Table 5.

Based on shielded shipments of 150 cubic feet of waste, transportation costs range $46 to $100 per cubic foot.

The shielded shipment will generally contain 10 to 2,000 curies per shipment. At the low end, the cost per curie will range from $700 to $1,500 per curie. At the upper end, the cost is reduced to $3.50 to $7.50 per curie which illustrates the value of reducing volume and increasing the specific activity of the waste.

DISPOSAL

Practically all of the radioactive waste generated to date at nuclear power plants have been disposed of at commercial shallow land burial sites. Of the six commercial burial sites that have opened in the U.S., only three remain in operation. A major effort is underway to form state compacts for the opening of additional regional disposal sites.

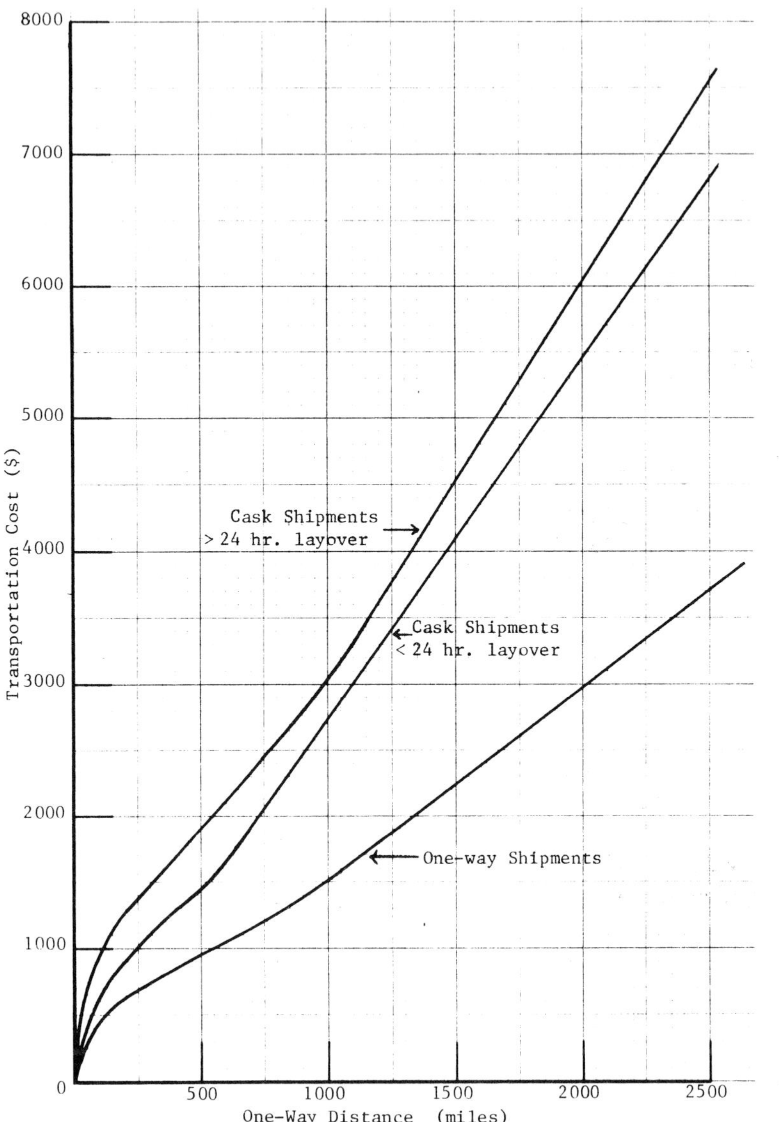

FIGURE 8. Current Transportation Costs (1983) for Radioactive Waste Shipments.

Lessons Learned

The three shallow land burial sites that have now been closed were closed primarily because of hydrogeological reasons. In at least two cases, the burial trenches were located over impermeable strata. Surface water continued to infiltrate and fill the trenches after they were covered. As a result, the potential

TABLE 5

Typical Radioactive Waste Shipping Costs
(1,000 miles, 2,000 miles round trip)

Transportaion and Permits	$3,000
Cask Usage and Services	$1,000 to $2,000
Disposable Container	$3,000 to $10,000
TOTAL:	$7,000 to $15,000

Based on shielded shipments of 150 cubic feet of waste, transportation costs range $46 to $100 per cubic foot.

for surface and subsurface migration was increased to such a point, it was no longer considered prudent to operate the sites as shallow land burial sites. The three sites have been closed and are in a caretaker status pending full decommissioning.

The experience from the operation of the original six shallow land burial sites has been studied in depth. Corrective measures have been incorporated into guidelines and regulations for selecting and operating future sites.

Present Burial Sites

Over the past few years, the prices for burial at the three remaining burial sites have increased dramatically. To a great extent, the increased cost of burial reflects the premature closure of certain of the sites and the need to accumulate reserves for the closure and decommissioning of the sites.

The burial costs at the three disposal sites are generally composed of the following types of charges.

    Basic volumetric burial charge

    Radiation surcharge

    Curie surcharges

    Weight surcharges

    Liner handling fees

    Perpetuity escrow funds

Figure 9 shows the burial charges for one of the commercial shallow land burial sites which are comparable to the other disposal sites. This curve shows the high burial cost for liners having high radiation levels. Table 6 gives a better picture of the economics of low-level radioactive waste disposal. Four typical shipments are shown having surface radiation levels ranging from 1 to 500 R per hour. This table also shows the higher burial cost at high activity levels. When costs are calculated on a $ per curie basis, the high activity shipments are much cheaper, again showing the benefits of reducing the volume and concentrating the waste.

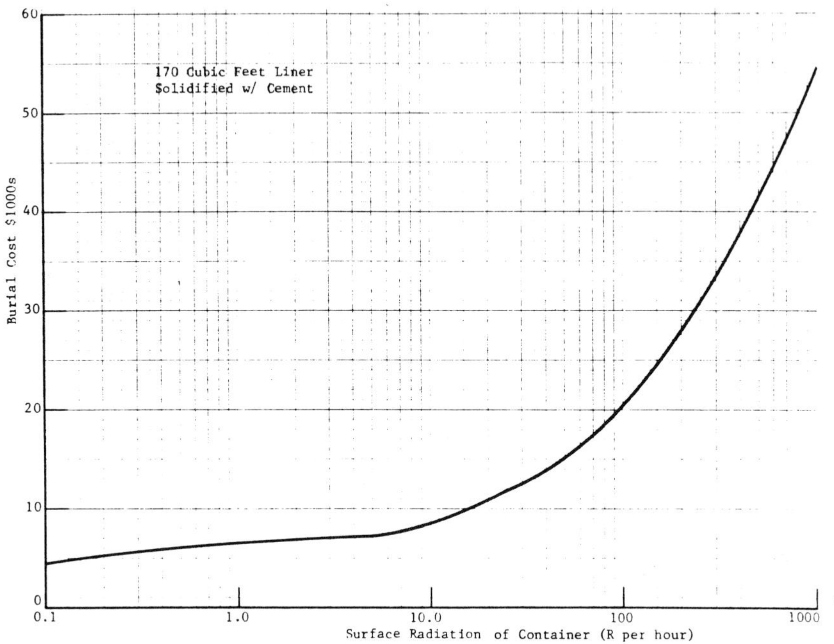

FIGURE 9. Typical Burial Costs (1983).

TABLE 6

Typical Burial Charges
(Solidified Waste)

Surface Radiation (R/hr.)	1	10	100	800
Type Shipment	L.S.A.	L.S.A.	Type B	Type B
Volume (CF)	170	170	100	80
Specific Activity (μCi/ml)	3.7	37	370	1830
Total Activity (curies)	18	180	1025	4100
Basic Burial Charges (at $15.45/CF)	$2,630	$2,630	$1,550	$1,240
Radiation Surcharge ($/CF)	13	20	70	200
($)	2,200	3,400	7,000	16,000
Weight Surcharge ($)	750	750	750	750
Curie Surcharge ($)	1,000	4,000	8,000	8,000
Cask Handling ($)	500	500	500	500
Total Cost	$7,080	$11,280	$17,800	$26,490
$ per curie	$ 393	$ 63	$ 17	$ 6.50

WASTE CONDITIONING

Waste conditioning encompasses those processes required to treat, handle, solidify and package low level radioactive waste to prepare it for shipment to the disposal site. Waste conditioning must place the waste in a form where it is suitable for both transportation and burial. As shown in the prior discussions, waste conditioning should include reducing the volume and concentrating the waste to reduce transportation and disposal cost and to make the best utilization of disposal facilities. A nuclear power plant will produce a certain number of curies of activity annually. The objective of waste conditioning should be to transport and dispose of curies in the most effective means available.

Dry Active Waste (Compactible)

This waste generally represents the largest waste type in terms of volume. Because of its very low specific activity, it can be reduced in volume without materially increasing the radiation levels. Incineration and high density compaction are the best available conditioning methods for handling of burnable and compactible dry active waste.

Non-Compactible Wastes

By definition non-compactible waste has a low potential for volume reduction. However, a large portion of the activity associated with non-compactible waste is surface contamination. Decontamination, ultrasonic cleaning and electropolishing represent conditioning methods by which radioactive non-compatible waste requiring disposal at shallow land burial sites can be reduced in volume.

Reactor Cleanup Resins

The resins generally represent a major fraction of the total activity that will be produced by a nuclear power plant. The activity levels of this material are such that further concentration and reduction in volume may not be worthwhile. Resins of this type should not be regenerated since this will actually dilute the wastes. The principal conditioning methods for this type of waste will be the use of high integrity containers and solidification methods that will not dilute the waste.

Filter Sludges

Many plants use powdered resin, diatomaceous earth and other filter media to remove radioactivity and other contaminants from liquid waste stream. The filter media is generally expended due to high pressure drop as solids are accumulated on and within the filter media. The filter media and solids, called filter sludges, are backwashed off the filter septum to a waste holding tank. Filter sludges represent a troublesome waste to handle because they are difficult to dewater and/or solidify. It is also difficult to further concentrate the radioactivity since the sludges generally contain significant quantities of extraneous materials.

Ion Exchange Regenerant

Many nuclear plants regenerate ion exchange resin, particularly boiling water reactor plants having deep bed demineralizers for condensate cleanup. With regeneration, the radioactivity is removed from the ion exchange resin and relative large quantities of liquid containing sodium sulfate result. This liquid is concentrated using evaporators and in some cases crystallizers, calciners and fluid bed dryers are used to reduce the volume of the waste and increase the concentrations of radioactivity. The degree to which radioactivity can be concentrated is limited because the majority of the solids will be the chemicals added for regeneration. A number of nuclear plants are no longer using regeneration and are disposing of the resins when they become expended. This can result in higher concentrations of radioactivity and more efficient disposal.

Concentrated Liquid Waste

The drains from equipment and the liquid collected in floor drums goes to liquid waste hold-up tanks. In the past most plants used evaporators to concentrate this waste for disposal. In the past several years, a number of plants have gone to demineralizers to treat this liquid. In pressurized water plants where the liquid contains relatively high concentrations of boric acid, the use of demineralizers can greatly reduce the quantities of water. By using weak base anion resins, the radioactivity can be removed without removal of the boric acid.

Future

The conditioning of nuclear waste for more effective transportation and disposal represents an area where future improvement can be made. Methods for the selective removal and concentration of the contained radioactivity will provide even better utilization of transportation equipment. More importantly, the reduced volume of low-level waste will make it economically feasible to utilize additional engineered safeguards in low-level waste disposal facilities which will facilitate decommissioning of the facilities and insure long-term internment of the waste.

Management of Radioactive Materials and Wastes: Issues and Progress. Edited by S. K. Majumdar and E. Willard Miller. © 1985, The Pennsylvania Academy of Science.

Chapter Twelve

HANDLING OF RADIOACTIVE MATERIALS AND WASTE IN HOSPITALS

Michael A. Vince, M.Sc.
Medical and Health Physicist
and
Radiation Safety Officer
St. Luke's Hospital
Bethlehem, PA 18015

Hospitals throughout the United States, not necessarily engaged in research activities but, rather, dedicated to providing the medical care needed by communities which they serve, often possess broad but specific assortments of radioactive materials which are useful in the diagnosis and treatment of a variety of conditions in patients. In general, the principle emissions from these radionuclides of medical importance are gamma rays, which are useful in both diagnosis and therapy, and beta particles, which find application in therapy. The sources of these radiations may be conveniently classified as *unsealed* and *sealed* sources.

Unsealed sources are used in nuclear medicine mainly for diagnostic and, occasionally, therapeutic purposes. They are administered to patients either orally or by intravenous injection. Unsealed radionuclides are also used in pathological studies such as radioimmunoassays, radiochromatography, dilution studies to determine volumes and flow of fluids and in investigations to diagnose specific functional disorders.

Sealed sources of radiation are used in the therapeutic management of cancer. As the classification implies, they are radioactive materials which have been encapsulated permanently in protective containers of various configurations. This class of radionuclides may be conveniently subdivided into three groups: encapsulated isotopes used in teletherapy, sealed sources used in temporary intracavitary and interstitial brachytherapy and sealed sources used for permanent interstitial brachytherapy.

These are not the only sources of radiation in hospitals but they are the sources that represent materials which must be controlled to lesser and greater extents in the clinical environment to prevent radioactive contamination and undesirable radiation exposure of patients and personnel. They require special handling and storage procedures and are, in some cases, potential sources of hazardous wastes which require disposal.

Before embarking on a detailed discussion of the methods of handling medically useful radioactive materials and their waste products, it may be instructive to contrast radiations emitted during radioactive decay with X rays which are generated by the familiar diagnostic X-ray machine. X rays are produced whenever a substance is bombarded by high speed electrons. When the current of energetic electrons is removed, the generation of X rays is terminated. The X-ray tube, which is the source of X rays in radiographic machines, functions essentially in this way and it is analogous to the electric light bulb. That is, light is only emitted while the bulb is "ON" and when it's turned off there is no residual light present. Likewise, the X-ray beam emitted from an X-ray tube is present only while the tube is energized but there is no residual radiation when the tube is turned "OFF".

On the other hand, radioactive decay is a continuous, random process in which unstable isotopes disintegrate to more stable configurations with the accompanying emission of radiation until the last unstable atom has disintegrated. There is no way to stop the process. However, the radiations being emitted can be blocked with appropriate absorbing and shielding material and, thus, in a sense, effectively "turned off." Nevertheless, behind the shielding, the process of radioactive decay continues. Consequently, care must be exercised in handling, storing and disposing of radioactive materials.

Ionizing radiations emitted by the nuclei of radioactive materials interact with human tissue by the rupturing of chemical bonds within the cells. This results in a traumatic disturbance of the normal function of the cell. The overall macroscopic effect will depend on the population and the nature of the cells which have been altered. In the case of acute exposures to radiation, that is, relatively large amounts of radiation delivered in a short period of time, these macroscopic effects can range from nausea, to small alterations in blood white cell counts, to genetic mutations and diseases like leukemia or even death, depending on the intensity of the exposure.[1] The radiation dosages normally encountered in the clinical environment will not cause the extreme harmful effects just cited and are generally considered to be below the level where harmful effects may be manifested. However, evidence that there is such a "threshold" level below which no biological effects can be observed, has not been unequivocably demonstrated. For this reason hospitals throughout the United States have adopted the ALARA concept recommended by the U.S Nuclear Regulatory Commission and is now a requirement for licensure for the possession of radioactive materials. Simply stated, the ALARA concept requires that radiation exposures to persons working with radioactive materials, as well as to the general public, be maintained *as low as reasonably achievable.*

In this chapter we shall discuss the various radionuclides and the forms in which they are found in the clinical environment. We shall then discuss in detail the procedures in providing protection for patients and personnel against undesirable radiation exposures and contamination, and methods of handling radioactive wastes.

CLINICAL APPLICATIONS OF RADIONUCLIDES

Unsealed Sources in Nuclear Medicine

In nuclear medicine, quantities of radioactivity present in unsealed radiopharmaceuticals and diluted radionuclides administered to patients range from tens of millicuries and less for diagnostic studies to a few hundred millicuries used in tumor therapy. Radionuclides used in nuclear medicine are carefully selected for specific characteristics which lead to the minimization of whole body radiation exposures of patients and of medical workers engaged in their care.

These radioactive materials possess relatively short physical half-lives, i.e., the time required for a radioactive substance to lose one half of its activity by radioactive decay, in the order of 1 hour to, at most, several weeks in duration. Radionuclides with half-lives shorter than about 1 hour decay too rapidly to be useful in conventional applications but are used with highly sophisticated equipment in special applications in which the practical aspects of short-lived isotopes can be circumvented.

Once the radionuclide is administered it must remain active in the patient long enough to accumulate sufficiently in the tissues of interest so that it may be adequately detected by external scintillation sensors and instrumentation to have diagnostic value, or so that the desired therapeutic effect may be realized. Radionuclides are chosen which are either organ-specific in themselves, i.e., they have an affinity for and accumulate in specific organs of interest, or they can be tagged to pharmacological compounds which are organ-specific. A good example of an organ-specific radionuclide is radioactive iodine which accumulates rapidly in the thyroid following administration.

Elimination of radionuclides from the body is dependent on the biological half-life of the substance, i.e., the time required for the body to eliminate one half of an administered dosage of any material by natural processes of elimination, i.e., urination, perspiraton and bowel movements. The biological half-life is approximately the same for both stable and radioactive isotopes of the same element.

In the vast majority of nuclear medicine applications, the radiation emitted from the administered radionuclide is used for diagnostic imaging. To be useful for this purpose, the energy of the emitted photons must be sufficient so that they emerge from the patient but such that they are absorbed efficiently by the solid scintillation detectors used with diagnostic imaging instrumentation. In the limited therapeutic applications of unsealed radionuclides in nuclear medicine the emitted radiation should be readily absorbed locally in the tissues to be treated.

The net result of the application of radionuclides, which meet the criteria outlined in the previous paragraphs, is the beneficial diagnosis or treatment of diseases and disorders with minimum whole body radiation exposure to the patient. Estimates of radiation doses imparted to patients as a result of nuclear medicine procedures are thoroughly detailed in the literature.[3-9] The characteristics and applications of radionuclides used in nuclear medicine are given in Table 1.A. Some of the ones of more importance or possessing peculiarities of interest shall be discussed below.

Before beginning, however, it should be noted that also included in Table 1.A. are unsealed radionuclides which are found in clinical laboratories and are used in pathology tests such as radioimmunoassays, radiochromatography, volumetric measurements of body fluids (referred to as body spaces) and investigations of bacterial activity. The radioactivities used in clinical pathological laboratories are, at least, three orders of magnitude less than the activities used in nuclear medicine, in the range of tens of microcuries and less. These minute quantities of radioactivity represent an insignificant risk to the environment and to the individuals working with them; therefore, they are exempt from most regulations controlling the handling of radioactive materials. Typically, the entire inventory of radioactive substances in clinical laboratories is in the order of 5-10 millicuries. A small glass vial used for a single test usually contains less than 10 microcuries. Radionuclides commonly used as pathological tracers are carbon-14, cobalt-57 and iodine-125. No special precautions are necessary in working with these very low level radioactive substances other than handling procedures dictated by common sense and housekeeping procedures common to clinical laboratories. However, work surfaces should be monitored regularly with survey instrumentation suitably sensitive to detect the radiation energy ranges and low emission rates which might be present. Work surfaces should also be wipe tested periodically (at least monthly) to assure that radioactive contamination is not being unknowingly allowed to systematically accumulate. Wipe testing procedures shall be discussed more fully later in the chapter.

In examining the contents of Table 1.A, the reader will be immediately struck by the large number of diagnostic imaging procedures performed using technetium-99m. Tc-99m is, in fact, currently by far the most popular radionuclide for nuclear medicine imaging studies. It approaches very closely what may be an ideal radionuclide for this purpose. It has a short half-life, it's readily tagged to organ-specific pharmacological compounds, it emits a gamma

TABLE 1

Characteristics of Common Radionuclides Used in Clinical Applications (10, 11, 12, 13)

A. DIAGNOSTIC NUCLEAR MEDICINE AND PATHOLOGY TESTING

Radionuclide	Half Life	Principle Photon (kev)	HVL Pb(cm)	Mode of of Decay	Imaging Applications		Non-Imaging Applications	
					Procedure	Usual Dose Administered (mCi)	Procedure	Usual Dose Administered (μ Ci)
Carbon-14	5732 y	none	N.A.	-β	Culture Studies		Bacterial Studies	Not administered-2 μ Ci)
Cobalt-57	270 d	122	0.02	EC			Hematology	0.5
Chromium-51	27.8 d	320	0.2	EC			Body space	50
							Hematology	100
Iron-59	45 d	1095	1.1	-β			Hematology	5
Iodine-125	60 d	35	0.00	EC			Renal	50
							Body space	5
Iodine-131	8.04 d	364	0.3	-β	Thyroid Scan	0.05	Thyroid Uptake	5
					Lung Scan	0.3	Renal	30
					Cardiovascular System	0.2-0.3	Body space	5
					Liver Scan	0.15-0.3	Gastro-intestinal	25
					Kidney Scan	0.2-0.4		
Indium-111	2.8 d	173	0.05		Cisternography	0.25-0.5		
Selenium-75	65 d	514	0.6	EC	Pancreas Scan	0.25		
Technetium-99m	6.02 h	140	0.03	IT	Brain Scan	15-20		
					Lung Scan	1-3		
					Cardiovascular System	15-20		
					Thyroid Scan	1-2		
					Liver Scan	0.15-0.3		

					Spleen Scan	3		
					Bone Scan	15-20		
					Kidney Scan	1-10		
Thallium-201	73 h	135	0.025	EC	Cardiovascular System	1.5		
Xenon-133	5.3 d	81	0.03	$-\beta$	Lung Scan	10-15	Blood Flow	1000
Ytterbium-169	31.8 d	198	0.07	EC	Cisternography	0.25-0.5		

Note: $-\beta$ = beta minus emission
EC = electron capture
IT = isomeric transition
NA = not applicable

B. BRACHYTHERAPY, EXTERNAL BEAM AND MOLD TREATMENTS OF TUMORS

Radionuclide	Half Life	Principle γ Photon (kev)	HVL Pb(cm)	Mode of Decay	Type of Source	Therapeutic Radiation Emitted	Source Configurations	Therapeutic Applications
Cesium-137	30.0 y	662	0.6	$-\beta$	sealed	γ-rays	Tubes and Needles	Interstitial and Intracavitary Brachytherapy
Cobalt-60	5.27 y	$\sim$ 1250	1.2	$-\beta$	sealed	γ-rays	Encapsulated Source in Teletherapy Machine	External Beam Radiation Therapy
Gold-198	2.7 d	412	0.3	$-\beta$	sealed	γ-rays	Grains or seeds Colloidal solution	Interstitial brachytherapy.
Iodine-125	60.3 d	35	0.00	EC	sealed	γ-rays	Sealed in hollow seeds	Interstitial brachytherapy

Iodine-131	8.04 d	364	0.2	$-\beta$	unsealed	β-particles	Administered orally in capsule or liquid form and by intravenous injection	Treatment of cancer of the thyroid and metastases
Iridium-192	74.2	136, 1060	1.0	$-\beta$	sealed	γ-rays	Seeds spaced in plastic ribbon tubes	Interstitial
Phosphorus-32	14.3 d	none	N.A.	$-\beta$	sealed	β-particles	Contact applicators	Contact mold brachytherapy
Radium-226	1620 y	many from 47 tc 2440	~ 1.4	α,β,γ	sealed	γ-rays	Needles and tubes	Intracavitary, Interstitial and contact mold brachytherapy
Radon-222	3.82 d	352-609	0.5	α,β,γ	sealed	γ-rays	Seeds	Interstitial brachytherapy
Strontium-90 /Yttrium-90	28 y /64	none /1750	1.1	$-\beta$ $-\beta$	sealed	β-particles	Contact applicators	Contact mold brachytherapy
Tantalum-182	115.0 d	43-1450	1.1	$-\beta$	sealed	γ-raus & β-particles	Wires	Interstitial brachytherapy

Note: $-\beta$ = beta minus emission
EC = electron capture
α,β,γ = alpha, beta, gamma emissions
N.A. = not applicable

photon energy which matches the sensitivity of solid scintillation crystal detectors and it is readily available. Tc-99m is a daughter-product of the radioactive decay of molybdenum-99 which has a half-life of 66.7 hours. Molybdenum-99 generators used for producing Tc-99m are available from radiochemical firms. It is not uncommon for hospitals to receive deliveries of molybdenum generators two or three times each week to support their requirements in nuclear medicine.

A radionuclide generator is simply a device consisting of a parent-daughter radionuclide pair contained in a lead-lined apparatus that permits separation and extraction of the daughter from the parent. The generation of the daughter is a consequence of the decay of the parent. A good description of radionuclide generators is given by Sorenson and Phelps[10].

There are other, less commonly used generators found occasionally in nuclear medicine, the daughter products of which are useful in highly specialized or experimental applications.[10] As in the case of molybdenum generators, with the passage of time their daughter products are eluted from them in diminishing quantities until they are no longer productive enough to keep.

Until the relatively recent development of short-lived radionuclides, such as Tc-99m, with convenient methods for handling them, liver, kidney and brain scans took about one hour and could achieve resolutions of about 2 cm. With the short-lived radionuclides, much larger doses can be used and improved scans can now be made in a few minutes, resolving objects of a few millimeters in diameter in the image field.

Iodine-131, until supplanted by technetium-99m, was broadly used in nuclear medicine studies (Table 1.A). It can be demonstrated that 100 μCi of Tc-99m delivers about the same radiation dose after complete decay as 1 μCi of I-131. The significance of this is that over 100 times more Tc-99m can be prescribed as I-131 with a much better diagnostic scan resulting.[11] I-131 is still used in nuclear medicine, primarily for the diagnosis of thyroid conditions. It is also widely used in the therapeutic treatment of cancer of the thyroid. The administration of the therapeutic dosage is given orally to the patient in either liquid or solid (pill) form. The latter is preferred since the chances for airborne I-131 to be present or for accidental spillage are eliminated. However, since the liquid form is considerably less expensive than the solid, it is often chosen in order to reduce the cost of the treatment to the patient. The I-131 compound used in liquid form is somewhat volatile. The stopper of the vial in which it is delivered should be removed under a protective hood operated at negative air pressure since there is always a partial pressure of iodine in the air above the liquid in the vial which is released when it's opened. In this way, uptake by respiration can be prevented among the medical care personnel administering the dosage.

Although the majority of the I-131 administered to the therapy patient is normally taken up by the thyroid and eliminated in the urine, some of it will be present in the patient's perspiration and other body fluids such as saliva. Bedclothes, articles of clothing and eating utensils can become contaminated.

Therefore, special precautions must be taken to prevent the spread of radioactive contamination from the patient. Disposable bedcovers, clothing articles and eating utensils are recommended and it is a good practice to cover with protective materials areas and items with which the patient is likely to contact. Protective materials such as plastic, paper, tape, etc. are recommended which can later be discarded after the radioactive contamination has decayed. Precautions in handling and disposing of other unsealed and sealed sources will be discussed subsequently in this chapter.

SEALED SOURCES USED IN RADIATION THERAPY

Radioactive materials are encapsulated in a variety of container configurations for teletherapeutic and brachytherapeutic applications in the management of cancer. Teletherapy is "long distance," external beam, radiation therapy as contrasted to brachytherapy which means "short distance" therapy[11]. Teletherapy, by definition, is delivered from sealed radioactive gamma-emitting materials, such as cobalt-60, and is similar to external X-ray beam radiotherapy delivered by high X-ray machines such as linear accelerators. Cesium-137 has also been used in the past for teletherapy (as well as radium prior to 1951) but have largely been supplanted by cobalt-60 units and now, more and more, by high energy X-ray machines which do not employ radioactive materials in their operation. Characteristics of materials used in teletherapy and brachytherapy are given in Table 1.B.

Almost immediately following the discovery of radium by Marie and Pierre Curie in 1898, sealed sources of this naturally occurring, long-lived, radionuclide were used in medicine. Until the evolution of artificially-produced radioactive materials following the development of the nuclear reactor, almost all temporary implant brachytherapy was done with radium sources. These sources are especially constructed with materials to give them strength and rigidity, as well as to absorb the undesirable alpha and beta particles emitted. Only the gamma rays are allowed to penetrate the walls of radium sealed sources. The radium is in equilibrium with its daughter decay products including radon gas, with its protective sealed sheath.

Radium sources must be tested periodically, usually semiannually, for leakage. There are many methods that have been developed for leak testing.[14] The recommended technique is to place a few needles or tubes in a test tube containing activated charcoal separated from the sources by cotton. After an "incubation" period of about 24 hours, the sources are returned to their protective storage container and the tubes are monitored in a well counter or with a sensitive Geiger counter. Detected radioactivity suggests the presence of a leaking source which may be identified by repeating the test for each source individually.[2] Leaking sources of radium and, for that matter, sources no longer in use, which previously could be returned to the supplier, are becoming increasingly difficult to dispose of by hospitals. The Environmental Protection Agency can be helpful in obtaining the appropriate permit and in contacting the appropriate brokers to arrange shipment to a hazardous waste disposal site.

Brachytherapy may be applied to the treatment of superficial lesions by employing molds or plaques made of wax or plastic which act to support the radioactive sources as well as to position them accurately and repeatedly with respect to the treatment site. Interstitial brachytherapy implants are accomplished with radioactive needles or small radioactive seeds. These are inserted into or in the vicinity of the lesion and are widely used for the treatment of intraoral and superficial tumors as well as tumors in surgically accessible organs such as the prostate.

Intracavitary implants are accomplished with specially designed metallic and plastic applicators which contain the radioactive sources during the treatment of body cavities. Most intracavitary implants are used in the therapy of gynecological lesions such as cancer of the cervix or uterus.

The execution of both permanent and temporary interstitial and intracavitary implants using *afterloading techniques* is of fundamental importance in providing protection from unnecessary radiation exposure of physicians and their assistants. Radiation exposure of medical personnel when using radium, in fact, was the major obstacle to the full development of brachytherapy until the evolution of short-lived sealed sources suitable for afterloading techniques which are described by Pierquin, et al[15] and Hilaris and Henschke.[16]

Implants with radium or radium substitutes such as cesium-137 or cobalt-60 (rare) are *temporary* implants because they must be removed a few days after the implantation. Implants with tiny seeds, grains or microspheres of short-lived radionuclides are implanted permanently. The radioactivity of these tiny sealed sources decays rapidly and after a short duration, weeks or, at most, months the remaining activity is inconsequential. The encapsulating material is inert biologically and chemically and remains in the patient with no adverse effects. Permanent interstitial implants have been made with a host of radioactive sealed "mini"-sources, listed in Table 1.B.

Certain opthalmologic conditions may be effectively treated with small opthalmic irradiators positioned on or near the cornea for a short duration. Most applicators now used for this purpose contain sources of Y-90 in secular equilibrium with the parent Sr-90. The front surfaces of such applicators absorb most of the low-energy betas emitted from Sr-90 but permit the high energy betas from Y-90 to enter the eye where they are absorbed.[17,18] Characteristics of this beta-source are given in Table 1.B.

The beneficial application of sealed radioactive sources is accompanied by the potential risk of radiation exposure to persons who handle them routinely. In the next section we shall discuss the underlying philosophy for controlling radiation hazards and briefly describe the accepted techniques for handling and disposing of radionuclides in the clinical environment.

PERSONNEL PROTECTION FROM THE HAZARDS
OF RADIONUCLIDES

Radiation sources located externally to the body pose the hazard of either whole body or local irradiation. The range of alpha particles (generally not found in the clinical setting) is in the order of a few centimeters in air. They can be stopped by a few sheets of ordinary paper. The range of beta particles is normally in the tens of centimeters and local irradiation from an external beta source would involve only the most superficial layers of skin (up to a few millimeters in thickness). However, the range of γ photons in air is measured in meters. Hence, it is unlikely that whole-body irradiation could occur except from γ rays (and X rays).

The deposition of radionuclides within the body, i.e., as a result of ingestion, inhalation, or absorption through the skin, poses an entirely different problem. In this case, short range particles, such as σ- and β-particles, are the most hazardous. They dissipate all their energy in a very short distance, causing dense ionization of atoms along their tracks. A restricted volume of tissue would be involved. An additional hazard from internal unsealed sources exists if the radionuclide is organically selective since it will produce a more intense local irradiation. Examples of this selective concentration are the aforementioned iodine in the thyroid and radium, strontium and phosphorus in the bones. Exposure of the body internally from a radiation source continues only as long as the source remains in the system. The most serious internal hazard is associated with ingested long-lived β-emitters that are selectively concentrated and not readily excreted. Strontium-90 is a prime example. Iodine-131 is considered moderately hazardous.

The three key words in providing adequate protection from *external* sources of radiation are *time, distance* and*shielding.* They apply equally well to sources of γ rays and β particles. Short working times and maximum working distances are effective in reducing radiation exposures from both γ photons and β particles. For example, it is not uncommon to measure an exposure rate of 60 millirems per hour at a distance of 1 meter from a patient who has received a radium intracavitary implant. Standing at that position for a period of 20 minutes, a talkative nurse would receive a whole body exposure of 20 millirems; however, if she would complete her tasks efficiently in 5 minutes she would receive only 5 millirems. Of even greater consequence is the distance factor since exposure falls off approximately as the square of the distance. The exposure of 20 millirems to the nurse would be only 5 millirems at 2 meters and about 1 millirem if she works efficiently. (See Appendix A for the SI conversion of units in radiation measurements.)

Much thicker shielding is required for gamma photons since they are considerably more penetrating then beta particles and the nature of their interactions with matter are quite different. Beta particles are directly ionizing and have a definite maximum range that is dependent on the energy characteristic of the radionuclide from which it is emitted. In other words, a specific thickness of absorbing material capable of stopping the beta particles of maximum energy from a particular radionuclide will stop all betas from that emitter regardless of the source strength. On the other hand, there is a probability that some fraction of gamma photons will always penetrate a thickness of absorber, since gamma rays are indirectly ionizing. The fraction absorbed increases with increasing absorber thickness but the quantity transmitted never reaches zero. A shield that is just thick enough to provide adequate protection at one source strength will not be sufficiently thick at significantly higher source strengths. Consequently, the protection offered from a gamma shield must always be evaluated in terms of the source strength.[19]

Rooms in which cobalt-60 radiotherapy units are located are built with wall thicknesses designed to adequately shield gamma rays with the highest source strength which might be installed in the unit. Higher source strengths than specified in the room design should never be used. These rooms are equipped with radiation monitoring devices having an audible signal that sounds if the source is not shielded in its housing as, for example, during irradiation of a patient. The door to the room is equipped with an interlock system which will not allow a therapy exposure to be made unless the leadlined door to the treatment room is closed. These measures provide adequate protection and, normally, medical personnel working with teletherapy units do not receive exposures significantly above the natural background radiation level.

Protection of personnel using small sealed sources of radium, cesium-137 and cobalt-60 (rate) is a more difficult task than providing protection from teletherapy units. For example, the lead shielding thickness required to reduce the dose rate to a safe level from 100 milligrams of radium (an amount not uncommon in intracavitary therapy) is 11.8 cm.[11] It's not difficult to understand why lead aprons commonly found in diagnostic radiology are not adequate against radium gamma rays!

To protect personnel from radiation arising from the use of sealed sources a few guiding principles are briefly outlined below. For specific details reference should be made to the appropriate handbooks.[20-22] To protect the hands from β and γ rays, sources should be manipulated with long handled forceps. It's a good idea to plan the moves in advance so that all items required for a particular procedure are on hand and the operation can be carried out expeditiously. The operation should be done behind a suitable barrier, such as a lead L-block equipped with a lead glass window for viewing the work. This block is superimposed between the sources and the body of the operator. Radium sources should be tested periodically for leakage (described previously). It's also a good idea to store radium in well-ventilated rooms to guard against the inhalation of radon gas should a radium source accidentally be broken.

Persons working with unsealed radioactive sources must be protected not only from radiation emitted by the sources but also from ingestion, absorption through the skin or inhalation of radioactive materials into the body. Neither eating nor smoking should be permitted in laboratories where radionuclides are used. In general procedures adopted by hospitals will depend, to a large extent, on the facilities available which will vary from one institution to another. Detailed and specific information is given in publications by individuals[19,24,25] and by advisory groups such as the ICRP and the NCRP.[26-33]

Procedures involving the preparation of radiopharmaceuticals using open sealed sources should be carried out in well-ventilated areas or, preferably, under a fume hood with an external vent. The spilling of activity on permanent work surfaces should be avoided as much as possible through the use of trays and water-proof, plastic-backed absorbant pads. Decontamination of unprotected work surfaces, especially if contaminated with long-lived isotopes, can be problematic. All work areas, sinks and items which persons working with radionuclides are likely to contact, such as doorknobs, telephones, etc., should be monitored periodically for activity and decontaminated when necessary.

To reduce the necessity for preparing radiopharmaceuticals in hospitals, eliminating the need for radionuclide generators on site, and, thus, reducing the radiation exposures to the hands and whole bodies of personnel in nuclear medicine, many institutions are eliminating their "hot" laboratories altogether and purchasing radionuclides as needed directly from pharmacies that handle radionuclides. Unused radioactivity is returned to the pharmacy for disposal by a contracted broker. This procedure is a great simplification, reducing the risks attendant to preparing radiopharmaceuticals, is economically attractive, and is highly recommended.

DISPOSAL OF RADIOACTIVE WASTE[10]

There are three general methods for disposing of radioactive wastes. The most important of these, especially if the waste materials have reasonably short-half-lives of a few weeks or less, is to *store and decay*. (After a decay period of ten half-lives has elapsed, only 0.1 percent of the initial activity remains.) A suitable holding area must be dedicated to this purpose and an appropriate bookkeeping procedure must be developed to assure that wastes are disposed of in a timely fashion. It is recommended that radioactive wastes be segregated into two categories: those having half-lives shorter than 3 days and those with longer half-lives. In this way, long-term accumulations of large quantities can be avoided. When the radiation level measured at the surface of the plastic storage bag is not distinguishable from background it may be discarded in the ordinary trash. Disposal by the store and decay method is often impractical for wastes with half-lives longer than a month if relatively larger quantities are involved. This is an economical and effective storage technique that is highly recommended for community hospitals.

Small quantities of radioactive wastes may be *diluted and dispersed* in the environment, e.g., radioactive gases may be vented to the atmosphere and liquid wastes may be disposed of in the hospital sewage system. This is an especially suitable procedure for disposing of patient excretions. Patients who have received diagnostic dosages of radionuclides should be advised to flush the toilet at least three times following a bowel movement, urination, or an episode of vomiting. In adapting this procedure assurances should be made that concentrations do not exceed the Maximum Permissible Concentration (MPC) values specified in 10 CFR 20 of the Congressional Federal Register. This technique is *not* recommended for routine disposal of laboratory wastes and unused radiopharmaceuticals.

Finally, frequently the only effective means of disposal of long-lived radioactivity, particularly if storage space is limited or if the source strength is high, is by *concentration and burial*. A permit is now required for the use of a hazardous waste disposal site for this purpose. Currently, the only site available for the disposal of radioactive wastes is located at the Hanford Waste Disposal Site in Richland, Washington. This situation will change in the future as the location of other sites in the nation is now under study. A permit to use the site in Washington must be obtained from that state. The procedures for applying for a permit can be obtained from state and federal Environmental Protection Agencies. A number of commercial companies are available that provide the transportation service for this method of disposal. They will also provide detailed instructions on the packaging procedures for these wastes required by the Department of Transportation.

APPENDIX A

In recent years, the Comité International des Poids et Mésures (CIPM) has adopted an international system of units with the abbreviation SI (Sisteme Internacional). These are slowly being introduced world wide and many countries have already adopted them. These units are slowly being incorporated into measures in the United States and, it is hoped, that conversion to the new system will be complete, at least in scientific textbooks and journals, by 1985. For the few instances in which units of radiation measurement were necessary in this chapter, the older, more familiar terms were employed. A short explanation of the SI units proposed for use in radiation and radiation protection measurements is offered here.

The problems of expressing the effects of radiation for the purpose of radiation protection are not simple. A radiation dose of one type of radiation may produce a much larger biological effect than the same dose from another type. To obtain a quantity that expresses on a common scale the damage occurring to an individual exposed to radiation the concept of dose equivalent was introduced.[11]

$$H \text{ (dose equivalent)} = D \times Q \times N$$

where D is the absorbed dose, Q is the numerical quality factor assigned to the type of radiation involved, and N is the product of other modifying factors that describe the radiobiological damage. In the older system of units, D is expressed in rads, H in rems and Q and N are dimensionless fractions. The relationship of these units to the new SI units is as follows:

SI Unit
$$1 \text{ gray} = 1 \text{ Gy} = 1 \text{ joule/kilogram} = 100 \text{ rads}$$
$$1 \text{ sievert} = 1 \text{ Sv} = 1 \text{ joule/kilogram} = 100 \text{ rems}$$

Exposure is a measure of the ability of radiation to ionize air. It is measured in units of coulombs (charge) per kilogram of air. The special unit of exposure used in the older system was the Roentgen (named for the discoverer of X rays).

$$1 \text{ Roentgen} = 2.58 \times 10^{-4} \text{ C/kg of air (exactly)}$$

or

$$1 \text{ C/kg} = 3876 \text{ R}$$

There is no special SI unit of exposure so C/kg is used.

For radiations described in this chapter and normally found in hospitals, the quality factor Q and modifying factor N are both equal to 1. Therefore, in the older units the rad and the rem are numerically equivalent. As a matter of fact, the Roentgen is almost numerically equivalent and, therefore, in the hospital environment 1 rad = 1 rem $\cong$ 1 Roentgen in most instances. This relationship doesn't hold true where alpha particles, neutrons, protons and other exotic radiations are involved in the irradiation.

The Curie (or millicurie, where 1000 mCi = 1 Ci), the unit of activity commonly used will be expressed in the SI system as the Bequerel.
$$1 \text{ Bq} = 1 \text{ disintegration per sec (1 sec}^{-1})$$
$$1 \text{ Curie} = 3.7 \times 10^{10} \text{ Bequerels (Bq)}$$

BIBLIOGRAPHY

1. Wang, C.H., D.L. Willis and W.D. Loveland. 1975. *Radiotracer Methodology in the Biological, Environmental, and Physical Sciences.* Prentice-Hall, Inc., Englewood Cliffs, New Jersey.
2. Hendee, W.R. 1979. *Medical Radiation Physics.* Second Edition. Year Book Medical Publishers, Inc., Chicago.

3. Webster, E.W., N.M. Alpert, and C.T. Brownell. 1974. Radiation doses in pediatric nuclear medicine and diagnostic X-ray procedures. In *Pediatric Nuclear Medicine*, ed. A.E. James, Jr., H.N. Wagner, and R.E. Cooke. W.B. Saunders, Philadelphia.

4. ICRP 1971. *Protection of the Patient in Radionuclide Investigations.* ICRP Publication 17. Permagon Press, New York.

5. NCRP 1977. *Medical Radiation Exposure of Pregnant and Potentially Pregnant Women.* NCRP Report No. 54. NCRP, Washington, D.C.

6. NCRP 1982. Nuclear Medicine—*Factors Influencing the Choice and Use of Radionuclides in Diagnosis and Therapy.* NCRP Report No. 70. NCRP, Washington, D.C.

7. Hine, G.J., and R.E. Johnson. 1970. Absorbed dose from radionuclides. *J. Nucl. Med.,* 11:468.

8. UNSCEAR 1977. *Sources and Effects of Ionizing Radiation.* United Nations Scientific Committee on the Effects of Atomic Radiation to the General Assembly, with annexes. United Nations, New York.

9. Roedler, H., A. Kanl and G. Hine. 1978. *International Radiation Dose in Diagnostic Nuclear Medicine.* Verlag H. Hoffman, Berlin.

10. Sorenson, J.D. and M.E. Phelps. 1980. *Physics in Nuclear Medicine.* Grune and Stratton, New York.

11. Johns, E.J. and J.R. Cunningham. 1983. *The Physics of Radiology.* Fourth Edition. Charles C. Thomas, Publisher, Springfield, Illinois.

12. Freeman, T.M. and M.D. Blaufox. 1979. Nuclear medicine, a summary of current techniques. *In: Physicians' Desk Reference for Radiology and Nuclear Medicine,* 9th Edition. Charles E. Baker, Jr., Publisher, Medical Economics Company, Oradell, New Jersey.

13. National Council on Radiation Protection and Measurements. 1978. A Handbook of Radioactivity Measurements Procedures. Recommendations of the NCRP, Report 58. Washington, D.C.

14. Wood, V.A. 1968. *A Collection of Radium Leak Test Articles.* U.S. Department of Health, Education, and Welfare Report MORP 68-1.

15. Pierquin, B., D.J. Chassagne, C.M. Chahhazian, and J.F. Wilson. 1978. *Brachytherapy.* Warrent H. Green, Inc. St. Louis.

16. Hilaris, S.B. and U.K. Henschke. 1975. General principles and techniques of interstitial brachytherapy. *In: Handbook of Interstitial Brachytherapy,* S.B. Hilaris, Ed. Publishing Sciences Group, Inc. Acton, Massachusetts.

17. Friedell, H., C. Thomas and J. Krohmer. 1954. Evaluation of clinical use of strontium 90 beta-ray applicator with review of underlying principles, A.J.R. 71:25.

18. Duggan, H. 1966. Results using strontium 90 beta-ray applicator on eye lesions. J. Can. Assoc. Radiol. 17:132. 1966.

19. Shapiro, J. 1981. *Radiation Protection, A Guide for Scientists and Physicians.* Harvard University Press, Cambridge, Massachusetts.

20. National Council on Radiation Protection and Measurements. 1968. Medical X ray and gamma-ray protection for energies up to 10 MeV. NCRP Report 33. Washington, D.C.

21. National Council on Radiation Protection and Measurements. 1976. Structural shielding design and evaluation for medical use of X rays and gamma rays of energies up to 10 MeV. NCRP Report 49. Washington, D.C.

22. National Council on Radiation Protection and Measurements. 1976. Radiation protection for medical and allied health personnel. NCRP Report 48. Washington, D.C.

23. Hendee, W. and S. Lohlein. 1968. Handling therapeutic doses of radioactive nuclides in a hospital. Radiol. Technol. 40:81.

24. Moore, M. and W. Hendee. 1977. *Radionuclide Handling and Radiopharmaceutical Quality Assurance*. Workshop Manual, Bureau of Radiological Health, USDHEW-FDA.

25. *Manual on Use of Radioisotopes in Hospitals*. 1958. American Hospital Association, Chicago.

26. International Atomic Energy Agency: Safety Series. No. 1: *Safe Handling of Radioisotopes,* 1963; No. 2: *Safe Handling of Radioisotopes — Health Physics Addendum,* 1960; No. 3: *Safe Handling of Radioisotopes — Medical Addendum,* 1960 (Vienna: IAEA).

27. International Commission on Radiological Protection. 1971. *Protection of the Patient in Radionuclide Investigations.* ICRP publ. 17. Permagon Press, New York.

28. National Council on Radiation Protection and Measurements. 1970. Precautions in the management of patients who have received therapeutic amounts of radionuclides. Recommendations of the NCRP, Report 37. Washington, D.C.

29. National Committee on Radiation Protection. 1964. Safe Handling of radioactive materials. Recommendations of NCRP, Report 30. Washington, D.C.

30. National Council on Radiation Protection and Measurements. 1972. Protection against radiation from brachytherapy sources. Recommendations of the NCRP, Report 40. Washington, D.C.

Chapter Thirteen

FEDERAL TRANSPORTATION REQUIREMENTS FOR RADIO-ACTIVE MATERIALS

Richard R. Rawl

Chief, Radioactive Materials Branch
U.S. Department of Transportation
DMT—223
400 Seventh St. S.W.
Washington, D.C. 20590

This chapter explains the basic Federal regulatory requirements[1] for transporting radioactive materials. Extensive changes to the transportation regulations were made in July, 1983 and these new requirements are described. There are many detailed requirements of the regulations that cannot be presented here due to space limitations. For this reason the contents of this chapter should serve as a guide and explanation of the basic requirements and not as a substitute for the detailed regulations themselves.

BACKGROUND

The transportation of radioactive materials has been regulated by Federal agencies since the early 1950's when there was concern expressed for the fogging of photographic film transported in the proximity of radioactive material. These earliest regulations, in their quest to protect radiosensitive materials, also provided a degree of radiation protection to transport workers and the public at large. As radiation protection philosophies and practices have evolved, their application in transportation has also changed. The International Atomic Energy Agency (IAEA) has been instrumental in fostering international agreement on what should be required as well as in ensuring that the latest international radiation protection principles are applied. The U.S. regulations have evolved in consonance with the international transport regulations as promulgated by the IAEA.[2] Current international and U.S. requirements reflect the continuing efforts to ensure that radiation exposures from transportation (both normal and accident conditions) remain as low as reasonably achievable.

CATEGORIES OF MATERIALS

For purposes of transportation, radioactive materials are defined as those which spontaneously emit ionizing radiation and which have a specific activity exceeding 2 uCi/kg. Materials with lower specific activities are not regulated in transportation. The materials which are regulated range widely in characteristics, from very low to large activities and with a range of concentrations (specific activities) as well.

In order to facilitate the specification of transport requirements which are commensurate with the hazards of the materials, a number of categories are used. The broad spectrum of radioactive materials are divided into the following categories for transportation purposes:

1. Excepted quantities—very low total activity;
2. Low Specific Activity (LSA)—limited specific activities;
3. Type A quantities—moderate activities;
4. Type B quantities—larger activities; and
5. Fissile materials—those capable of undergoing nuclear criticality.

Each category has specific requirements, the stringency of these being based on the hazard of the materials being transported. In all cases the requirements are designed to provide suitable:

1. Containment of the material;
2. Protection from radiation exposure;
3. Rejection of decay heat; and
4. Prevention of criticality.

Because of the wide range of radiotoxicities presented by the radionuclides which are shipped, a system for ranking them has been developed. The system used incorporates the metabolic data and dosimetric modeling of the International Commission on Radiological Protection which is then factored into a series of transportation-related models. The results are two values for every radionuclide; an A_1 value for "special" form and an A_2 value for "normal" form. "Special" form refers to radionuclides which are essentially non-dispersible, usually by virtue of being contained in high integrity welded capsules. This non-dispersibility for special form materials must be proven by subjecting the material to: a 9 m drop onto an unyielding surface; heating in air to 800°C for 10 minutes; the impact of a 1.4 kg steel bar dropped 1 m; and if capsule is long and slender, a bending test; all without any significant release of activity. These tests are described in detail in Title 49 Code of Federal Regulations, Section 173.469 (referred to as 49 CFR 173.469). Special form materials generally present only an external radiation hazard.Normal form materials are all forms other than special form and are comprised of the powders, liquids, gases, etc., that are shipped. Normal form materials may present an external radiation hazard and a contamination or internal hazard if inhaled or ingested. Since A_1 (special form) values are limited by external radiation they are generally higher than the A_2

(normal form) values which are limited by internal organ considerations or A_1, whichever is least. An artitrary ceiling of 1000 Ci is placed on the less radiotoxic nuclides.

Examples of A values are:

Radionuclide	A_1 (Ci)	A_2 (Ci)
241_{Am}	8	.008
14_C	1000	60
60_{Co}	7	7
99_{Mo}	100	20
239_{Pu}	2	.002

The complete list of radionuclides and their A values are shown in 49 CFR 173.435. Multiples and submultiples of the A values are used throughout the regulations to specify limits since they provide reference values for each radionuclide based on its radioactive properties and relative hazard in transportation.

EXCEPTED QUANTITIES

At the lower end of the hazard spectrum are those materials which contain a very limited amount of total activity. Within this category there are three subdivisions of materials:

1. Limited quantities—any form of material;
2. Instruments and articles—manufactured goods with radioactive materials as a component part; and
3. Manufactured articles of uranium or thorium metal.

Excepted quantities take many forms and include diagnostic kits, samples, electron tubes, smoke detectors and small check sources. These materials present a minimal hazard in transportation and hence have minimal requirements applied to them. Basic requirements are that the radiation level at the surface of the package must not exceed 0.5 mrem/h, the surface of the package must be free of significant removable contamination and the materials must be retained in their package under conditions encountered in normal transportation. Consignments of excepted quantities must be described on a document which is placed in or on the package or forwarded with it (49 CFR 173.421-1). This description identifies the material by category and indicates its radioactive nature in case it becomes lost or involved in an accident.

Limited quantities (49 CFR 173.421) must also have their inner packaging marked "Radioactive" as an additional precaution against uniformed opening. Packages of limited quantities must not exceed the limits given in column 4 of Table 1. Instruments and articles (49 CFR 173.422) have their activity limited on a per item and per package basis as given in columns 2 and 3 of Table 1,

TABLE 1

Activity Limits for Limited Quantities, Instruments and Articles

Nature of contents	Instruments and Articles		Limited Quantity Materials package limits
	Instrument and article limits[1]	Package limits	
Solids:			
Special form	$10^{-2}A_1$	A_1	$10^{-3}A_1$
Other form	$10^{-2}A_2$	A_2	$10^{-3}A_2$
Liquids:			
Tritiated water:			
<0.1 Ci/liter	—	—	1000 Curies
0.1 Ci to 1.0 Ci/1	—	—	100 Curies
>1.0 Ci/liter	—	—	1 Curie
Other Liquids	$10^{-3}A_2$	$10^{-1}A_2$	$10^{-4}A_2$
Gases:			
Tritium[2]	20 Curies	200 Curies	20 Curies
Special form	$10^{-3}A_1$	$10^{-2}A_1$	$10^{-3}A_1$
Other forms	$10^{-3}A_2$	$10^{-2}A_2$	$10^{-3}A_2$

[1]For mixture of radionuclides see §173.433(b).
[2]These values also apply to tritium in activated luminous paint and tritium adsorbed on solid carriers.

above. The radiation level from any unpackaged item is restricted to 10 mrem/h.

Manufactured articles whose sole radioactive contents are natural or depleted uranium or natural thorium (49 CFR 173.424) are treated similarly to limited quantities. One additional requirement placed on them is that the outer surface of the uranium or thorium must be enclosed in a durable protective sheath.

LOW SPECIFIC ACTIVITY (LSA) MATERIALS

Another category of materials with a limited potential hazard are those materials with low specific activities. Unlike the other categories of materials, restrictions are not placed on the *total* activity within a package but are placed on the *specific activity* of the material which is put into the package. The allowable specific activities were derived using a transportation accident model and relating the resulting internal exposures (inhalation) to the ICRP recommendations.

Within the category of LSA materials (49 CFR 173.403(n)) there are five subcategories of materials, some of which are LSA by the very nature of the radionuclides involved and some of which must have their concentrations limited to below established values. Materials which are inherently LSA are:

1. Uranium or thorium ores and physical or chemical concentrates of these ores; and

2. Unirradiated natural or depleted uranium and unirradiated natural thorium.

All other LSA materials must be evaluated to meet one of the following:

3. Tritium oxide in aqueous solutions with a concentration that does not exceed 5.0 millicuries per milliliter;

4. Nonradioactive objects externally contaminated with radioactive material—contamination must not be readily dispersible—when averaged over 1 square meter must not exceed 0.0001 millicurie per square centimeter for radionuclides with an A_2 value of not more than 0.05 curie and must not exceed 0.001 millicurie per square centimeter for other radionuclides; and

5. Materials in which the activity is essentially uniformly distributed and in which the estimated average concentration does not exceed—
   - 0.0001 millicurie per gram for radionuclides with an A_2 value of not more than 0.05 curie,
   - 0.005 millicurie per gram for radionuclides with an A_2 value of more than 0.05 curie but not more than 1 curie, or
   - 0.3 millicurie per gram for radionuclides with an A_2 value exceeding 1 curie.

In order to determine if a given material is LSA or not, the last subcategories require some investigation and calculation. First, the specific activity of the material by radionuclide(s) must be ascertained as well as the A_2 value (49 CFR 173.435) for each radionuclide. Then one must compare the specific activity of the material by radionuclide with the specific activity limit according to the A_2 value of the radionuclide. If any radionuclide is present in a specific activity exceeding the specific activity limit given for its A_2 value the material cannot be classified as LSA. Additionally, in order for a material to be classed as LSA under these criteria the radionuclides present must be grouped by A_2 values in accordance with the above listed break points. The specific activities for all of the radionuclides in each group must be summed and the total specific activity cannot exceed the limit for the group. As a final step the specific activity present per group must be ratioed to the group limit and summed to be sure it is less than unity, as follows:

$$\frac{APG_1}{0.0001} + \frac{APG_2}{0.005} + \frac{APG_3}{0.3} \leq 1$$

where:

APG_1 = the total specific activity in millicuries per gram of material for all radionuclides in the material with an A_2 value not exceeding 0.05 curie;

APG_2 = the total specific activity in millicuries per gram of material for all radionuclides in the material with an A_2 value of more than 0.05 curie but not exceeding 1.0 curie; and

APG_3 = the total specific activity in millicuries per gram of material for all radionuclides with an A_2 exceeding 1.0 curie.

While the computation necessary for a mixture of radionuclides may appear somewhat involved it is necessary in order to ensure that the material will not pose a significant internal hazard if inhaled under accident conditions. In return for this "built in" safety certain packaging and labeling requirements are waived. This allows LSA materials to be shipped more economically and in some cases in bulk. Any material which meets the definition of LSA may be shipped in either of two ways. The first is in regular commerce, tendered to a carrier in a routine manner (49 CFR 173.425(a)). This method requires the use of a Type A package. Type A packages (49 CFR 173.403(cc)) must be capable of withstanding the normal conditions of transport and rough handling as demonstrated by meeting certain test requirements (see next section). When shipped this way packages of LSA material must also be marked, labeled and described on detailed shipping papers in the same manner as a Type A consignment.

The second method for shipping LSA is the most widely used and involves "exclusive use" consignments of LSA packages. Exclusive use conditions require the sole use of the vehicle by a single consignor (shipper) and all loading or unloading must be under the direction of the consignor or consignee (receiver). In return for exclusive use conditions of carriage the shipper is allowed to use "strong, tight" packaging instead of the Type A. The strong, tight packaging is not required to meet specific tests such as a free drop but it must be designed and constructed so that there will be "no leakage of radioactive materials under conditions normally incident to transportation." This allows the shipper considerable freedom in package design yet it must be sufficient to completely contain the contents even after the shock and vibration of normal transport.

When shipped as exclusive use in strong, tight packages there are a number of additional conditions (49 CFR 173.425(b)) which must be fulfilled, including:
- packages must be free of significant removable external contamination;
- the vehicle and packages must be loaded so that the radiation levels do not exceed (49 CFR 173.441(b)):
    - 1000 mrem/h at the package surface. Packages exceeding 200 mrem/h at the surface must be shipped in a closed transport vehicle;
    - 200 mrem/h at any accessible external surface or edge of the vehicle;
    - 10 mrem/h at any point 2 m from any accessible surface or edge of the vehicle; and
    - 2 mrem/h at any normally occupied position in the vehicle.
- there must be no loose radioactive material in the conveyance;

- the shipment must be braced to prevent shifting;
- except for unconcentrated ores, the conveyance must be placarded;
- each package must be marked "Radioactive—LSA"; and
- specific instructions on the maintenance of exclusive use must be provided by the shipper to the carrier and included in the shipping papers.

Some LSA materials may be transported in bulk, that is without individual packaging (49 CFR 173.425(c)). The materials that can be carried this way include ores and ore concentrates, uranium and thorium metals and alloys. Materials and contaminated objects which are restricted to lower concentrations and contamination limits may also be shipped in bulk (49 CFR 173.425(c)).

Licenses of the U.S. Nuclear Regulatory Commission (NRC) should be aware that the NRC requires that packages containing LSA material with a total activity exceeding A_2 must meet the Type A packaging standards (10 CFR 71.52).

TYPE A QUANTITIES

Quantities of materials which would not present a serious hazard if involved in an accident are allowed to be shipped in packagings designed for normal conditions of transport and rough handling. These quantities are restricted to activities not exceeding the A_1 value for special form materials nor A_2 for normal form materials. The A values are derived from a hypothetical accident scenario and the limited contents provide a limitation for expected consequences. Experience has shown the models to be quite conservative.

Any form or specific activity material can be shipped as a Type A quantity, provided of course, that the total activity in a package does not exceed the Type A quantity (A_1 or A_2 as appropriate for the material form). Special provisions and requirements are placed, however, on oxidizing radioactive materials (49 CFR 173.419) such as thorium nitrate and pyrophoric radioactive materials (49 CFR 173.418) such as finely divided uranium metal. There are also special tests for Type A packages designed for liquids and gases (49 CFR 173.466) because of their greater dispersibility.

The most widely used Type A packaging is known as a DOT Specification 7A. This is a performance oriented package specification in that its ability to withstand certain test conditions is required; the actual construction design is not specified, leaving the shipper free to design a packaging to meet his needs. This provides the shipper with a great degree of freedom in package design but also places on him the responsibility to test or evaluate each package design to the required tests and to maintain adequate documentation of this (49 CFR 173.415(a)).

The tests and environmental conditions which the packaging must be capable of withstanding include:

1. Temperature range of $-40°C$ to $70°C$;

2. Acceleration, vibration and vibration resonance that may occur during normal transportation;
3. Reduced pressure of 0.25 atm;
4. Water spray test simulating exposure to rainfall of 5 cm/h for one hour;
5. Free drop of 1.2 m (graduated down to 0.3 m for packages with mass exceeding 15 Mg);
6. Corner and rim free drops for certain fiberboard and wooden packagings;
7. Compression of 5 times the package weight or 1.3 Mg times the vertically projected area in meters; and
8. Penetration of a hemispherical ended 3.2 cm diameter steel bar with a mass of 6 kg dropped 1 m.

Additionally, packagings designed for liquids and gases must be subjected to a free drop of 9 m and the penetration test with the bar dropped 1.7 m.

In order to determine if the package design successfully passes the tests, two acceptance criteria are specified: no loss or dispersal of the radioactive contents; and no significant increase in the recorded or calculated radiation levels at the package surfaces. These acceptance criteria are designed to ensure that packages which have been subjected to rough handling can still be safely transported to their destination.

TYPE B QUANTITIES

Quantities of material with a total activity exceeding the A_1 or A_2 values could pose a significant hazard if the contents were released from the packaging. Activities exceeding the Type A values are known as Type B quantities and must be transported in accident resistant packagings.

In addition to meeting all of the Type A packaging test requirements a Type B packaging must be subjected to:
1. Free drop of 9 m onto an unyielding surface;
2. Thermal environment of 800°C for 30 minutes;
3. Free drop of 1 m onto a 15 cm diameter steel bar penetrator.

In order to successfully pass these tests the packaging must prevent loss of the radioactive contents (except for limited amounts of contaminated coolant or gases) and must limit radiation levels to a level less than 1000 mrem/h at 1 m from the package surfaces.

All Type B packagings used domestically must either be specifically approved by the NRC or Department of Energy (packagings restricted to DOE use) or be listed in Title 49 as specification packages (49 CFR 173.416). The allowed contents for each package design are specifically given in either its approval certificate or Title 49. For more details on Type B packaging requirements the NRC regulations should be consulted[3].

FISSILE MATERIALS

Because of the potential criticality hazard of fissile materials packages, they must be evaluated to ensure criticality safety under both normal and accident conditions of transport. Individual packages must remain subcritical by an adequate margin and arrays of the packages must also remain subcritical. The packages are evaluated in both the undamaged and damaged conditions with damaged meaning after being subjected to the Type B tests.

The criticality evaluation determines the number of packages of a particular design which may safely be transported together. The number of packages in a shipment is then limited to these numbers either directly through a provision in the package approval or indirectly through labeling.

In order to indicate the degree of control which needs to be exercised to ensure criticality safety, fissile packages are divided into three categories, as follows:

Fissile Class I—Packages which may be transported together in unlimited numbers.

Fissile Class II—packages which may be transported together only in limited numbers.

Fissile Class III—packages which must be transported under exclusive use conditions with particular actions being taken to ensure criticality safety.

The assignment of the proper fissile class is given either in the DOT regulations (for specification packages) or in the NRC or DOE approval certificate for the package design. The number of packages which may be transported together or the required labeling information is also specified.

HIGHWAY ROUTE CONTROLLED QUANTITY (HRCQ)

At the upper end of the spectrum of radioactive materials transport are those packages which contain a HRCQ of radioactivity (49 CFR 173.403(e)). These activities are well above the Type A quantity and comprise the upper range of Type B quantities. They must be packaged in Type B (and fissile, if appropriate) packaging and are subject to specific routing controls.

The HRCQ is defined as a quantity of material which exceeds:

1. 3000 times A_1 (for special form materials);
2. 3000 times A_2 (for normal form materials); or
3. 30,000 curies, whichever is least.

Carriers that transport HRCQ's must provide special training for the driver (49 CFR 177.825) and provide to the shipper a copy of the routing that was used for the shipment. The shipper in turn must provide this to DOT (49 CFR 173.22). The actual routes which are used for transporting HRCQ's must be "preferred routes." Preferred routes consist of the interstate highway system and/or any route designated by a State as being preferred. In order to designate a segment of highway as preferred in lieu of an interstate segment the State must quantitatively evaluate the selection from a radiological risk perspective.

OTHER SHIPPING REQUIREMENTS

The preceding sections have dealt mainly with the packaging requirements for transporting radioactive materials. This is fitting because the underlying regulatory philosophy is that most of the safety provided during transport should result from the packaging requirements, including of course, materials limitations in the less stringent packaging categories. Such an approach minimizes the need for any human action during transport to ensure safety.

There are some conditions, however, when actions during transport may be necessary such as for minimizing external radiation exposure from the accumulation of packages or responding to an accident in which the packages may have been damaged. In order to support such actions there are a number of requirements directed at both the shipper and carrier.

* Shipping Papers

  Each consignment of radioactive material (other than excepted quantities) must be described (49 CFR 173.200) in detail on the documentation for the shipment. These "shipping papers" must include a package by package description which provides the following:
    1. The proper shipping name as selected from the 49 CFR 172.101 table of proper shipping names;
    2. The hazard class "Radioactive material" unless these words are contained in the proper shipping name;
    3. The identification number given in the 49 CFR 172.101 table for the selected proper shipping name;
    4. The name or symbol of each radionuclide;
    5. A description of the physical and chemical form of the material or the notation that it is special form;
    6. The activity of each package in Ci, mCi, or uCi; and if the package contains a HRCQ the words "Highway Route Controlled Quantity";
    7. The category of label applied to the package;
    8. The Transport Index (for yellow category-II and -III labels);
    9. Fissile information; and
    10. Package approval identification markings.

  Each shipping paper must also contain a standardized certification (49 CFR 172.204) by the shipper that the packages are properly prepared, described and meet the DOT regulations. Shipments via passenger-carrying aircraft are restricted to radioactive materials intended for use in, or incident to, research, or medical diagnosis and for such shipments a certification to this effect is required (49 CFR 172.204(b)(4)).

The Transport Index (TI) referred to in item 8 above is the highest radiation level at 1 m from the surfaces of the package (except for fissile packages). For fissile packages the TI is the highest of either the radiation oriented value or one derived from for criticality safety considerations (49 CFR 173.403(bb)).

* Marking

    Each package of radioactive materials (other than excepted quantities and exclusive use consignments of LSA materials) must be marked with certain information (49 CFR 172.300), including the;
    1. Proper shipping name;
    2. Identification number;
    3. Name and address of the consignee or consignor;
    4. Gross weight of the package (only for packages exceeding 50 kg);
    5. "Type A" or "Type B" as appropriate;
    6. Package certification or specification marking (for Type B and fissile packages only)

* Labeling

    Each package of radioactive materials (other than excepted quantities and exclusive use consignments of LSA materials) must be labeled with one of the three radioactive materials labels (49 CFR 172.400 and 172.403). The labels are known as the White-I, Yellow-II and Yellow-III and appear as follows, with the upper half of -II and -III labels being yellow and the -I, -II and -III stripes being red:

The choice of the correct label is dependent upon the Transport Index, the maximum radiation level at the package surfaces, and the fissile characteristics (if any) of the package. These criteria are applied using Table 2.

In choosing the correct label category the shipper must select the highest category required for *any* of the three criteria given in Table 2 above (TI, RL or Fissile Criteria). Category White-I is the lowest and Category Yellow-III is the highest.

Each label has either two (for White-I) or three (for Yellow-II and Yellow-III) entries which must be filled in by the shipper. All labels must have the "contents" entry filled the in with symbols or names of the radionuclides in the package and the "activity" entry must be filled in with the contained activity. Yellow labels must also have the "Transport Index" box filled in with the appropriate TI for the package.

TABLE 2

Radioactive Materials Packages Labeling Criteria

Transport Index (TI)	Radiation Level at Package Surface (RL)	Fissile Criteria	Label Category[1]
N/A	RL ≤0.5 millirem per hour (mrem/h	Fissile Class I Only No Fissile Class II or III	White-I
TI ≤1.0	0.5 mrem/h < RL ≤ 50 mrem/h	Fissile Class I, Fissile Class II with TI ≤1.0, No Fissile Class III	Yellow-II
1.0 <TI	50 mrem/h < RL	Fissile Class II with 1.0 < TI, Fissile Class III	Yellow-III

- Placarding

   Certain vehicles must display a large (10" x 10") square-on-point placard which shows the word "Radioactive" and the internationally recognized trefoil symbol (49 CFR 172.556). The upper half of the placard is yellow. Only those vehicles transporting a package bearing a Yellow-III label or an exclusive use shipment of LSA materials need to be placarded radioactive. Vehicles transporting a HRCQ must display a large square white background outlined by a black border behind the "Radioactive" placard (49 CFR 172.507).

CARRIER REQUIREMENTS

   There are few actions needed to be taken by the carrier during transport under normal conditions. The actions which are required are directed at minimizing exposure of personnel to radiation and ensuring criticality safety for fissile materials. These objectives are met by either limiting the accumulation of packages based on their transport index or in special situations by requiring maintenance of exclusive use conditions.

   Carriers must normally limit the accumulation of yellow labeled packages to less than 50 TI's per vehicle (49 CFR 177.842) and per group of packages when in storage. Groups of packages must be separated from each other by a distance of at least 6 m. Minimum separation distances are also specified for the cargo-to occupied space distance on the vehicle. These separation distances are based on the number of TI's actually present and establish an upper limit on the dose rate in occupied areas.

   Under exclusive use conditions these goals may be met through other means such as formalized radiation protection programs for personnel in lieu of the

50 TI maximum per vehicle rule. Likewise, critically safety may also be provided by ensuring that no additional radioactive materials will be loaded onto the vehicle (49 CFR 173.457).

Carriers are also required to report incidents involving any release of material and to ensure that vehicles are decontaminated before being returned to service (49 CFR 171.16).

INTERNATIONAL SHIPMENTS

Provisions are contained in the regulations to allow import and export shipments in essential conformity with the international transport regulations as promulgated by the International Civil Aviation Organization and the IAEA. There are a small number of differences between the U.S. and these international regulations and those U.S. requirements which must be met when in the U.S. are listed in these sections (49 CFR 171.11 and 171.12).

FUTURE REGULATORY ACTIVITY

The IAEA has revised the international transport regulations and should publish the new requirements in 1985. While the changes which have been proposed are not wide ranging they may be significant in certain areas such as for LSA materials. In order to maintain continuity with the international requirements and to ensure that all safety concerns are addressed, the U.S. Department of Transportation and NRC will most probably be proposing amendments to the domestic regulations following the IAEA's publication of its regulations. This action could be reasonably expected in 1986.

CONCLUSION

While it is impossible to inform the reader in this short chapter of all the details involved in making a safe and legal shipment of radioactive materials, the major areas have been covered. It is hoped that the reader (and potential shipper) now has a basic knowledge of the numerous aspects of making a shipment.

Anyone planning to make a shipment needs to obtain a copy of the DOT regulations[1] and may need the NRC regulations[3] as well. Periodically the DOT publishes guidance material which shippers find useful and which may be obtained from: Department of Transportation, Materials Transportation Bureau, Washington, D.C. 20590.

REFERENCES

1. Title 49 Code of Federal Regulations, Parts 100-199, Superintendent of Documents, U.S. Government Printing Office, Washington, D.C. 20402.
2. *Regulations for the Safe Transport of Radioactive Materials,* Safety Series No. 6, 1973 Revised Edition (As Amended). International Atomic Energy Agency, Vienna, Austria.
3. Title 10 Code of Federal Regulations Parts 71; Superintendent of Documents, U.S. Government Printing Office, Washington, D.C. 20402.

PART 3

Requirements, Socio-Political Considerations and Recommendations

As the radioactive waste materials have increased, there is a growing recognition by the public that there are many social and political problems that must be addressed. One of the most critical of these is the location of the waste sites. Almost, without exception, there is public opposition to having a low-level radioactive waste disposal site in a nearby neighborhood.

The initial chapter provides an analysis and evaluation of the problems of siting low-level radioactive facilities. Such aspects are discussed as public attitudes toward waste and waste disposal, public perception of radioactive risks, and problems of public participation. To resolve the opposition, such possible incentives must be offered to abutting communities as assurances that property values will be maintained, local agricultural products will not be affected, early and periodic health screening will be available, local services such as roads and bridges will be maintained, local site monitoring will occur on a regular basis, and the local community will have a role in long-term management including a voice in deciding various site operation options.

The resolving of political issues is also crucial to the disposal of nuclear wastes. Foremost is the prevention of polarization of opinion regarding nuclear power into irreconcilable camps and concomitant difficulty in separating issues of nuclear armament, nuclear energy and nuclear waste. In nuclear waste disposal, popular perception is critical in a political system in which democratic decision making is valued. The second chapter considers polarization, perception and risk, distribution of benefits and costs, the importance of crises and fortuitous events, and finally, an analysis of political considerations in the establishment of a nuclear waste policy.

The final chapter of this section presents the findings of the International Council of Scientific Unions on the disposal of radioactive wastes. Their major conclusion is that nuclear wastes may be safely disposed of using current technology. It is recognized that interim storage of wastes for 50 to 100 years would reduce the problem of thermal loading at the final disposal sites. Additional research, however, is needed to assure the success of interim storage in radioactive waste disposal.

Chapter Fourteen

PROBLEMS IN SITING LOW LEVEL RADIOACTIVE WASTES: A FOCUS ON PUBLIC PARTICIPATION

Richard J. Bord, Ph. D.
Professor of Sociology
The Pennsylvania State University
201 Liberal Arts Tower
University Park, Pa. 16802

A major obstacle to the establishment of waste sites of all kinds has been vigorous public opposition[1]. Attempts to site low level radioactive waste (LLRW) are almost certain to confront similar, if not more vociferous, levels of public resistance. A recent report by the General Accounting Office [2] documents the snail's pace at which states have moved to implement the Low-Level Radioactive Waste Policy Act of 1980. That act gave the States responsibility for LLRW and encouraged the formation of interstate compacts and regional disposal facilities. However, public officials are reluctant to do anything that might identify them as favorable to LLRW siting. In Pennsylvania politicians are especially wary of pushing LLRW siting in the wake of the anti-nuclear sentiment generated by the Three Mile Island power plant accident of 1979.

Faced with the need to safely dispose of the State's radioactive wastes it becomes increasingly necessary to establish effective ways of eliciting the public's cooperation rather than encouraging their opposition. The mechanism that is supposed to foster cooperation between the public and government agencies in cases involving potentially risky technologies is the public participation program. Unfortunately, public participation programs associated with waste siting often result in greater conflict and more intransigent opposition rather than cooperation and conciliation. This unconstructive outcome of public participation programs is a function of (a) public attitudes toward wastes and waste disposal in general, (b) their fear of radiation risks in particular, and (c) shortcomings in the structure of the participation programs themselves which includes public distrust of those agencies managing the programs.

A better understanding of each of these issues should provide guidelines which promote the construction of more effective public participation programs. While it is probably impossible to design a public participation program that will guarantee siting, there is much that can be done to deal effectively with public fear, hostility, and distrust. The second part of this paper will suggest options to improve the public participation process.

PROBLEMS IN ELICITING PUBLIC SUPPORT FOR LLRW DISPOSAL

Public Attitudes Toward Waste and Waste Disposal

It is clear from public opinion data and from the numerous unsuccessful attempts to site solid and chemical wastes that waste of any kind is viewed as an unwelcome intrusion into communities asked to bear the burden of its presence.[3] Communities generally resent being tagged as a dumping ground for wastes they did not create. They deeply resent being saddled with wastes from outlying areas. This is especially true of rural areas which are asked to accept urban wastes. The feeling is that other people are getting all the benefits while the community hosting the waste site is getting all the costs. This is the equity problem which will have to be successfully dealt with in any waste siting program.

Perceived risks play a role in generating public opposition to local waste siting attempts. The mass media, particularly television, has been unrelenting in its attention to waste dump problems ranging from the infamous Love Canal incident to recent exposes of Dioxin contamination
in the Midwest and of New York City solid waste facilities causing unanticipated chemical pollution problems. This steady diet of fear arousing communications can hardly inspire confidence that any potentially risky substance can be safely contained in the ground. In fact, the Sierra Club[4] has already taken an unequivocal stand against storing LLRW in shallow land burial facilities. They support their argument with examples of problems which characterized the sites at Maxey Flats, Kentucky, Sheffield, Illinois, and West Valley, New York.

An effective public participation program must incorporate information, education, and feedback mechanisms that first, provide convincing evidence that long term shallow land burial can be done safely, and second, that allows the public some voice in deciding which site integrity options they are willing to live with.

Public Perception of Radioactive Risks

Modern attitude theory views beliefs as the foundation of attitudes and behavior.[5] When applied to health related behavior a model is generated that links behavior to (a) beliefs about one's perceived susceptibility to a particular health threat, (b) beliefs about the severity of that threat, and (c) beliefs about the cost-benefit outcome of engaging in that particular type of behavior.[6]

The health risk that is most frequently linked to radiation is cancer. Cancer is a serious threat and one that has a relatively high probability of occurrence. Therefore, perceived seriousness and perceived susceptability are high for this particular issues. When a community is asked to accept a LLRW site it perceives that it is being asked to accept, amount other things, an increased cancer risk. It should come as no surprise that anxiety and outright opposition are the results.

A scientist may try to counter these fears with data which attempts to demonstrate that the increased cancer risk from a LLRW site is minute and not at all comparable to the many other risks a person assumes during the course of his or her daily routine. This information is an attempts to change the threatened individual's beliefs and thus his or her attitudes and behavior. However, the outcome of these information campaigns is seldom greater receptivity toward radioactive risks.

Various risk management scholars point out the limits of such information campaigns as applied to nuclear energy or radioactive waste. Slovic and his colleagues[7] have been involved in a program of research designed to investigate the factors underlying decisions on various risk-benefit dilemmas. Their research results relating to radioactive risks are illuminating. Fear of radioactivity is apparently a result of the public's perception that its risks are involuntary, unknown to those exposed or to science, uncontrollable, unfamiliar, and severe.

That people need to feel in control of their environment has been argued and documented by a number of theoretical and research streams: psychological reactance theory,[8] the learned-helplessness phenomenon,[9] and anomie-alienation theory[10]. The basic idea is simple. People are willing to endure costs if they perceive them to be self selected or if they feel there is something they can do to deal with the costly phenomenon should problems arise. The unknown and unfamiliar are, by definition, impossible to control. These kinds of events induce a range of responses varying from depression-apathy to outright rebellion. Slovic, et. al.,[11] go on to argue that these negative reactions to radioactive risks need not be viewed as irrational. Lay people do not evaluate risks using statistical data, they rely on inferences based on what they have heard or observed. These inferences involve a number of judgmental characteristics. First, the relative availability of information is important since people will judge an event as probable if instances of it are easy to imagine or recall. Second, since instances that are easy to imagine or recall will play an important role in probability assessment then the relative intensity of the event will also be important. That is why a few dramatic failures are remembered much more vividly than a chain of routine successes. Third, people are usually very confident about these kinds of judgements. Finally, there exists a desire for certainty so that if risks cannot be totally ignored their source may be rejected outright.

The implications of these judgemental characteristics for radioactive waste are reasonably straightforward. Much of what people hear or read about radiation is in a negative vein. Risks from nuclear power, radioactive waste, nuclear accidents, government irresponsibility in past radiation incidences, and disagreement among experts about radiological impacts are the kinds of topics that are found most often in the printed media or on television. This means that the most available and dramatic information concerning nuclear events reinforces perceptions of uncertainty and danger. Since radiological impacts are quite long term and probabilistic it is difficult to unambiguously demonstrate the safety of certain dose levels, one can only argue from a particular interpretation applied to a limited set of data. Furthermore, there are few examples of successes in the siting of radioactive or hazardous waste that can be pointed to as evidence that the technology works. Even whose sites generally viewed as safely and responsibly managed have their detractors.[12] Unfortunately, there are more negative than positive examples. The mass media has expressed little interest in well run waste sites of any kind.

One further problem is that very few members of the general public are aware or understand the distinctions between low level and high level radioactive waste. Furthermore, many of those who are informed of the differences are unimpressed with the distinctions made by the regulatory agencies.[13] The degree of antipathy that has come to be associated with high level radioactive waste is likely to transfer somewhat uncritically to radioactive wastes of all types. This would be especially likely in lieu of effective public education programs or in the absence of public trust in the agencies which have responsibility for regulating this material.

Again, intensive public information and opportunities for feedback are essential if public support for a LLRW site has any realistic chance of occurring. However, given Slovic and his colleague's arguments public education alone will be insufficient to generate support. If public fear is primarily a result of their inability to detect the risk or to be aware of its impact then means must be provided to help insure radiation detection should the off-site migration of nucleides occur. Public involvement in the site monitoring process could help overcome some of these fears.

Problems with Public Participation Programs in General

Citizen participation programs have been instituted to fulfill four basic functions:[14]

1. Participation as Policy—From this perspective citizen participation is a fundamental element in realizing democratic ideals. The rationale is that people have a right to influence policy decisions that may affect their lives.
2. Participation as Strategy—Since the early 1950's it has been a social science "truism" that one way to enhance acceptance of any program is to involve those affected by it in the decision making process.[15] Such participation is supposed to result in enhanced perceptions of personal control and greater commitment to the subsequent decision.
3. Participation as Communication—From this perspective the public is viewed as a resource that may be able to improve the quality of the decision being made. In decisions regarding risky technologies the public may

have a point of view which encourages considerations overlooked by those whose major orientation is technological.

4. Participation as Conflict Resolution—Citizen participation can be a process that results in reduced tensions and more controlled conflict. The intersection of ideas that ideally occurs in public meetings can hopefully moderate dogmatic opinions and lead to greater understanding and trust.

The crucial issue is whether past public participation programs involving waste siting, and related issues, have fulfilled these functions. If they have fallen short of their proponents' expectations what reasons can be offered and what steps can be taken to ameliorate these defects.

It is apparently very difficult to construct a public participation program on waste siting that is viewed by the public as an exercise in democratic decision making.[16] Agencies are criticized for involving the public too early when plans are still vague or too late when the decision is a foregone conclusion, for not allowing meaningful input, or for involving a very limited segment of the affected public.[17] Even more disturbing, participation programs administered by governmental agencies are often viewed as attempts to mislead or manipulate the public. Distrust of governmental agencies tends to be increasingly the norm.[18]

The assumption that participation in decision making results in greater support for the proposed program must be qualified by two other assumptions. First, if the public does not trust the siting agencies or the proposed technology, and the participation program fails to build trust and understanding, then participation will be viewed as simply a skirmish in a larger protracted conflict. Second, public participation programs tend to disproportionally attract those who strongly oppose the proposal. From their perspective the participation program can be viewed as a forum to publicize their discontent and help mobilize further opposition.

These is also considerable evidence that public participation programs dealing with issues of nuclear power and waste may actually result in greater negative attitudes than existed prior to the public discussion process.[19] Agencies apparently have difficulties in effectively communicating need and assuring public safety. The public involvement-discussion process can actually heighten citizen awareness of risks, convince them that either the resolve or the technical means to neutralize these risks do not exist, and result in well-formed negative attitudes where previously there may have been honest ambivalence or no attitude at all. It is significant to note that the Swedish government's attempt to educate some 80,000 citizens about nuclear power resulted in increased confusion and uncertainty about the issue because participants could not resolve the conflicting opinions of nuclear experts.[20] A successful public participation program must integrate expert testimony in a way that promotes consensus rather than ambiguity or in a way that indicates that the sponsoring agency does not fear openness and discussion.

Clearly, participation programs may enhance destructive conflict rather than moderate it. There is some evidence that the existence of public controversy per se increases opposition to a technology, even when proponents and critics are equally vocal.[21] This outcome is especially likely when the technology is poorly understood or when the public perceives a lack of expert consensus about the risks of that technology. Apparently, when faced with potentially high risks plus ambiguity the public tends to choose rejection as the safest option.

Finally, although it is possible that the public's non technical viewpoint may provide valuable insights to decision makers, few scholars of the public participation process have noted the problems involved in trying to get meaningful public input on poorly understood technological issues. Very few citizens understand the nature of the risks from radiation in general or LLRW in particular, the issues involved in successfully locating a LLRW site, or the extent and urgency of the need to provide such sites.

Limited public understanding means that the public can provide very little useful information concerning site selection, site design, waste handling, or any other technical issues. Furthermore, the public is quite likely aware of their knowledge limitations. In effect, the public input is limited to mundane expressions of concerns about water, air, and soil contamination, asking that trucks stay away from populated area, and similar kinds of very general concerns. The public is assured that their input will be seriously considered but their input is limited to general expressions of concern that are already being dealt with in greater detail in the technical specifications, so the public may legitimately wonder what utility their input has. The participation process can become a formalistic, but somewhat nonfunctional, ritual. Effective public participation on LLRW siting will have to incorporate techniques which permit the public to make meaningful judgements about specific issues. Furthermore, public judgements will have to be explicitly incorporated into the decision process on site selection and site management.

Past public participation programs have had problems fulfilling the four functions attributed to them by Wengert.[22] These programs suffer from the following defects: they are generally administered by agencies which inspire little public trust; they allow insufficient or somewhat useless public input; they tend to involve those elements for the public that are most vociferous in their opposition; they fail to effectively educate the public about the technical aspects of the proposed technology; expert testimony tends to be used in ways that increase public confusion and hostility.

Summary of Part I

The above discussion highlights the following issues which must be dealt with in designing an effective public participation program in the case of LLRW siting:

1. The program must be administered by trusted people.
2. People affected by the site must be allowed to shape the decisions that will affect their lives. This can be done by actively involving abutting communities in negotiating issues such as:

    a. incentives that may help restore perceptions of equity;
    b. site monitoring techniques and the role of local involvement;
    c. long-term management options;
    d. specific aspects of site design and operation.
  3. The participation program should include an effective information, education, and feedback process.

Each of these issues are elaborated in the next section.

DESIGNING AN EFFECTIVE PARTICIPATION PROGRAM

Deciding Who Shall Administer the Participation Program

A participation program could be run by a state agency, a private agency hired by the state, or some third party such as an appointed board of citizens representing different areas of the state. Given the degree of public distrust of government agencies care must be exercised not to politicize the siting attempt. This is especially important in Pennsylvania with its relatively large number of organized anti-nuclear protest groups. Walsh's warning is appropriate: "Public hearing processes...may become more an instrument of protest mobilization than of social control, especially when an organized protest ideology is available."[23]

The Minnesota experience in attempting to site hazardous waste has demonstrated the ability of a citizen's board to inspire trust and confidence.[24] The board can be made up of citizens chosen to represent various geographic, demographic, and socio-economic segments of the state. This board can be charged with responsibilities such as: operating the citizen education, information, and feedback program; responding to reasonable requests for information; communicating with interested or concerned communities; and perhaps playing some role in the site selection process.

The problems with a citizen board arrangement are (a) the cost of reimbursing board members for their time and labor plus the necessary supporting staff, and (b) the possibility that even a citizen board might not solve the credibility problem since it is always possible, in situations of intense conflict, to perceive them as allied with state or industry interests.

Involving Those Most Affected by Siting

Standard public meetings, which comprise the bulk of most citizen participation programs, are generally held in various regions and those who are interested attend the meetings. This format can be retained for the public information-feedback aspect of a LLRW participation program. Regions of the state that meet preliminary geological, hydrological, and demographic siting criteria can be divided into manageable units and public meetings and workshops held as are deemed necessary.

However, modern public participation scholars strongly advocate involving potential host communities in deciding their fate relative to a LLRW site.[25] Determined opposition from potential site abuttors is a strong possibility. it is at precisely this point that many waste siting programs grind to a halt. If this outcome can be avoided or moderated at all it may be by (a) early, consistent, and total involvement of those communities in the participation process, and (b) granting those communities some power to negotiate risk factors and incentives. Among the options that might be offered for community discussion, judgement, and negotiation are deciding acceptable incentives, deciding acceptable site monitoring techniques, and deciding acceptable long-term management procedures and various site operation factors.

1. Allowing Abutting Communities to Negotiate Incentives—Meeting the Equity Problem.

The only way to offset the unequal burdens imposed upon a community by waste siting is to provide tangible or incentives. Safety considerations can certainly be regarded as the most important incentive in promoting LLRW waste siting. However, this discussion assumes safety and focuses on other kinds of benefits and assurances. While the economic benefits provided by the inclusion of a LLRW site into a small community may be fairly substantial, NRC and DOE estimates vary from 60 to 80 direct jobs plus multipliers, they are probably insufficient to foster perceptions of equity.

There are a number of possible incentives that can be offered abutting communities. The following appear to be among the most important:

    a. Assurances that property values will be maintained. Bealer[26] and others have noted that one of the main reasons for community opposition to waste siting is the fear that property values may be adversely affected. A guarantee of property value would ease local fears and probably not be very costly since there is no evidence that well run waste sites lower property values.

    b. Local agricultural products pricing can be guaranteed. Attempts to site hazardous waste in rural areas have met determined opposition from farmers. Price guarantees could reduce this opposition and would also probably not be expensive since there is little evidence that waste sites adversely affect local agricultural prices.

    c. Local services but especially road and bridge repair can be provided to offset truck traffic damage. A limited access road to the site may be the most desirable option from an abutting community's point of view.

    d. Early and periodic health screening can provide a running medical profile of community residents which can be used to check on possible health problems created by site failure.

The costs of these incentives can be met by a surcharge on the waste. Again, the relative cost of such a program must be weighed against the possibility of not siting at all.

2. Allowing Abutting Communities a Role in Site Monitoring.

Since the basis of much of the public fear of radiation is the difficulty in detecting possible environmental contamination, local citizens should be permitted to participate in site monitoring procedures. The following options are possible: routine reports from the site operator to the community; local community representatives directly involved in site monitoring; a liaison appointed by the site operator to be available to concerned citizens; funds provided to communities so that they may hire their own monitoring personnel and buy their own equipment.

Each of these options involve more expense and diminishes the site operator's autonomy somewhat. However, these costs are minute if they result in public confidence.

3. Allowing Abutting Communities a Role in Deciding Long-Term Management Options.

From the public point of view all other issues in LLRW disposal may pale in comparison to the issue of who will be responsible for the long term management of the site.[27] The most exacting regulations and technical specifications are meaningless unless they are put into practice by a knowledgeable and responsible site operator. It will be essential that whomever is chosen as the site operator have absolutely impeccable credentials. The operator should have a spotless record of effective, relatively long-term chemical or radioactive waste management along with no demonstrable connections to past or recent public scandals.

The site operator could be chosen from private industry, from an existing state agency, or from a new state authority set up to specifically manage a LLRW site. Sites operated by private industry would be regulated by the state DER or comparable agency.

Given the degree of mistrust of both government and industry that has apparently been generated in Pennsylvania since the TMI incident, it will be important to allow the public some voice in deciding who the site operator should be. Funds can be provided abutting communities so that they may hire a private engineering firm to do initial site screening and so that they may execute a thorough search for a trusted site operator. However, the option can continue to exist for an interested operator to contact a potential host community and engage in good faith bargaining.

4. Allowing Abutting Communities a Role in Deciding Various Site Operations Options.

The criteria for site design and operation developed by NRC start with a base case and include a number of increasingly expensive options which can upgrade safety to some degree. These options involve things such as trench construction, waste packaging and stacking, and trench cover design.

These various options, including their cost and relative contribution to site safety, should be carefully explained to community members. Abutting communities can be permitted to negotiate for acceptable risk factors. For example, a community may decide that the only way they would feel safe given a LLRW site is if the trenches were reinforced with concrete and the wastes grouted. Being allowed a voice in setting such standards can give the local community a real sense of meaningful participation and help insure feelings of confidence and safety.

Designing An Effective Information and Education Program

Much of the public fear of radiation risks, which one scholar has labeled a "phobia",[28] can be attributed to public ignorance. Very little has been done by responsible scientists to educate the public, at all age levels, about the nature of radioactivity and its risks relative to other risks in the environment. In an atmosphere of partial information and a lack of expert consensus on crucial questions high fear levels should be expected and not viewed as a mental aberration. Given a great deal of misinformation and ignorance public education on radiation is something long overdue.

After the state has undergone preliminary screening on geological, hydrological, and demographic criteria those regions that are identified as holding potential sites can be involved in an intensive public education-feedback campaign.

Information packets can be prepared which thoroughly discuss the following kinds of issues:

1. the benefits of nuclear science and the use of radioactive materials—the need for a LLRW site;
2. the risks associated with various kinds of radioactivity and a comparison of these risks to others;
3. the distinction between low-level and high-level radioactive wastes;
4. site safety criteria specified by the regulatory agencies and options available to minimize risks;
5. site management issues and how they are to be dealt with;
6. incentives that can accrue to host communities;
7. assurances available to protect host communities;
8. the public participation process and the opportunity for local involvement in decision making.

These information packets can be distributed in public places, mailed to known interest groups of all persuasions, discussed on television, radio, or in newspapers, and given to local community leaders. The information in these packets can provide the discussion material for public meetings and workshops.

However, it is not sufficient to simply present information. The public must be provided a means to comment on the information and offer their perception of the most desirable options. While it is routine to accept testimony at public meetings this kind of format biases public feedback in the sense of favoring the opposition. Besides testimony taken at public meetings carefully constructed surveys could be done which allow systematic feedback on specific issues. The public can be asked to evaluate options related to: population density and distance from site; site proximity to water supplies; the appropriate transportation networks; the relative importance of agricultural land, land with suspected minerals, recreation-scenic areas, or other value; waste form preferences; various site enhancement factors; long term assurances and incentives. The data gathered in this fashion can be made public and also used in public meetings and workshops.

Finally, a public education program will only be as credible as the experts who provide the information. Numerous public opinion polls indicate that government and industry experts are not highly trusted by the general public. University affiliated scientists and radiologists tend to inspire greater trust. Since experts representing regulatory agencies must be included in the siting process consideration should be given to constructing an advisory team of experts which includes independent scientists and even scientists representing environmental concerns.

Such a team would be responsible for providing technical information and responding to public inquiries on safety considerations and other issues. Since the public dislikes ambiguity in expert testimony the team should try to come to some consensus before publicly addressing issues. However, if consensus cannot be reached it is probably better to publicize the various dissenting opinions, and the reasons behind them, than to act as though differences do not exist and have them "leaked" later. This kind of forthrightness can do much to inspire trust.

SUMMARY AND CONCLUSIONS

Attempts to site wastes of all kinds have foundered because of strong public opposition. LLRW siting attempts are certain to meet similar problems. Local communities tend to see little or no benefits but high costs in hosting waste sites. Fear of pollution, the unknown aspects of radiation risks, a lack of confidence in governmental agencies, are all factors promoting public resistance. Compounding these problems has been the failure of citizen participation programs to fulfill the functions for which they were designed. Instead of fostering more open communication, regulating conflict, and generating better ideas, participation programs dealing with waste siting tend to generate more conflict and mobilize determined opposition.

Recent approaches to this problem emphasize the intense involvement of affected communities in decisions that will affect their lives plus offering incentives to help restore equity. Local communities can be involved in site monitoring, in approving certain site design and operation options, in deciding which incentives they feel justly compensates them for assuming the burden of a LLRW site, and even in helping select the site operator. While this approach entails some monetary cost and loss of autonomy for governmental agencies and waste industry representatives it does offer the hope of enhancing feelings of confidence, trust, and control among those who must live with the site.

Furthermore, there is a need for more effective public education about waste sites and radiation risks. The only information about these issues that reaches the public tends to be negative exposes presented by the mass media.

Finally, public participation programs must develop a means to enlist meaningful public feedback. Besides the traditional public meetings, which generally mobilize the opposition, careful surveys can be done of a representative spectrum of the public getting their judgements about specific criteria. These judgements can be used to shape siting policy. Such an approach would help avoid the criticism that public input is not taken into account.

While the suggestions included in this paper go far in dealing with public fear and distrust they cannot guarantee siting success. There are a number of uncontrollable contingencies that can affect any siting program. Another energy crises, for example, may increase the prestige of the nuclear industry and make LLRW siting less onerous. Or, news broadcasts of waste site failures or of nuclear accidents could make LLRW siting more problematic. The problems of waste siting will not disappear nor are the solutions easy ones. They demand serious consideration by talented scientists of all kinds. Waste siting difficulties certainly rank near the top of challenges facing advanced industrial societies.

REFERENCES

1. Bealer, R.C., K.E. Martin, and D.M. Crider. 1982. *Sociological Aspects of Siting Facilities for Solid Waste Disposal.* Department of Agricultural Economics and Rural Sociology. Publication 158. The Pennsylvania State University, University Park, PA; Environmental Protection Agency. 1979. *Siting of Hazardous Waste Management Facilities and Public Opposition.* Report SW-809.

2. General Accounting Office. 1983. *Regional low-level radioactive waste disposal sites — Progress being made but new sites will not be ready by 1986.* RCED-83-48.

3. Environmental Protection Agency, op. cit.3-23; Massey, D.T. 1978. *Attitudes of nearby residents toward establishing sanitary landfills.* Report

No. ESCS-03. Natural Resource Economics Division; Economics, Statistics, and Cooperative Service, U.S. Department of Agriculture; Bealer, et.al., op. cit. 1-50.

4. Sierra Club Newsletter. 1982. *Insecure landfills: The West Valley Experience.*

5. Fishbein, M. and I. Ajzen. 1975. *Belief, Attitude, Intention and Behavior.* Addison-Wesley Publishing Company, Reading, Mass.

6. Maiman, L.A. and M.H. Becker. 1974. The health belief model: origins and correlates in psychological theory. *Health Education Monographs* 37: 332-347.

7. Slovic, P. 1978. Judgment, choice and societal risk taking. in K.F. Hammond (ed.) *Judgment and Decision in Public Policy Formation.* Westview Press, Boulder Colorado; Slovic, P., B. Fishhoff, and S. Lichtenstein. 1979. Rating the Risks.*Environment.* 21: 13-20, 36-39.

8. Brehm, J.W 1966. *A Theory of Psychological Reactance.* Academic Press, New York, NY.

9. Seligman, M.E.P. 1975. *Helplessness: On Depression, Development, and Death.* W.H. Freeman, San Francisco.

10. Seeman, M. 1954. On the meaning of alienation. *American Sociological Review.* 24: 783-791.

11. Slovic, op. cit., 15.

12. Shapiro, F. 1981. *Radwaste.* Random House, New York, NY.

13. Sierra Club Newsletter. 1983. A "low-level" nuclear waste primer.

14. Wengert, N. 1976. Citizen participation: practice in search of a theory. *Natural Resources Journal.* 16: 23-40.

15. Coch, L., and L.R.P. French, Jr. 1948. Overcoming resistance to change. *Human Relations.* 1: 512-532.

16. Pogell, S.M. 1979. Government-initiated public participation in environmental decisions. *Environmental Comment,* 6: 3-15; Kasperson, R.E., G. Berk, P. David, A. Sharaf, and J. Wood. 1979. Public opposition to nuclear energy: retrospect and prospect. in C. Giseld, et.al., (eds.)*Sociopolitical Effects of Energy Use and Policy.* National Academy of Science.

17. Abrams, N.E. and J.R. Primack. 1980. Helping the public decide; the case of radioactive waste management.*Environment,* 22: 14-40.

18. Walsh, E.J. 1981. Resource mobilization and citizen protest around Three Mile Island. *Social Problems,* 29: 1-21.

19. Orr, D.W. 1979. U.S. energy policy and the political economy of participation. *The Journal of Politics,* 41: 1027-1056; Deese, D.A. 1982. A cross national perspective on the politics of nuclear waste. in E.W. Colglazier, Jr. (ed.) *The Politics of Nuclear Waste.* Pergamon Press, New York.

20. Slovic, et. al., op. cit., 17.

21. Mazur, A. and B. Conant. 1978. Controversy over a local nuclear waste repository. *Social Studies of Science,* 8: 235-243.
22. Wengert, op. cit., 25-27.
23. Walsh, op. cit., 9.
24. The Izaak Walton League of America, Inc. 1981. Waste alert bulletin. 9: 1-2.
25. Howell, R.E. and D.E. Olsen. 1982. *Citizen Participation in Nuclear Waste Repository Siting.* Department of Rural Sociology, Washington State University for Battelle Project Management Division. ONWI-267; Hurley, M. 1982. *Social and Economic Issues in Siting a Hazardous Waste Facility.* Citizens for Citizens, Inc., Fall River, Mass.
26. Bealer, et. al., op. cit., 16.
27. Shapiro, op. cit., 168.
28. Du Pont, R.L., M.D. 1980. Understanding fear of nuclear power. Presented at the International Conference of the Atomic Industrial Forum, Inc., Washington, D.C.

Chapter Fifteen

POLITICAL CONSIDERATIONS OF NUCLEAR WASTE DISPOSAL POLICY

Robert S. Friedman

Professor of Political Science
Department of Political Science
The Pennsylvania State University
University Park, Pennsylvania 16802

A substantial segment of the scientific and engineering community believes that the technical issues associated with nuclear waste disposal have been resolved and all that remains to create effective public policy is resolution of political questions. Unfortunately, even if they are correct regarding technical issues, political problems are complex and tend to defy easy solution. Four primary political considerations need to be resolved in order to implement a workable nuclear waste disposal program.

Perhaps foremost among these is the polarization of opinion regarding nuclear power into irreconcilable camps and the concomitant difficulty in separating issues of nuclear armament, nuclear energy and waste disposal.[1] The heterogeneous and diffuse policy making system that prevails in the United States usually requires a middle-of-the-road consensus to produce and implement public policy in any issue area. The consensus that existed during the halcyon days of the 1940s and 1950s in favor of nuclear development has evaporated.

Technical issues surrounding the problem of nuclear waste disposal emphasize risk analysis. Scientists, engineers, economists, and statisticians are comfortable with cost-benefit studies that deal with probabilities of actual risks connected with various alternative paths. However, politicians must deal not only with reality but with the perceptions that individuals and groups in the society have about reality.[2] In nuclear waste disposal matters, as in a number of other areas, it is popular perception that is critical in a political system in which democratic decision making is valued. Fear of catastrophic consequences from nuclear incidents is widespread. This perceived reality, whether accurate or not, pervades all decision making involving nuclear issues.[3]

A third problem closely linked to the issue of perception of reality has to do with the distribution of benefits and costs of particular approaches to public policy. Distributive decisions are those in which some or all of the participants in the system derive benefits equal to or greater than their costs. Redistributive decisions are those in which some individuals are net beneficiaries and other net losers or payers.[4] In cost-benefit terms there are few distributive decisions since all programs have a price tag and someone must pay. Even a program maintaining the beauty of our national park system, for example, might seem redistributive to the older taxpayer who is not likely ever to use such facilities. However, in political decision making, the key is perception, and if a decision can be made to look distributive to a wide segment of the population, chances for achieving agreement are enhanced. The difficulty in the nuclear waste disposal issue is that for the localities selected as possible disposal sites there are few mechanisms available to sell the activity as a distributive program.

Finally, establishment of public policy with respect to nuclear waste disposal requires constraints on the economic behavior of segments of the marketplace in order to assure health and safety for future generations. The history of governmental regulation of this kind in the United States is replete with failure until a major crisis occurs or is perceived to have occurred with regard to the issue. Illustratively, current policy with respect to the regulation of the safety and efficacy of prescription and over-the-counter drugs was established only after revelations regarding birth of deformed babies whose mothers had used the drug thalidomide.[5] Despite the publicity associated with the Three Mile Island accident in 1979, there has not been a nuclear accident or event with health consequences comparable to the thalidomide affair since the bombing of Hiroshima and Nagasaki.

The discussion that follows places these four political consideration into the context of the efforts to develop and implement both low level and high level nuclear waste disposal policies in the United States.

POLARIZATION

Despite death and destruction at Hiroshima and Nagasaki, the period following World War II witnessed a substantial consensus in the United States for the maintenance of a nuclear weapons program as a deterrent to the spread of communism and for the development of nonmilitary atomic energy uses as part of an expanding economy.[6] The federal government subsidized university research and private sector use of nuclear power. By the 1960s the consensus began to erode, partially as a result of nuclear related events, but at least equally because of increasing distrust of the political and scientific leadership of the country. The latter was triggered by allegations of misconduct, duplicitous behavior and immorality of the Viet Nam War effort. The epithet "military-industrial complex" tarnished the scientific and technical community as well as the civilian and military leaders of the nation. The distrust was further exacerbated by the Watergate scandal that culminated in the resignation of President Nixon and the criminal convictions of some of his associates.

Erosion in support of nuclear energy cannot be disengaged entirely from these events. However, independently, criticism has grown not only of nuclear armament, but also of nuclear energy as a safe alternate energy source. Environmental organizations and anti-nuclear groups have flourished since the 1960s and have garnered media attention in expressing their fears of health and safety problems connected with the generation of nuclear power.[7] The accident at Three Mile Island on March 28, 1979 has provided them with ammunition in their campaign to dismantle or at least to arrest the growth of the nonmilitary nuclear industry.

Supporters of nuclear power see it as an economically sound nonpolluting reliable alternative to the uncertainties of Middle Eastern oil and "dirty" domestic fossil fuel sources. They cite data to demonstrate that no civilian lives have been lost in the generation of nuclear power since its inception. They believe that through careful planning nuclear waste can be disposed of safely and permanently probably by geological disposal in mined cavities. However, one group of detractors has said of them, "the unthinking optimism of the nuclear mindset continues unchecked."[8]

Nuclear power advocates would be quick to respond that there is an antinuclear mindset of which the following statement by Resnikoff, a staff associate of the Sierra Club, might be characteristic:

> We are skeptical, distrustful. We feel the need to protect ourselves. Given this distrust, the means of protecting ourselves will not be through consensus, which is a very one-sided arrangement, but through the standard means of the democratic process: the vote, the courts, and civil disobedience. Working out how safe is safe in the backrooms of the NRC, DOE, EPA, Battelle or Keystone is almost impossible given past history, unfolding events and our view of the "sides" in this controversy. The solution is not through consensus bodies, but via the standard democratic processes which will carry out the public will.[9]

Such sharply divergent statements are indications of a growing polarization of positions about nuclear power. Evidence of polarization is also found in opinion polls. Results of surveys undertaken since the accident at Three Mile Island demonstrate bimodal responses among the general population as well as among those in the impacted area regarding the desirability of encouraging nuclear power development with perhaps 40 percent supporting nuclear energy, and 40 percent in opposition and the remainder uncertain. As Table 1 suggests, when respondents were asked whether they favored the restart of the undamaged reactor at Three Mile Island, if the Nuclear Regulatory Commission attests to its safety, 44 percent of those in the immediate area remained adamantly opposed and 32 percent in the remainder of the State of Pennsylvania persisted in their opposition.

TABLE 1

Respondent Support for Restart of Undamaged Reactor by Location

Location	Favor Restart	Oppose Restart	Don't Know/Unecided
Three Mile Island Area	46%	44%	9%
Remainder of Pennsylvania	44%	32%	24%

Source: Survey by Cambridge Reports Inc., July 1982

PERCEPTION OF RISK

Sometime after the accident at Three Mile Island a researcher interviewing an official of a trade association in Florida casually mentioned that he had driven within a mile of the site four days after the accident. The official inquired, "What was the sky like?" Presumably she was linking her recollections of the mushroom clouds associated with nuclear explosions and the events at Three Mile Island. Wide disparity between reality and perception is not unusual in matters of health and safety risks. For example, in a recent study, a group of "psychologists found that when the death rates between two causes of death differed by less than a factor of 2, their subjects could not pick out the more likely cause with any reliability. When the two death rates differed by more than a factor of 2, their subjects could usually get the order right. But they did not do very well on the ratios. For example, for a series of questions for which the true ratio was 1,000 to 1, the average values of the answers to the individual questions ranged from less than 2 to 1 to roughly 5,000 to 1."[10]

Unfortunately for proponents of nuclear power even risk assessment projections which indicate very low probability of public danger do not allay fears. Zinberg has stated her view of the problem well:

Nuclear power and nuclear wastes are unique in the fears that they arouse. Pronuclear scientists often do not understand this concern; indeed, they express surprise that the public is so concerned about safety when so few people have been killed in nuclear accidents. They point out that although more than 100,000 miners have been killed in coal mining accidents since the turn of the century, —this record has provoked little public outcry. In part, the difference in public concern results from the recognition that mining has killed only miners, whereas the release of radiation from nuclear accidents is likely to harm many others not directly involved in nuclear power generation. The same argument can be used with airplane and automobile fatalities: the majority involve individuals who have chosen the risk, whereas radiation like the less lethal but more extensive effluents from coal-fired power plants is socially imposed and perceived as a restriction on individual freedom by those who oppose it.

In addition radioactivity—though an accepted part of medical treatment—is popularly associated with destruction. The nuclear holocausts of 1945, the secrecy of AEC activities, and the deceptions about radioactive fallout during the above-ground nuclear tests in the 1950s taints nuclear power's present image. The continued escalation of nuclear weapons manufacture maintains for nuclear power its lethal reputation. And not insignificantly, fears about that which is invisible but capable of penetrating people, buildings, and the ecosphere may well contribute to the public's wariness of nuclear power.[11]

In a democratically run polity perceptions of reality are ignored by office holders at the risk of their positions. In February of 1980 General Public Utilities Company, the Nuclear Regulatory Commission and the City of Lancaster, Pennsylvania signed an agreement that no waste-water from the damaged reactor at Three Mile Island would be discharged into the Susquehanna River before December 31, 1981 or until the NRC completes an environmental impact statement regarding the discharge of such water. Despite the fulfillment of the necessary requirements, GPU has no immediate plans to discharge TMI-2 waste water into the river. The mayor of Lancaster, as a member of the NRC Advisory Committee on the Three Mile Island Unit 2 Clean-Up, has made his opposition to dumping wastewater into the Susquehanna River very clear. Whether his opposition derives from his own fears of health and safety risk or are in response to community pressures upon him are not discernible. However, opinion in his community in opposition to dumping is widespread among the citizenry and he empathizes.

Nuclear power proponents urge improved communications as one means of reducing the fears of the public toward their industry. They believe that by expressing relevant statistics in easily understood terms such as reduction in life expectancy from various risk sources public fears will diminish, because of what they believe are data favorable to their industry.[12] More likely it will take a combination of accident free operation of existing nuclear facilities, reduction of concern about nuclear warfare, and increasing fears of risks from the use of other energy sources to change the negative perceptions about the safety of nuclear power.

DISTRIBUTION OF BENEFITS AND COSTS

Most of the discussion of polarization and negative perceptions of the nuclear industry have centered on power generation as opposed to those of waste disposal. However, while the issue of distribution of benefits and costs can be raised with respect to all phases of the process of generation and disposition of nuclear energy, isssues involving distribution of benefits and costs in waste disposal are especially troublesome. One important key to the politics of distribution is perception. Programs that are perceived as distributive—providing benefits to individuals or groups at minimal or no cost to others—are likely to be popular and easily enacted and implemented. Programs perceived as redistributive—benefiting some at substantial cost to others who do not necessarily benefit—will certainly engender the opposition of those who must support the programs unless some means can be found to demonstrate to them that they accrue valuable benefits in compensation for the costs.

For many years public education and highway building in the United States were widely perceived as distributive activities. Those in our society with no children to educate were convinced of indirect benefits to themselves of improving the educational level of the society so as to make it more productive. Similarly, road building was seen as a boon to all segments of the economy. Increasingly, both of these areas have encountered difficulties in maintaining support in recent years because of controversies resulting from a redefinition by some groups of the benefits and costs to themselves. In the case of public education, issues such as school busing to assure racial desegregation and rising costs for those who prefer to use private and parochial schools have led many to challenge existing educational policies. Similarly, as Interstate Highways have threatened to alter and even destroy neighborhoods, affected parties have opposed their construction with considerable success.

Nuclear waste disposal can only be seen as redistributive to the residents of the locale selected as a repository. Fear of radiation leaks and the uncertainties of the future are likely to generate opposition in any community selected as a disposal site. To be sure, an argument can be made that the general interest requires ultimate selection of sites leading to the achievement of the general welfare. Unless there is a demonstration, however, of compensation through direct payments of money or increased economic activity in the form of jobs, the community selected is likely to oppose the choice. In effect, they will argue, "Why us?" As the discussion that follows concerning existing nuclear waste legislation demonstrates, the probable results are likely to be avoidance of specific site selection as long as possible and when sites are chosen to pick sparsely populated loci and/or those with minimum political clout. The net result is likely to be susceptible to two criticisms; first, a location that is suboptimal in scientific and technical terms; and second, a location that exacerbates political inequities because the process has victimized the politically weak.

PROGRAM CHANGE AND CRISIS

The American political system is highly fragmented. Federalism, separation of powers, and checks and balances, the core elements of the system, are designed to make certain that government will be limited and that governmental action will not be arbitrary. The ideological position of the Founding Fathers was based upon a fear of the excesses of absolute monarchy. Twentieth century conditions are far different from those at the time of the founding of the republic, in that social and economic interdependence has developed as a result of the industrial revolution. Nonetheless, the prevailing ideology in the United States continues to oppose government intrusion upon the marketplace except where failures have occured or where health and safety of the polity is deemed to be at considerable risk. Even in cases where substantial majorities appear to favor concerted public action, government has been loathe to move or has acted in largely symbolic ways. Illustratively, only after mass unemployment and a deepening depression had persisted for more than three years did the nation in 1933 agree to intensive governmental action to try to restore confidence in the economy. Similarly, despite numerous warning signals and the full power of the presidency, Congress barely approved military conscription in the summer of 1941, just four months before Pearl Harbor was attacked.

In some instances crises or fortuitous events have led to action where stalemate had previously prevailed. The 1962 amendments to the Food, Drug and Cosmetic Act are a case in point. Some congressional leaders had been pressing for changes for years. They had argued that FDA examination only of the safety of drugs was insufficient to safeguard the citizenry from chicanery and price excesses in prescription drugs. Some wanted emphasis placed on assurance that drug products were efficacious—capable of improving the health of a substantial proportion of the users. The drug industry and many conservatives were opposed to the legislation as an unnecessary restriction on the market. They argued that if a drug did not work, physicians and patients would eventually discover the fact and the drug would disappear from the market-place. President Eisenhower was opposed to the legislation as were the Republican leaders in Congress. President Kennedy, when he entered office, although not opposed to such legislation, offered no assistance to supporters. Then a tragic event "broke the dam." An official of the Food and Drug Administration found that thalidomide was unsafe and the drug was not cleared for use in the United States. Information providing horror stories concerning deformed babies was disseminated by the news media. The opposition to legislation to include efficacy within the meaning of marketable drugs evaporated as the Republican leaders in Congress offered their support and President Kennedy embraced the proposal. Even the drug industry gave up the fight. Ironically, whatever the merits of the efficacy provision, thalidomide had already been barred from the market by existing law. No changes were needed in the law to deal with the problem.[13]

Conditions with respect to nuclear waste disposal decision making are ripe for inaction. Political fragmentation in decision making is especially notable regarding the issue. Responsibilities on high level waste are seen as primarily national government problems and responsibilities on low level waste disposal as state and local problems. Even, however, in matters of high level waste, members of Congress from localities that are potential disposal sites have been vocal in protecting their interests. Polarization on the issues of nuclear power, perceived fears about radiation hazards and the redistributive inprint on any decision made add to the likelihood of inactivity.

In addition, despite the impact of the Three Mile Island accident on the nuclear power industry, there has not been a crisis or fortuitous event of sufficient proportion to break the "log jam." There have, of course, been legislative enactments and some implementation of these in matters both of high and low level waste disposal, but as the next section demonstrates, they establish a framework for decision making without providing any guaratee of implementation.

POLITICAL CONSIDERATIONS AND NUCLEAR WASTE POLICY

Waste disposal was not a major issue during the early years of the nuclear age. After World War II, considerable emphasis was placed on peaceful development of a bountiful new source of energy while the military continued to add to its nuclear arsenal. "Assured by scientists that a safe technical solution to the nuclear waste disposal problem was at hand, government authorities permitted wastes to collect in temporary storage facilities. Only after temporary facilities became pressed for space, and the public's concern for the environment spread to nuclear waste, did the government accelerate its search for a permanent answer to the nuclear waste question."[14]

At first high level wastes were stored at three primary sites: Hanford, Washington; Idaho Falls, Idaho; and Aiken, South Carolina. A number of events, however, led to reexamination of the issue of storage. Several leakages occured at the Hanford site, restiveness developed among Idaho political leaders after a fire at a plutonium plant in Colorado necessitated unexpected waste shipments to their state and, most importantly, a National Academy of Science Report condemned existing practices.[15] The AEC undertook a project to explore the possible use of salt vaults near Lyons, Kansas as a permanent repository. Initial expectations within the AEC were that a superior site had been found and that the waste problem may have been settled. Unfortunately, the Kansas Geological Survey raised a number of technical questions that induced the Kansas congressional delegation to challenge the plan and it was soon abandoned. In the aftermath, the AEC and its successors turned to the concept of the Retrievable Surface Storage Facility as a solution. Like earlier proposals, it too was challenged on technical grounds. This time the Environmental Protection Agency issued an environmental impact statement giving the solution its lowest possible rating.[16] More recently, the Department of Energy has returned to exploration of the salt bed concept at a Carlsbad, New Mexico site as well as at sites in Kansas, Louisiana, Mississippi, Nevada, Texas, and Utah. Because of the uncertainty of salt domes as the solution to the problem, government researchers are also probing other geological formations as potential solutions.[17]

For a time it appeared that a solution to at least part of the nuclear waste disposal technology problem might be chemical reprocessing of nuclear fuels. In 1956 the AEC agreed to share information on the subject with the private sector. The State of New York, in an agreement with the Getty Oil Company, began construction of such a plant at West Valley in 1963. Three years later it began operation. Throughout its life the enterprise was plagued with technical, economic, and political problems. Finally, it was shut down by the operator in 1972 to alter its capacity. High modification costs prompted Getty to withdraw from the business and after the Three Mile Island accident, the governor of New York disavowed the initial agreement. Other reprocessing plants have fared little better.[18]

Low level waste disposal has been described by critics as "haphazard."[19] A considerable amount has been dumped in containers at sea. Low level solids have been buried at one of a number of commercial or Department of Energy sites. As in the case of high level radioactive waste, the AEC was given full control of disposition of low level waste. However, in 1959, to arrest growing concerns from states in which primary storage was taking place, Congress amended the act so that the governors of states were authorized to enter into agreements with the AEC to regulate low level waste disposal under regulations and standards set by the AEC. As a result existing low level waste disposal sites are licensed and regulated by host states, since all are in agreement states. Any proposed site in non-agreement states would be regulated by the federal government.[20]

By the late 1970s events began to converge in a manner that was to compel Congress fo the first time seriously to consider comprehensive policy with respect to both high and low level waste disposal. President Carter had acted to abandon chemical reprocessing eliminating that as an option for handling of spent fuel. Environmentalists and antinuclear activists were using the waste issue as an argument for restricting if not terminating the development of nuclear power. Utilities were rapidly using up their storage space. And finally, states that had hosted disposal sites were threatening to cut off access of shipments from outside their borders. Indeed, Governor Riley of South Carolina announced, "I would be adamantly opposed to an AFR [away-from-reactor] in South Carolina in the absence of a commitment to a true plan for permanent disposal of nuclear waste. There is a basic law of nuclear waste often overlooked—all waste remains where it was first put—just look at the 23 million gallons of temporarily stored (military) waste at the Savannah River Plant."[21]

The Riley threat apparently was instrumental in the passage of low level waste legislation at the close of the Ninety-sixth Congress in December 1980. Efforts to include high level waste disposal fell apart when the Senate could not agree upon the role of the states in the disposal of waste produced by the production of nuclear weapons. The low level waste legislation mandated decentralized responsibility making the disposal of commercial low-level waste a state responsibility. States were free to build their own dump sites or could form regional compacts with the approval of Congress to establish burial sites. After 1985, regional groups could refuse waste from noncompact states.[22]

The legislation created no waste disposal sites. It placed decision making in the hands of individual states and their representatives. Predictably the host state for any regional site would have to be rewarded quite handsomely to counteract the perceived costs of risking the presence of a disposal site. In light of the perceived fears of nuclear radiation the probabilities of the leadership of any state accepting the responsibilities as host state remains in doubt.

Efforts to enact high level waste disposal legislation were renewed in the Ninety-seventh Congress with both the Reagan administration and the nuclear industry pressing for action. The key issues that needed resolution in order to obtain passage of a bill were disposition of military waste and the assurance that no state could be compelled to provide a disposal site without participation in the process. Both issues were resolved by agreeing to the demands of the forces that had blocked previous action. First, military waste disposal was, for the most part, exempted from provisions of the act; and, as a result, the supporters of the nuclear defense program lent their support to the act. Second, an amendment proposed by Senator Proxmire, Democrat of Wisconsin, provided that a state could veto a federal government decision to place a repository within its borders unless both houses of Congress voted to override it. In view of historic experience with Congressional decision making, the Proxmire amendment was tantamount to protecting any unwilling state from receiving a repository against its wishes.[23]

The action taken by Congress directed the Secretary of Energy to study five potential sites for location of a permanent underground repository for high level nuclear waste and to recommend three of them to the President by 1985 for further consideration. It also required that another five potential sites be studied for a second repository, three of which were to be recommended to the President by 1989. The President was required to submit his recommendation for the first site by March 31, 1987 and for the second by March 31, 1990. The legislation also dealt with interim storage facilities and required the Secretary of Energy to report to Congress by June 1, 1985 on the need for and the feasibility of a monitored retrievable storage facility; specifying that the report was to include five different combinations of proposed sites and facility designs involving at least three sites.[24]

Superficially the enactment of the two bills in successive Congresses appears to have overcome the handicaps associated with the four political considerations discussed in this chapter. In fact, the legislation has created the outlines of a plan to overcome these considerations but there remains the need to identify actual waste disposal sites. Low level waste disposal interstate compacts are being debated and personnel in the Department of Energy will expend resources on high level waste disposal site selection. The test will come when real locations are identified and efforts to veto them are upheld or overridden.

One additional political event has further compounded the nuclear arena. The Supreme Court in April 1983 upheld a California statute permitting the state to ban future nuclear power plants until the federal government creates permanent disposal sites for radioactive waste.[25] The decision could encourage the identification of sites, but it might also serve to halt all new plant development if sites are not agreed upon by the dates established in the 1982 legislation.

SUMMARY AND CONCLUSION

The framework for a program of high and low level nuclear waste disposal has been provided by 1980 and 1982 acts of Congress, but until that program is in place, problems related to polarization of the issues, perception or risk, distribution of benefits and costs, and the absence of immediacy linked to crisis remain obstacles to success. Both acts of Congress pushed into the future the crucial issue of where sites were to be located and, as a result, delayed the furor. However, an indication of the difficulty likely to arise is found in a provision of the 1982 legislation which barred construction if "a repository of its surface facility would be adjacent to a one-square mile area with a population of 1,000 or more."[26] This provision was authored by Representative Trent Lott, Republican House Whip, and was presumed to be designed to block identificaiton of a site as a possible location for a repository within Lott's Mississippi district.

The granting of responsibility to the Secretary of Energy to identify sites for high level waste disposal finessed the redistribution consideration for the time being. The location of authority for low level waste disposition in the states not only deflected the redistribution issue by pushing the decision into the future, it was also designed to placate many of the opponents of nuclear power who believe that DOE and NRC make arbitrary decisions at the expense of the popular will.[27]

However, as the second half of the decade of the 1980s approaches, decisions will have to be made regarding disposal sites. The confidence that once existed that science and technology can provide safe repositories no longer exists and the bureaucracies that will recommend to the president and governors policy solutions are not adequately trusted.

In order to create a program for the establishment of nuclear waste repositories several conditions must prevail. Perhaps foremost is the need to alter the public perception of risk. In short, there will need to be recognition that cigarette smoking and automobile driving, acts of volition, are potentially more dangerous to one's health than radiation leaks from nuclear power plants or waste repositories. Second, the process of repository site selection will have to include wide public participation in the process in order to obtain legitimacy. Without it Congress and the state legislatures are certain to override any proposal no matter how widely accepted by scientists and engineers. Finally, states and localities selected as sites for repositories will need to be compensated adequately in exchange for accepting the onus of serving as host.

Political scientists have not been notably successful forecasters of policy outcomes. However, the evidence of American history does not provide encouragement that maximization of control at the state and local level and oversight by Congress of administrative actions, as meritorious as they might appear in terms of democracy, are harbingers of success for unpleasant policy decisions. States rights and Congressional intervention to block executive action were used to maintain second-class citizenship status for Black Americans until the judicial process was resorted to as a device to alter policy. Most likely, a major policy breakthrough will occur only after a mishap or tragedy, the final product involving either a waste disposal program in the context of continued use of nuclear power or one premised on its abandonment.

REFERENCES

1. An interesting exposition of competing viewpoints is found in Colglazier, E.W., Jr., 1982, editor. *The Politics of Nuclear Waste*, Pergamon, New York, NY, xvii-xxii.
2. King, M.R. and R.S. Friedman. 1981. *An Approach to Understanding Conequences and Change in the Regulatory Process,* Institute for Policy Research and Evaluation, University Park, PA.
3. Douglas, J. March 1982. Toward better methods of risk assesment. *EPRI Journal* 7:23-28; and Morgan, M.G. November 1981. Probing the question of technology induced risk. *IEEE Sprectrum* 18:58-64.
4. Friedman, R.S. 1971. *Professionalism: Expertise and Policy Making,* General Learning Press, New York, NY; Lowi, T.J. 1964. American business, public policy, case-studies and political theory. *World Politics* 16:677-715; Salisbury, R.H. 1968. The analysis of public policy: a search for theories and roles. In *Political Science and Public Policy*, A. Ranney, editor, Markham, Chicago, IL.
5. Harris, R. 1964. *The Real Voice.* Macmillan, New York, NY.
6. Davis, David. 1978. *Energy Politics.* 2nd Edition. St. Martin's, New York, NY.
7. Lipschutz, R.D. 1980. *Radioactive Waste: Politics, Technology and Risk.* Ballinger, Cambridge, MA; and Murray, R.L. 1982. *Understanding Radioactive Waste.* Battelle, Columbus, OH.
8. Hilgartner, .S, R.C. Bell and R. O'Connor. 1983.*Nukespeak*. Penguin, New York, NY, 164.

9. Colglazier, E.W. Jr., op. cit., 190.

10. Morgan, M.G., op. cit., 63.

11. Colglazier, E.W. Jr., op. cit., 183-184.

12. Douglas, J., op. cit., 28.

13. Harris, op. cit.

14. Editorial Research Reports. 1982. *Energy Issues: New Directions and Goals.* Congressional Quarterly, Washington,D.C., 27.

15. Lipschutz, R.D., op. cit., 113-138 and Willirich, M. and R.K. Lester, 1977. *Radioactive Waste: Management and Regulation.* The Free Press, New York, NY, 59-87. See also, National Academy of Sciences, National Research Council, Division of Earth Sciences, Committee on Geologic Aspects of Radioactive Waste Disposal. *Report to the Division of Reactor Development U.S. Atomic Energy Commission,* Washington, D.C.,1966.

16. *Ibid.*

17. Lipschutz, op. cit., 139-159 and Murray, op. cit., 65-76.

18. Lipschutz, op. cit., 122-124.

19. Ibid., 125.

20. Jordan, J.M. and L.G. Melson. 1981. *A Legislator's Guide to Low-Level Radioactive Waste Management.* National Conference of State Legislatures, Denver, CO, 15.

21. Colglazier, E.W. Jr., xv.

22. *Congressional Quarterly Weekly,* December 20, 1980, 3623-3625.

23. On June 23, 1983 the U.S. Supreme Court declared the Congressional "veto" device to be an unconstitutional violation of separation of powers, presumably negating the effect of the Proxmire amendment and clouding the picture with respect to the implementation of the 1982 legislation on high level nuclear waste disposal site selection.

24. *Congressional Quarterly Weekly,* December 25, 1982, 3103-3105.

25. *New York Times,* April 21, 1.

26. *Congressional Quarterly Weekly,* December 25, 1982, 3105.

27. Walsh, E. March 1983. Three Mile Island: meltdown of democracy? *The Bulletin of the Atomic Scientists* 39:57-60.

Management of Radioactive Materials and Wastes: Issues and Progress. Edited by S. K. Majumdar and E. Willard Miller. 1985, The Pennsylvania Academy of Science.

Chapter Sixteen

DISPOSAL OF RADIOACTIVE WASTES*

J.M. Harrison

J.M. Harrison is a former president of the International Council of Scientific Unions. Currently, he is chairman of the Science Advisory Board of the Northwest Territories, Canada, and president of the Canadian National Commission for Unesco. His address is 4 Kippewa Drive, Ottawa, Ontario, Canada K1S 3G4.

Summary: Scientists appointed by the International Council of Scientific Unions have concluded that nuclear waste may be safely disposed of using current technology. Interim storage for 50 to 100 years greatly reduces the problem of thermal loading at the final disposal sites, but more research devoted to such interim storage is needed.

The president of the National Academy of Sciences, Washington, D.C.,wrote to the president of the International Council of Scientific Unions (ICSU) on 29 August 1977 to suggest that ICSU enter the discussion of nuclear waste management. The letter stated that "the tentative scope of the proposed effort would be to obtain the considered opinions, and consensus where available, of an international body of scientists (which combines the prestige of the respective national academies and scientific unions and the knowledgeability of a wide range of experts from several relevant fields) on technical issues involved with the management of the back end of a nuclear fuel cycle, including reprocessing (whether and what), waste solidifications, and isolation (disposal)." The proposal was made jointly with the president of the National Academy of Engineering and with the encouragement of the International Atomic Energy Agency (IAEA).

A small preparatory committee, appointed by the president of ICSU, was asked to prepare a report on the issue for circulation at the 19th General Assembly of the ICSU in September 1978. The matter was discussed informally with senior officers of the IAEA and of the Nuclear Energy Agency (NEA) of the Organization for Economic Cooperation and Development. All expressed interest and indicated they would provide information and data where available.

Members of the preparatory committee were J.M. Harrison, Canada, convenor, consultant: H. Lacombe, France, Laboratoire d'oceanographie physique; A. Preston, United Kingdom, Directorate of Fisheries Research; Gilbert White, United States, president of ICSU's Scientific Committee on Problems of the Environment; Y. Yamamoto, Japan, professor emeritus, Tokyo University; V. Yemelyanov, U.S.S.R., corresponding member, Academy of Sciences.

The committee's report suggested that ICSU might undertake to appraise existing research efforts on the safe disposal of high-level radioactive wastes (HLW), which are highly radioactive materials containing long-lived radioactive isotopes, generally resulting from spent nuclear reactor fuel. ICSU is responsible to no government, has no vested interest in nuclear power (or any other kind of power), has access to the world's scientific and technical community, and can come to conclusions purely on scientific grounds.

In 1978 the General Assembly of the ICSU authorized a review of the research being conducted on disposal of high-level radioactive wastes. The review was to be performed by three working groups whose chairmen would be nominated by the appropriate international scientific union or scientific committee. W.S. Fyfe was nominated by the International Union of Geological Sciences as chairman of Working Group No. 1, Terrestrial Disposal; C.D. Hollister, by the Scientific Committee on Ocean Research as chairman of Working Group No. 2, Marine Disposal; and F. Morley, by the Scientific Committee on Problems of the Environment as chairman of Working Group No.3, Environmental Pathways. The general committee of ICSU appointed a steering committee to provide overall guidance to the working groups, to assist in selecting their members and to review their reports. The steering committee was composed of J.M. Harrison, geologist, Canada (chairman); H. Lacombe, physical oceanographer, France; F.B. Staub, biochemist, Hungary; T. Watanabe, mineralogist, Japan; V.S. Yemelyanov, nuclear specialist, U.S.S.R., and the chairmen of the three working groups.

A preliminary report was presented to the 18th General Assembly in 1980 and the working groups were asked to continue their reviews and to prepare a final report for consideration by the General Assembly in 1982.

During the period of review, the steering committee was informed that the U.S.S.R. preferred to make its international presentation through the IAEA, and V.S. Yemelyanov attended only one early meeting of the commiteee. Also, after preparation of the preliminary report, F. Morley withdrew as chairman and member of Working Group No. 3; he was succeeded by R.H. Clarke, a colleague and member of Working Group No. 3.

Thousands of scientists in laboratories around the world are working on the problems of HLW disposal. The accessibility of their reports and publications varies widely, but the quantity of this material is enormous. Faced with the problem of what could be achieved with the means available, the committees sought aspects of the problems where there might have been insufficient research on subjects of significance. Provision of special funds from the Science Council of Japan and from the Japanese Nuclear Safety Research Association greatly helped the study.

REVIEW OF ONGOING ACTIVITIES

Four international organizations are particularly involved in coordinating programs of research into disposal of HLW, and to these programs particular attention was paid. The four are the IAEA and NEA, which were referred to earlier, the Commission of the European Communities, and the Council for Mutual Economic Assistance (CMEA). The last organization named coordinates research in the U.S.S.R. and most of the Eastern European countries.

In general terms, these international agencies coordinate and promote technological developments, provide safety guidelines, initiate research programs, and coordinate specific activities concerning seabed disposal and terrestrial disposal. All have a large publications program, and they have supported numerous technical reports and conferences. Their publications are freely available.

In reviewing national programs of research on the disposal of HLW, recent publications were consulted, as well as papers given at international symposia and scientists from the various countries. Argentina, Brazil, China, Mexico, and Norway, all users of nuclear power, had not published details of research programs on disposal of HLW up to the time the review ended. Countries with broad programs on terrestrial or seabed disposal, or both, include Belgium, Canada, Federal Republic of Germany, France, Japan, the Netherlands, Sweden, Switzerland, the U.S.S.R. (and other CMEA member countries), the United Kingdom, and the United States, with the largest of all programs described. The total field of research on both geologic and subseabed disposal options is very extensive, covering the topics of interest in varying degrees of detail.

In addition to the foregoing, the following countries have comparatively small programs of research into HLW disposal: Australia, Austria, Denmark, Finland, Italy, and Spain. While small, however, their contributions can be important because they are directed toward specific aspects of HLW disposal.

ICSU SUMMARY AND RECOMMENDATIONS

Summary: This is a brief account of the review by the ICSU's steering committee and of the reports prepared by the working groups on terrestrial disposal, marine disposal, and environmental pathways. The conclusion and recommendations are based on several postulates that were agreed to by the steering committee and the working groups.

1) Disposal is taken to mean the sequestering of material with no intention to retrieve it (although it may be technically possible to do so), whereas storage is intended to be temporary.

2) Criteria for site selection should be determined by safety and science and not by economics and politics, although these factors will inevitably be involved for the actual selection.

3) Nuclear wastes are with us now, so safe means must be found for their disposal. Programs are already under way in different countries to investigate various methods of disposal.

4) High-level wastes are potentially hazardous for at least 10^5 years and the behavior of the nuclides contained therein must be considered over that period.

5) The system of dose limitations developed by the International Committee on Radiation Protection is rational, and the dose limits are set at a level so low that risk is comparable to other risks of everyday life.

6) Primary attention should be given to international agencies and the research they support, sponsor, or coordinate; but national programs should be reviewed where necessary.

7) The review should be limited to disposal using technology now available.

8) The working group on environmental pathways should consider migration of nuclides in the biosphere and the other two working groups should review data concerning the migration of nuclides from the repository to the biosphere. Some overlap would be inevitable and probably useful.

The steering committee and the working groups are confident that HLW can be safely disposed of provided that some specific lines of research are intensified (see Recommendations that follow). Since it is impossible to eliminate all risks from any activity, safe disposal means their reduction to an acceptable level.

Disposal on land or beneath the sea demands a systems approach and close cooperation, beginning with early planning, and must involve all concerned. This approach must include the design of waste containers, the planning and building of the repository, and the transportation, emplacement, and monitoring of the filled container.Care must be taken to ensure that procedures for encapsulating the HLW do not create more dangers to any part of the population than the procedures would avoid in disposal.

The steering committee and working groups agreed that the HLW would be more safely disposed of if solid wastes were retained in appropriate storage facilities for 50 years or more beforehand. Hence, more research is needed on temporary, long-term storage.

Although both the proposed research and research under way are directed to HLW, their successful conclusion will ensure safe disposal of all toxic materials, including mercury and arsenic.

RECOMMENDATIONS

What follows are the annotated recommendations made by the working groups. These are extremely condensed versions of extensive reports (1). The data from which this article was prepared were those available up to about the first third of 1982. Since then new programs have been initiated in several countries; and, as is usually the case, some of the recommendations have been acted upon.

1) More effort is needed to ensure secure storage of solidified HLW for periods of up to 100 years.

Comment. The heat content of the HLW decreases rapidly with time, reducing thermal effects in the disposal medium and making it possible to use less underground space for disposal. Present data are inadequate for optimum choice of disposal facilities. It follows that more attention should be given to studies on specific sites, so that optimum choices may be made in due time. The first disposal sites should be small so that any unexpected results can be better controlled.

2) Development of underground laboratories in all proposed host rocks should be accelerated.

Comment. Such laboratories could be used for interim storage. Materials could be tested there, and fluids monitored for the next 50 years or more. Some might even become small disposal sites.

3) Much more information is needed about fracture systematics and resultant permeability, not only for the immediate host rocks but for all those providing containment.

Comment. Techniques are needed to predict fracture systems and their relation to permeability on all scales. For example, large cracks in rocks such as granite are cause for concern, but a myriad of small fractures in rock such as shale might actually inhibit the movement of fluids. Methods are needed to locate fractures and estimate flow volumes through them over long periods.

4) Specialists in the study of the geological deposits that have formed during the last 5 million years or so should be more involved in research on HLW disposal.

Comment. Studies on these deposits can determine their tectonic stability and the effects of changes in climate, variations in water levels, and so forth. Seismic studies are also necessary, but they would not allow predictions over the periods of time under consideration.

5) Mining methods and technologies should be carefully studied to ensure the most favorable conditions for HLW emplacement.

Comment. Minimum disturbance to the rock during excavation is an important requirement for the repository. The employment of new giant machines for boring large underground passageways, for example, would greatly reduce the use of explosives, which would probably mean fewer fractures in the rock.

The engineers for the repository, those who will construct the canister, and the scientists who will set the specifications for containment must work as a team to achieve the best procedures and methods for containment.

6) The disposal of HLW and other toxic wastes through the process of subduction should be carefully analyzed.

Comment. If further research shows that the sedimentary load is rafted under the crust in some areas of subduction, and if the speed of subduction can be predicted, ready-made disposal sites are available. All materials would be carried into the interior of the earth.

7) The effect of thermal perturbances on the containing medium and its resultant effects on fluid flow, fluid chemistry, rock and sediment alteration, and other physical properties should be studied so that they can be predicted with assurance.

Comment. This is important for all types of disposal. In the terrestrial environment, these studies would help determine the degree of fracturing of the rock, the subsequent permeability, and the hazard of lost containment. In the marine environment, the effects of porefilled, fine-grained sediments must be better understood. For example, will the heat source cause a miniature eruption and eject the canister? Or will it cement the sediments to produce a "rock" that could be fractured?

8) A pilot survey should be conducted based on cores of deep-sea sediments, which are tens of meters long, from several locations.

Comment. It is essential for site selection that the cores be carefully studied, for they would relate the past history of the surveyed area, the rates of sedimentation, the composition of the sediment, the possibility of geological disruption, and other geologic information.

9) Much more information is needed about both the sorption and mechanical characteristics of deep-sea sediments.

Comment. Information needed about the sediments includes grain size, composition, ion-exchange capacity, ability to absorb radionuclides, and contained biota and consequent bioturbation.

10) A more careful assessment of the adsorption of radionuclides by sediment systems is needed.

Comment. Radionuclides can occur in various oxidation states that may behave differently from related elements. They could move erratically, especially as the strata will be disturbed by emplacement of canisters. Moreover, the sorption characteristics of the containing media at elevated temperatures appear to be little known.

11) More research is needed on the processes that occur at the boundary between sediment and overlying sea water.

Comment. Results from the release of contained material into the water at the seabed surface will depend on such factors as mixing and advection at the boundary and exchange between the boundary layer and the overlying, stratified, deep water.

12) Physical and chemical models of diffusion processes in the seabed are needed.

Comment. Diffusion processes in undisturbed sediments are not well understood and much more verification is needed on disturbed sediments, such as what may happen as a result of emplacement of heat source, the effects of convection and adsorption, effects on the biota, and the potential escape of radionuclides.

13) Physical oceanographers need to determine how those radionuclides that reach the water would be distributed.

Comment. What aspects of dispersion and dilution are most significant? The local, transient effects of radionuclides in the large discrete plumes or blobs of deep water that have recently been recognized might lead to higher doses to the biota. The probabilities of occurrence and the effects of such events, such as upwelling, need to be evaluated.

14) Major research is needed on all aspects of engineering related to emplacement.

Comment. Scientists must ensure that engineers are involved in all aspects of planning the disposal of HLW. Only by so doing can optimum designs and techniques be assured. This collaboration should begin with techniques for coring and drilling of holes and continue through the design, construction, and emplacement of canisters to the plugging of boreholes, both on land and in the seabed.

15) More data are needed on rates of removal of radionuclides from the ocean by sedimentation and their return to the ocean from the seabed.

Comment. When fine-grained sediment descends through water containing radionuclides, it will certainly adsorb some of them and carry them to the bottom. Conceivably, the seabed sediments will lose some radionuclides to the water. Predictions on relative rates of those processes are needed.

16) Better understanding of how radionuclides concentrate in marine organisms is essential.

Comment. Different radioactive elements may be concentrated in different parts of an organism, so it is especially important to distinguish the factors that lead to concentrations of radionuclides in the edible parts.

17) The factors that determine dose per unit of intake for radionuclides need much more study.

Comment. Rates of metabolism of radionuclides, and hence dose per unit intake, depends on the chemical form of the radionuclides that is ingested or inhaled. Further work on this topic is needed.

18) More effort should be given to investigating the chemical speciation of radionuclides in ground water and to research the methods for retarding radionuclide migration.

Comment. Radionuclides are adsorbed at different rates by different media.

Little is known about the kinetics of sorption reactions, so the transport may not be as slow, relative to ground water, as is thought.

19) Research should be intensified on the mechanisms involved in movement of radionuclides in the terrestrial environment, including freshwater, over long time periods.

Comment. Considerable work has been done on transfer rates over relatively short term (decades), but the time scale of HLW is very long and, in general, there is insufficient knowledge to make adequate long-term predictions.

APPENDIX

Terms of reference for the working groups. The working groups will:

1) Review relevant research completed, under way, or planned for the purpose of: (i) ensuring that the proposed methods of disposal of high-level radioactive wastes, whether into geological formations on land, below the seabed, or on the seabed, provide the degree of containment necessary to protect the biosphere from undue risks of radiation originating from the wastes; (ii) estimating with sufficient accuracy any radiation exposure to man that may result from such disposal; and (iii) assessing any harm to ecosystems from such disposal.

2) Conduct the reviews so that account is taken of the relevant behavior of nuclides following loss of any man-made containment of the wastes.

3) Principally base their reviews on the relevant activities of the IAEA, NEA, the Commission of the European Communities, and the Council for Mutual Economic Assistance, but extend these to national agencies where necessary to complete the reviews.

Membership of the working groups. The membership of Working Group No. 1, Terrestrial Disposal, included Dr. V. Babuska, Czechoslovak Academy of Sciences; Dr. W. S. Fyfe (chairman), University of Western Ontario; Dr. D. L. Norton, University of Arizona; Dr. N.J. Price, Imperial College of Science and Technology, United Kingdom; Dr. E. Schmid, Anglo-American Corporation of South Africa Ltd.; Dr. S. Uyeda, University of Tokyo, Bunkyo-ku; and Dr. B. Velde, University Pierre et Marie Curie.

Working Group No. 2, Marine Disposal, included Dr. Kurt Bostrom, University of Lulea; Dr. Egon T. Degens, Universität Hamburg; Dr. E. K. Duuersma, Delta Institute for Hydrobiological Research, Netherlands; Dr. Charles D. Hollister (chairman), Wood's Hole Oceanographic Institution; Dr. Ronald Pusch, University of Lulea; Dr. John C. Swallow, National Environmental Research Council, United Kingdom; and Dr. Gleb Udintsev, U.S.S.R. Academy of Sciences.

Working Group No. 3. Environmental Pathways, included Dr. B. G. Bennett, Environment Measurements Laboratory, New York; Dr. Y. Inoue, Kyoto University; Dr. R. H. Clarke (chairman, 1982) National Radiological Protection Board, United Kingdom; Dr. P. Jumans, University of Washington; Dr. J.P. Massue, Council of Europe; Dr. F. Morley (chairman, 1980), National Radiological Protection Board, United Kingdom; Dr. I. Nerethnicks, Royal Institute of Technology, Sweden; and Dr. J.B. Robertson, U.S. Geological Survey.

REFERENCES AND NOTES

1. The full reports can be obtained from the International Council of Scientific Unions, 51 Boulevard de Montmorency, 75016, Paris, France.
2. A list of the references consulted by the working groups and steering committee would fill several pages. Moreover, this article provides only the conclusions of a study that directly involved about 30 individuals. It is suggested, therefore, that readers interested in specific details refer to the ICSU report with its 174 citations. The most comprehensive coverage is contained in the various reports and proceedings of the International Atomic Energy Agency in Vienna and of the Nuclear Energy Agency. Probably the best place to begin is *Underground Disposal of Radioactive Wastes* (International Atomic Energy Agency, Vienna, 1980). The Council for Mutual Economic Assistance also published material of relevance to the countries of Eastern Europe, but mainly in the Slavic languages. Many papers from the region, however, are included in several of the publications from The International Atomic Energy Agency. In addition, various regional groups have sponsored studies, seminars, and workshops on different aspects of waste disposal. From all of these it is relatively easy to investigate the literature of any aspect of radioactive wastes.

PART 4
Emergency Plannings and Preparations

The safety of humans who handle radioactive materials is a fundamental concern in the industry. The basic practice is to design facilities that permit the handling of radioactive materials by use of barriers placed between the worker, the environment, the general public and the radioactive materials. These barriers are normally dictated by laws and regulations and are intended to control the geographic dispersion of radioactivity. The initial chapter of this part reviews the concept of barrier failure as the primary cause of most radioactive spills. The basic conclusion is that the best way to prevent a spill is the management of radioactive spill prevention.

Because of the sensitivity of the public to the possible problems of a nuclear disaster, many false impressions persist. A basic question then is, what is the risk? As a result of many government and industry studies, it has been concluded that nuclear energy is inherently safe. Procedures to evaluate release of radioactive materials into the atmosphere in a nuclear accident were developed more than 20 years ago. While these procedures were extremely conservative, it was recognized as a consequence of the Three Mile Island accident that the emergency guidelines should be updated. The NRC has now formulated new criteria for evaluation of the release of radioactive materials into the atmosphere.

The management of radioactive spills is a broad category covering a variety of events. The discussion in Chapter 19 on waste management is limited to spills that are localized. The basic goal in localized management of spills is first to assure the health and safety of personnel, and then to control the spread of contamination, assess the extent of the spill, conduct clean-up operations, and finally, verify that the area is no longer contaminated. These procedures must be swift and effective to reduce or mitigate any adverse impacts on public health and safety. Because every spill is unique, all plans must be flexible to meet a variety of situations.

Although the possibility of a radioactive waste accident is extremely remote, hospitals must develop plans to handle a possible emergency. Chapter 20 proposes such a plan. To be practical, a geographic hierarchy of preparation is advocated. Within regional hierarchies, those hospitals nearest nuclear facilities should maintain the highest levels of preparation and planning. The plan must require facilities for the regional management of contaminated victims, a system of communications between hospitals, procedures for the public to minimize radioactive hazards and the handling of radioactive patients, and finally the means of containment of the radioactive contamination.

Management of Radioactive Materials and Wastes: Issues and Progress. Edited by S. K. Majumdar and E. Willard Miller. © 1985, The Pennsylvania Academy of Science.

Chapter Seventeen

MANAGEMENT OF RADIOACTIVE SPILL PREVENTION

Stephen Marchetti

Manager, Radiological and Decommissioning Operations
West Valley Nuclear Services Co.
(A Subsidiary of the Westinghouse Electric Corporation)
West Valley, New York 14171

Facilities designed to process and/or permit handling of radioactive materials are, in the simplest terms, designed to place barriers between the worker, the environment, the general public, and the radioactive material. These barriers which can take many forms are often dictated by regulation and are intended to provide the means to maintain control over radioactivity by confining it to within the smallest practicable volume.

As a general statement, the level of barrier required is directly proportional to the level of hazard presented by the material being handled. In some cases, these barriers may be physical and redundant such as the barriers present in commercial nuclear reactors (see Figure 1); in other cases where the hazard level is low (or perceived to be low), the barriers provided may be administrative rather than physical. The author proposes that in nuclear or any facilities for that matter, barriers against accidents are task or job related and consist of both engineered (designed-in) controls and administrative controls. It is further posed that these barriers are important in the prevention of accidental loss of control over radioactive material confinement.

Regardless of the durability of the barriers, sound work practice requires assuming that a barrier will ultimately fail and allow the release of radioactive material, thus plans for responding to and managing these events are essential.

BARRIER

Radioactive materials by their nature emit radiation, some forms of which can penetrate material and deposit their energy in human tissue by a process called ionization. The principal means of protection against radiation exposure

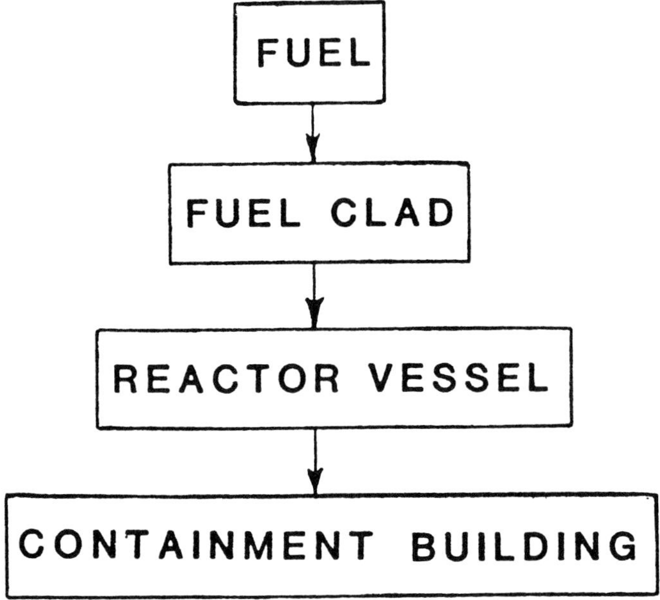

FIGURE 1. Physical barriers in a commercial nuclear reactor.

is the installation of shielding or a dense material between the worker and the material which absorbs the energy of the radiation and minimize the workers' exposure to it. In addition to the potential hazard of direct exposure, radioactive materials can exist as finely divided solids or in other physical forms and if not properly confined can be spread within the work place or to the environment. In facility design and operation, these events are prevented by installing containment walls or barriers to prevent the spread of contamination.

In any situation where a worker is exposed to potential hazards in the work place, barriers to accidents are provided. In the case of radioactive materials, these barriers may consist of engineered controls such as glove boxes, shield walls, ventilation, etc. which are designed into the facility and require no overt action on the part of facility personnel to utilize them. Administrative barriers may also be provided and could include procedures, policies, personnel controls, and operating specifications. Training may also be considered a barrier since it is intended to produce desirable actions by personnel and to discourage undesirable or hazardous activities in the work place.

For any task or job, there are three key elements which are always involved; a man, a machine, and a procedure. "Man" refers to any individual or group of individuals assigned to perform a task. "Machine" refers to the task or equipment or machinery necessary to get the job done. The "procedure" is the in-

FIGURE 2. The concept of barriers.

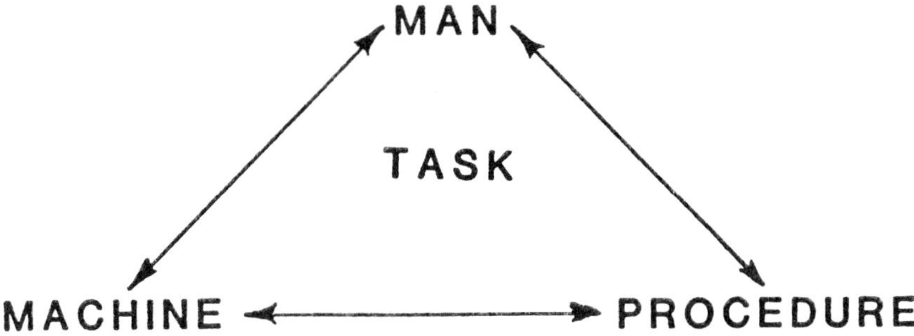

FIGURE 3. Interfacing required for task completion.

struction or method used to operate the machine. The interrelationships between these elements are shown in Figure 3 and illustrate the fact that for any task to be successful, all these elements must be in good working order. Additionally, the three elements are interfaced by engineered and/or administrative controls as discussed in the following paragraphs.

Man-Procedure

This interface deals largely with planning and completing training necessary for the man to successfully and correctly perform the task; the man *must* fully and correctly understand how to perform his duties. Since the technical knowledge is not inherited and is not a natural ability or instinct, these instructions must be communicated by written procedures. Procedures must be clearly written in understandable language, and it must be confirmed that the man understand the procedures and the proper way to use them before starting work. Training provides a means to ensure that the man not only understands the procedure but is capable of performing it in the desired manner. The principal objective in training is to effect this interface as a means of modifying the man's behavior to produce the desired result which is performance of the procedure in an accurate manner.

Procedure-Machine

This interface requires that the written procedure be specific for the machine being operated or the equipment involved and that the procedure identifies unambiguously any part of the machine and its controls. Procedures require a level of detailed review to ensure both their clarity and operability.

Machine-Man

The machine-man interface deals with ergonomics or what is popularly referred to as today as "human factors." Bruce Hertig writing in the *Fundamentals of Industrial Hygiene* defines ergonomics as "the application of the human biological sciences in conjunction with engineering sciences to achieve the optimum mutual adjustment of man and his work, the benefits being measured in terms of human efficiency and well being." This philosophy simply implies that the machine must be designed and engineered such that a man can do what he is required to do to perform the task given the constraints posed by the operation.

It should be apparent to the reader at this point that man-machine-procedure interface is integrally related to the concept of barriers discussed previously as indicated in Figure 4. Furthermore, it should be obvious that a breakdown, weakness, or failure of any element or interface shown in Figure 4 indicates the potential for an incident, an accident, or in our case, what is commonly referred to as a spill.

In utilization and/or production facilities where radioactive materials are handled, these barriers provide the principal means of radioactive material control or confinement. Failure of any barrier or element can lead to a loss of confinement or what is commonly called a "spill." Webster defines spill as to cause

FIGURE 4. Barriers related to interfaces.

or allow accidentally or unintentionally to fall, flow, or run out so as to be lost or wasted! This definition clearly fits with the concept of barrier failure as an initiating mechanism.

MANAGING SPILL PREVENTION (PRE-PLANNING)

Prevention is the most important phase in the management of radioactive spills. Typically, facility managers place an inordinately high level of attention on facility design as the principal means of spill prevention. Although facility design is important, the previous discussion shows that it does not and cannot provide the only barrier against the accidental loss of control of radioactive material. This has been demonstrated in a number of cases, where in spite of thorough facility design, failure of administrative and procedural controls and less than adequate training have resulted in spills of radioactive material. This serves to illustrate the need for a balanced approach to developing barriers to loss of radioactive material confinement. More specifically, this concept endorses the need to evaluate work (and accident potential) in terms of tasks or what is commonly known as "job safety analysis" to ensure that proper barriers against undesirable results are in place. Periodic and thorough reviews by facility management are important in assuring that the task elements and the barriers to prevent undesirable performance are in place and operating effectively.

ASSESSMENT

In spite of the number and durability of the barricades in place to prevent accidents, or in this case spills, it is always prudent to assume that these events will occur and to plan accordingly for their management. In order for planning to be effective, a thorough knowledge of the material being handled and the potential routes for interaction with man and the environment is important. In general, the assessments routinely performed as part of the licensing process are "worst case;" that is, conditions which would produce the greatest impact on man and his environment. Similar assessments should be prepared as part of the operating phase of the facility and, in the area of spill prevention, require a more detailed knowledge of both the material being handled and the potential consequence of accidents involving spills of that material. Such assessments are generally performed in terms of dose or consequence analysis. In assessing the potential impact of radioactive material spills, a detailed knowledge of the chemical and physical properties of the contaminant, its toxicological properties, and radiological properties is required.

Chemical and Physical Properties of Contaminants

Radioactive contaminants can exist in all three physical states (solid, liquid, and gas) and may exist in all of the states within any given process or facility. Factors such as particle size and solubility of the solid form can dictate the amount of material taken into a worker's body and the distribution of that material for the purpose of determining worker dose. Further, material particle size determines the potential for spreading radioactive material within the facility by either personnel tracking or air currents (particle resuspension).

For contaminants in liquid form, volatility, flash point, and boiling point are factors which bear on both the consequence and control of radioactive material spills. In the gaseous form, reactivity and flash point are important considerations.

Toxicological Properties

Many compounds used in the processing of radioactive materials present a level of risk from the industrial hygiene viewpoint and merit consideration in assessing the potential consequences of radioactive material spills. While not a major problem at hospital facilities, processes such as the dissolution of uranium metal which uses concentrated nitric or hydrofluoric acid can present a level of industrial hygiene concern as great as if not greater than the radioactive material itself. Factors such as the threshold limit value (TLV) and the potential for metabolic uptake must be considered where there is a potential for chemical spills.

Radiological Properties

Since the radiological properties of materials bear most heavily on the consequence of exposure, they must be thoroughly characterized. Factors to be considered include the potential for metabolism (which is as much a function of chemical and physical form), the type of radiation produced and its quality factor, the energy of the radiation and its range in air, and in the material spilled, its daughter products and the maximum permissible concentration of the material in air and water. Table 1 summarizes the essential factors requiring characterization during the planning and assessment phase for spill management. Collectively, these factors dictate the consequence of a spill by determining the prospective pathways to man and the environment and ultimately the consequence to man in terms of dose.

Pathways Analysis

Obviously, a spill has initial boundaries but due to factors such as resuspension, evaporation, atmospheric dispersion, soil migration, and waterborne transport, the consequences can be made less local and may become regional in scale. A spill with potential regional impact can involve a greater population and hence require a more detailed assessment of the pathways to man. For plan-

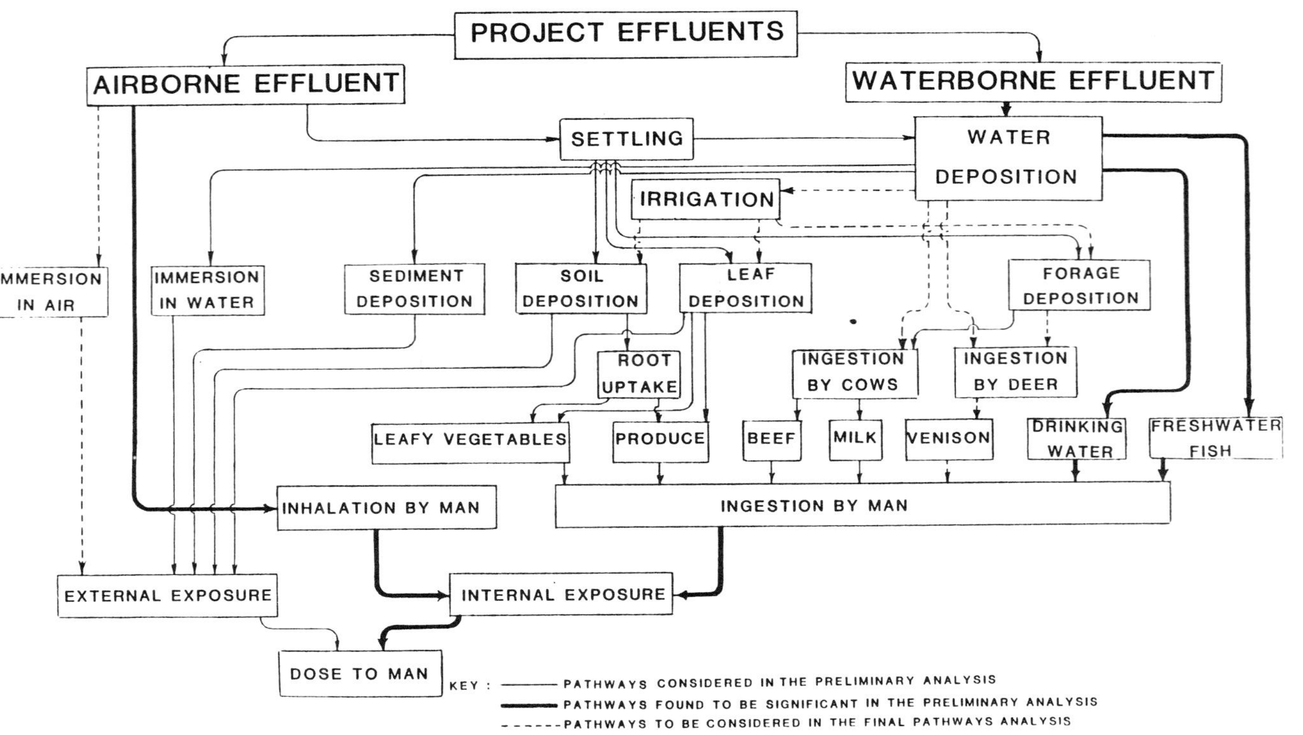

FIGURE 5. Significant pathways at a typical nuclear facility.

TABLE 1

Material—Dependent Factors Required to Perform Pre-Spill Characterization and Planning

Properties	Factors Requiring Characterization	General Information Required	Information is Used to Determine:
Chemical and Physical	Solid	• Particle Size • Solubility	• Metabolic Potential • Potential for
	Gas	• Volatility • Boiling Point	Resuspension or Dispersion
	Liquid	• Flash Point • Pyrophoricity	• Human Uptake Potential
	Multi-Phase		
Toxicological	TLV Metabolism	• General Toxicity • Uptake Potential	• Disposal Options • Human Uptake Mechanisms
	Dangerous Properties	• Metabolic Potential	• Recovery and Cleanup Plans
Radiological	Type of Radiation Energy of Radiation Range in Spilled Media Daughter Radiation	• External Exposure • Internal Exposure	• Recovery and Cleanup Plans • Potential for Human and Pathway Exposure
	Maximum Permissible Concentration		• Potential Consequence of Release
	Quality Factor		• Significance of Pathways Involved

ning purposes, these assessments may be generic. However, in performing actual post-spill dose assessments to determine the need for either protective or corrective action, site-specific pathways data is certainly required and may be complex in nature. Figure 5 shows a compartment pathways model for an actual nuclear facility which potentially produces effluents in both solid (fine dispersible particulate form) and liquid form. The complexity of the model is to determine the transport of radioactivity through the compartments to determine the dose to man and thus the consequence of the event. The calculations performed utilize a release component, a pathway transport, and a dose component.

The release component determines the amount of radioactive material available to a given pathway for transport. The solution for the release component has the general form of:

$$R_i = T_i \, f_{ia} \, A_i$$

where:

R_i is the material released and available to enter a given pathway for i th nuclide at a specific location.

T_i is the atmospheric or water dilution term for the i th nuclide.

f_{ia} is the fraction of the i th nuclide removed by other removal mechanisms (a) such as fallout and rainout.

A_i is the activity of the i th nuclide released.

The general form of the pathway transport equation is:

$$I_i = U_a \, C_{ia} \, f_{ia}$$

where:

I_i is an individual intake of radioactivity from a given pathway for the i th nuclide,

C_{ia} is the concentration of the radioactivity for the i th nuclide in a specific compartment of the pathway (a),

U_a is the annual intake of the foodstuff or material from an individual compartment or pathway (a)

F_{ia} is the fraction of total material ingested which is ingested from the area of interest (a) containing nuclide (i).

The dose is then simply:

$$D_T = I_i \, F_i$$

where:

D_T is the annual dose to the organ of interest,

I_i is the concentration of nuclide (i) in the organ of interest,

F_i is a concentration to dose conversion factor for nuclide (i) in a given organ.

The above equations are in their simplest form and are in fact substantially more complex when multiple pathways and nuclides are involved. They are further complicated when relatively short half-life nuclides are involved and corrections must be made for exponential decay within compartments. The calculations often involve simultaneous solution of a series of equations. Fortunately, computer models are available to perform these complex assessments.

Clearly, spills of material which can be confined to a facility or within site boundaries do not have a regional level of impact. This stresses the importance of proper spill management within the facility boundaries and serves to illustrate the need to prevent dispersion of spilled materials and thus to prevent a regional dose consequence.

RECOVERY ACTION PLANNING

The assessments described above have a much broader utility than simply predicting the consequence of releases. They are also important in planning response and recovery actions which may be categorized as either protective or corrective in nature. For example, a thorough pathways assessment can detect those pathways which are significant with respect to consequence and can aid in determining optimum placement of detection instruments to provide early warning of a release and to monitor any continuing residual releases. Additionally, an understanding of these pathways in terms of their significance can aid in determining which corrective action would provide the greatest benefit in terms of reducing consequence from a spill. Consider the complex pathway model shown in Figure 5. The airborne effluent pathway has a direct effect on consequence to man through the inhalation route. This route potentially permits inhalation of radioactive materials and a long-term internal exposure depending on the chemical and radiological characteristics of the effluent. The waterborne effluent pathway has its greatest significance in the drinking water compartment followed by ingestion by man. This pathway also results in a potential long-term internal exposure. Although other compartments in the waterborne effluent pathway such as the cow-milk-man and the ingestion by game animals and forage deposition with ultimate ingestion by food animals may have a long-term significance, these are generally not thought of as having an immediate impact on man.

It is obvious that the consequence to man through many of these pathways can be mitigated by taking preventive measures such as condemnation of crops, condemnation of milk products, and temporary use of alternate drinking water supplies. Such actions are called protective actions, because they protect the population against receiving unnecessary radiation exposure. Other protective actions may be taken further into the pathway and may be administered on an individual basis. For example, potassium iodide tablets can be issued to individuals prior to receiving a dose of radioactive iodine. The potassium iodide, which contains no radioactive iodine, acts as a thyroid block preventing thyroid uptake of radioactive iodine. Another form of protective action is evacuation. All of these protective actions require extensive preplanning and coordination in order to be effective. Generally speaking, when protective actions are taken, corrective actions are required. Consider the waterborne-effluent mediated pathway involving deposition of radioactive materials onto forage with subsequent uptake in cows by ingestion of forage. This pathway ultimately leads to man through the cow-milk-man pathway. By condemning forage available in pastures and placing animals on store feed, the cow-milk-man pathway is blocked, and the potential for dose is eliminated. While this may also be considered a protective action, it is in fact corrective because the source of radioactive material which could enter the pathway is eliminated. Followup work would be required to assure that contaminated forage was harvested and properly disposed of. Corrective actions may thus be considered to be restorative in that the spilled material is removed and the area returned to the prespill condition.

SUMMARY

The author has reviewed the concept of barrier failure as the primary cause of most spills. Further, it is concluded that management of radioactive spills is actually management of radioactive spill prevention. The paper further stresses the need for proper preplanning and assessments in order to determine the most effective means of radioactive materials monitoring and consequence assessment. The development of protective and corrective action plans in terms of pathway analysis was also discussed. It is important for the reader to recognize that correlations may be drawn between the actions taken in the event of radioactive spills and the actions taken in the event of chemical spills. Both are potentially hazardous materials, and have potential regional impact in terms of consequence to both man and his environment. In terms of actual consequence to man, the hazard presented by chemical spills can be far greater than the risk presented by spills of radioactive material. In either case, it is important to use sound judgment in investigating the circumstances surrounding the event to ensure that:

The specific problem and its potential hazard to man are mitigated, and Adequate analysis/assessment is directed at evaluating the potential cause and preventing occurrence.

BIBLIOGRAPHY

Olishifski, J.B., *Fundamentals of Industrial Hygiene,* National Safety Council, Chicago, 1979.

Management of Radioactive Materials and Wastes: Issues and Progress. Edited by S. K. Majumdar and E. Willard Miller. © 1985, The Pennsylvania Academy of Science.

Chapter Eighteen

EMERGENCY PLANNING AND PREPAREDNESS FOR A NUCLEAR ACCIDENT

E. Preston Rahe
Manager, Nuclear Safety Department
Westinghouse Electric Corporation
Box 355
Pittsburgh, Pennsylvania, 15230

Many scientists and engineers who have the responsibility for nuclear safety sometimes feel that they are living in the world of Chicken Little. Everytime something doesn't go as planned, the cry is heard "the sky is falling; the sky is falling." This is a false assumption, for the "sky is not falling." However, meltdown has become a household word synonymous with ultimate disaster. The word, which refers to the complete melting of the fuel in a nuclear power disaster, was brought to public attention in the movie, The China Syndrome, and its implications were underlined by the accident at Three Mile Island. Fear of a meltdown has encouraged the viewpoint that dependency on coal and oil to produce electricity should prevail.

A basic question thus is, how rational is this viewpoint? According to estimates of government scientists, for nuclear power to be as dangerous as coal, there would have to be a reactor meltdown every two weeks somewhere in the United States. While many scientists would question the accuracy of this government statement, the Union of Concerned Scientists, a decidedly anti-nuclear group, have indicated that nuclear risks become equal to coal if a meltdown were to occur every six months.

There are 79 operating reactors in the United States providing 13 percent of the nation's electricity. In the development of present-day nuclear technology, is there a relationship between the fear of nuclear energy today and the introduction of new technologies in the past? In general, a study of technological progress experienced in American society throughout history can be characterized by a pattern of progress, setbacks, followed by new creative efforts. Many scientists and engineers believe that this is the cycle nuclear energy is facing today.

Emergency preparedness is not a new endeavor, but is one that has been successfully faced in the past. For example, Benjamin Franklin wrote about the need to prepare for emergencies during his time. In his day, efficient saw mills made it possible to build frame houses more versatile and economical than log cabins. Advances in urban living had made it possible to crowd these frame houses together in cities. These changes in lifestyle, as well as cleverly conceived fireplaces, stoves and lamps, made life comfortable, but raised the possibility of catastrophic fires. Franklin was seriously concerned with the fire hazards, but realized that avoidance of progress, such as the banning of cities or modern conveniences, was impossible. Instead, he worked to make progress serve man. Franklin proposed architectural modifications to make houses more fireproof, the licensing and supervision of chimney sweeps and he conceived one of the earliest coordinated emergency planning and preparedness concepts—the volunteer fire company.

The nuclear industry faces a similar challenge today. While society is quite willing to accept the benefits of nuclear technology, it would be desirable if those benefits came risk-free. While this is not possible, there are two choices. The first is to abandon nuclear technology and the second is to reduce the risk to the lowest possibility and at the same time prepare to deal with an emergency if it materializes. As Franklin did, the second option is the only viable one.

The question must then be asked, what is the risk? Risk is the concept used to assess the probability of the adverse impact of a nuclear emergency on human beings. Nothing in life is risk free. Recognizing this, people tend to respond in a predictable fashion to different levels of risk. To illustrate, a risk level of one in one-thousand is normally considered unacceptable, and great effort is spent to reduce it. These risks don't exist today. At a level of one in ten-thousand, resources are still expended to avoid these risks. However, this type risk is accepted if the benefits are sufficient. An example is the use of the automobile. While the danger of the automobile is recognized, the risks are reduced by such measures as seat belts, guard rails, and speed limits.

At a level of one in one-hundred-thousand, the public is generally tolerant of the risk. It is recognized that a risk exists and a nominal amount of resources will be committed to reduce it. An example of this type of risk is accidental drowning. In order to prevent it, lifeguards are provided. Another example is accidental child poisoning and to prevent it the child proof container has been invented.

When the risk approaches one in one million, it is known as an "act of God" and there is hope that it will never be experienced. An example of this is a hurricane.

When the safety of nuclear energy is discussed, the question is, what is its risk category? Repeatedly, independent studies, as well as government and industry reports, have concluded that nuclear energy is inherently safe. Often times though, the perception of risk, either through lack of information or inaccurate information, is what the public awareness is of the risk at a given time.

There have been many reports, including newspapers, that have perceived nuclear energy as a very high risk undertaking. For example, the Washington Post wrote about the Sandia Report on reactor safety indicating that "The most detailed government study of potential consequences of accidents at atomic power plants has concluded that the worst-case death toll could exceed 100,000 persons and damage could top $300 billion at certain locations." This news story was based on a very broad *interpretation* of the Sandia Report, and not what the report actually concluded. Both Sandia and Nuclear Regulatory Commission officials disavowed this interpretation, noting the extreme over-conservatism in assumptions necessary to reach such a conclusion. As estimated by the NRC, the chance of such a consequence from an accident at a single reactor in one year is one in one billion. Even though the risk is infinitesimal—small enough to be ignored in a general risk hierarchy—the possible disaster is feared by the general public.

Nuclear scientists believe that even if a hypothetical accident did occur, the consequences would be less severe than generally envisioned. The basis on which the risk is determined, that is, the amount of radiation that would be released to the environment following an accident, is overly conservative to the extent that it bears little resemblance to reality. The method which is used to predict high releases of radioactive materials from a nuclear accident has remained unchanged and, until recently, unexamined for the past 20 years. In this model the release of radioactive materials for a postulated accident is overstated. Thus the consequences of the accident will be overstated as well. Twenty years ago, the method to predict radioactive releases was based on the assumption that a large fraction of the radioactive material would escape the core. This was intentionally assumed to ensure conservatism since actual plant data did not exist at that time.

The most important lesson of Three Mile Island may prove, in time, to be what did not happen during the emergency. According to the conservative model, at least a thousand times more radioiodine, the fission product isotope of most concern in human effects, should have been released from the plant as a result of the accident. In reality, the release was so small that a person standing naked at the plant boundary would have received a dose of radiation equivalent to a few chest x-rays. The residents of Middletown, Pennsylvania received less radiation for the duration of the accident than if they had vacationed in certain areas of Brazil where there is a high concentration of radioactive elements in the soil.

Analyses of radioactive releases at Three Mile Island and other past accidents provide an insight to the conservatism in current regulatory assumptions. Accident records show that the presence of water and steam in the accident environment plays a key role in limiting the release of radioactive products, as well as in delaying the release of iodine. In a light water reactor, water is always present in the core under normal operating conditions and would be present along with wet steam even under damaged core conditions. In all light water reactor accidents to date, as well as supporting experimental data, no more than

0.3 percent of the available iodine has ever been released to the atmosphere. The current regulatory models assumes that 25 percent of the available iodine will be released during an accident.

There are an array of research projects in progress in many countries to study the release of radioactive materials during an emergency. Excellent progress is being made, and the results confirm that the prior release estimates are too high. These new data are thus providing a basis for more realistic emergency planning guidelines. Additionally, the NRC has issued a tentative schedule for incorporating research results into reactor licensing. NRC staff recommendations for revised final release values of radioactive materials are expected in the near future.

As a part of on-going research, utilities have adopted emergency guidelines to protect the public in the event of an accident. The effect of continuing research on these plans will be to aid in determining at what point in time, if ever, during an emergency, the population should be evacuated. This is an extremely important consideration since, if the preliminary conclusion is correct that the traditional model is overconservative, the health and safety of the public would be better served by improvements in shielding methods such as staying indoors and closing all outside air access to the home, than by evacuation. Evacuation is very stressful for people and involves real physical dangers.

Emergency procedures for nuclear accidents were discussed as early as 1947, long before the emergency of Three Mile Island. From the early years of commercial reactors, regulations required operators to submit emergency response plans, not only for the site area, but typically for an area extending 10 miles from the reactor. There was also the requirement that efforts would be coordinated with the local authorities at the time of an emergency.

The radiological emergency plans today normally consist of two categories of activities: (1) there are the emergency plans, which apply to the plant and its site, which are under the jurisdiction of the utility and the NRC, and (2) there are the emergency plans that apply to the offsite situation which are under the jurisdiction of the state and local authorities, with the Federal Emergency Management Agency coordinating the efforts. The NRC and FEMA work together to assure that their plans are complementary.

The TMI accident revealed some deficiencies in existing emergency plans. It was evident that the plant operators should be able to recognize more clearly the inception of an abnormal event, immediately diagnose the difficulty, and follow its progress more easily. Further, the personnel should be better trained and equipped to evaluate and control the developing situation. It was also apparent that communications with the media and the public should be improved.

In response to the Three Mile Island incident the NRC developed criteria for emergency response. These included such items as the need for an onsite technical support center from which the operator of a reactor can obtain immediate technical advice, an operational support center from which onsite emergency response can be coordinated, an emergency operations facility from which the offsite emergency response can be coordinated, and a safety parameter display system which the operator can use to access major plant functions to determine the plant condition. This final requirement was implemented because, during an accident, the operator must analyze a host of information on which to base actions. At TMI this information overload was a major factor in delaying necessary actions which could have lessened the severity of the accident.

The safety parameter display system developed by Westinghouse is unique in the industry. The idea for it came from medical use in intensive care units where continuous monitoring of essential human functions is vital. The industrial system monitors eight important plant parameters such as temperature and pressure. These parameters are displayed to the operator in an easily recognized geometric shape. As the parameters change in response to a plant malfunction, the operator can observe several dials and gauges. This is an important diagnostic tool which enables an operator to rapidly respond to malfunctions.

In addition to the actions by the NRC the nuclear industry has also responded. The Nuclear Safety Analysis Center and the Institute for Nuclear Power Operations emerged from the experiences at Three Mile Island. The Nuclear Safety Analyses Center functions to collect and analyse information on equipment malfunctions at all nuclear power sites. In this manner, all nuclear plant operators are made aware of the situation at all other plant sites so that information can be used to evaluate individual plants for similar circumstances. The Institute of Nuclear Power Operations serves the industry by developing criteria for the training of nuclear power plant operators and the management of nuclear power plants under construction and in operation.

It is now believed that if an emergency should occur in the future, the nuclear power industry is for better prepared than prior to TMI. An hierarchy system for emergency response has been developed based on specific plant conditions. This hierarchy is divided into four classes of emergency actions levels, each with its own requirements for onsite and offsite response.

The first class is called an unusual event and indicates a potential degradation of the plant safety level. State and local authorities would be informed of the unusual condition as soon as it is discovered and would provide fire or security assistance if necessary. Unusual events can generally be handled by plant workers and do not affect the public's safety.

If the condition of the plant should worsen, the plant will be taken to an alert status. In this case there may be an actual or potential substantial degradation of the plant safety level. In addition to the actions in response to an unusual event, the licensee would activate the on-site technical support center, the on-site operational support center as well as bring emergency operation facility and other key onsite and offsite emergency personnel to standby status. There is still no adverse affect on public safety, but state and county officials begin preparing emergency operation centers in case the situation becomes more pronounced.

The next class of events is called a site area emergency and involves actual or likely major malfunctions of plant equipment or systems needed for public protection. Any radioactive releases are not expected to exceed EPA protective guidelines except near the sight boundary. The offsite emergency operations facility would be activated, and local authorities would begin public notification of emergency status, as well as mobilizing evacuation resources.

In the unlikely event that the plant safety level should further degrade, a general emergency would be declared. A general emergency occurs if there is actual or imminent core melting, with potential for loss of containment integrity. Releases can be reasonably expected to exceed EPA guidelines offsite beyond the immediate site area. This requires a full mobilization, sheltering of population out to at least five miles, and consideration of evacuation. Sheltering and evacuation are management decisions, and would be up to the governor of the state.

Based on current regulations, FEMA approves each site-specific plan of state and local governments for each power reactor site after 1) formal review offsite preparedness, 2) holding a public meeting at which the preparedness status has been reviewed, and 3) a satisfactory joint exercise has been conducted with both utility and local participation.

Annually, each state, within any position of the 10-mile emergency planning zone, must conduct a joint exercise with the utility to demonstrate its preparedness for a nuclear accident. While it is unlikely that these extreme measures will be needed as a result of an accident at a nuclear power station, the fact that these plans have been well thought out and implemented have already proven their benefit to society.

The preparedness for a nuclear accident can be of great advantage in other types of emergencies. For example, on December 11, 1982, a non-nuclear chemical storage tank exploded at a Union Carbide plant in Louisiana shortly after midnight. More than 20,000 people were evacuated from their homes. They were evacuated under the emergency response plan formulated for use in the event of a nuclear accident at the nearby Waterford Nuclear plants. Clearly, this illustrates how a plan conceived for one purpose is appropriate to handle other types of accidents that occur in a modern industrial society.

Management of Radioactive Materials and Wastes: Issues and Progress. Edited by S. K. Majumdar and E. Willard Miller. © 1985, The Pennsylvania Academy of Science.

Chapter Nineteen

THE MANAGEMENT OF RADIOACTIVE MATERIALS SPILLS

Michael T. Ryan,[1] Ph.D. and David G. Ebenhack[2]

[1]Director, Environmental and Dosimetry Laboratory
Chem-Nuclear Systems, Inc.
Barnwell, S. C. 29812
and
[2]Vice-President, Regulatory Affairs
Chem-Nuclear Systems, Inc.
Columbia, S. C. 29210

The management and handling of a radioactive materials spill is a broad category covering a variety of events. A radioactive materials spill can be defined as a release of radioactive material from a location where it is controlled to an area where it is not controlled or where it is controlled in a less than desirable way. A radioactive materials spill can be as simple as a dropped beaker in a radiochemistry laboratory or as complicated as a transportation accident with failure of shipping containment on public access highways. While a whole variety of radioactive material spills can be envisioned, the principles of spill management can be applied with a certain degree of uniformity. It will be the focus of this chapter to present and discuss the principles of radioactive materials spill management and give examples of how these principles are implemented in practical situations. The discussion will be limited to spills that are localized. Large distributed releases of radioactive materials such as major accidental releases from fixed nuclear facilities are beyond the scope of this chapter and are covered elsewhere in this book. The Nuclear Regulatory Commission has published guidance for major releases of this type[1,2].

Evaluations of accidents involving radioactive materials spills have all indicated that prior planning and preparedness for radioactive materials spills is essential for proper and prompt response.

The reactor accident at Three Mile Island is the most vivid example of the need for emergency preparedness. In fact, the Kemeny Commission which investigated the accident cited emergency preparedness as a key issue in their recommendation[3]. Whether a large release from a nuclear power station or a more localized spill of radioactive materials, preparedness and planning are essential for proper management of the incident.

BASIC PRINCIPLES OF RADIOACTIVE SPILL MANAGEMENT

All nuclear facilities should be prepared to deal with spills of radioactive materials within the confines of the facility and in areas of public access. No radioactive materials spill management plan can be generic, but by review of some basic principles a plan could be developed for a particular application. These principles are discussed and exemplified below. The severity and consequences of a radioactive materials spill can range from minor spills in nuclear facilities that are usually handled with normal contamination control procedures up to accidents involving spills of larger quantities of radioactive materials with release to areas of public access. No matter what the severity of the radioactive materials spills, the same principles of spill management apply. These are:

1) Assure the health and safety of personnel immediately involved in the spill.
2) Control to the extent possible the immediate spread of contamination without risking further injury to personnel.
3) Make an initial assessment of the spills so that the proper elements of the applicable spill management plan can be implemented.
4) Implement the response plan. The major elements of the response plan include proper notification of civil authorities and parties responsible for the radioactive materials involved in the spill and initiation of spill containment and stabilization.
5) Conduct clean up operations. During this phase whatever actions are necessary to clean up the spill should be completed. Generally, after initial spill emergencies such as personnel injury, fire and public safety hazard have been controlled, a less urgent approach can be taken for complete spill clean up. The time taken to effectively complete spill clean up is dictated by the specific conditions of a particular spill.
6) Verify, through radiological assessment survey, that spill clean up is complete.

These principles seem simple in concept but can be complicated in terms of application to radioactive materials spills. Each principle will be discussed in detail with emphasis on its application in various situations.

It is first necessary to discuss the role of a radioactive materials spill emergency response plan. Emergency response plans at nuclear facilities address the basic question "What if we had a release of radioactive materials?" The response plan should provide specific guidance regarding who should respond to the spill, what initial actions should be taken and who should take them, who is in charge of the response effort and which authorities and individuals should be notified that the spill has occurred. Prompt response by qualified personnel often prevents the radioactive materials spill from becoming worse.

In the management of any radioactive materials spill, clear communication through the designated chain of command is essential for proper response. As in any emergency, someone has to be responsible and take charge of the situation. As an emergency progresses, the responsibility may shift from one individual to another, but the authority in charge and the lines of communication must be clear for effective emergency management.

Facility personnel and public safety personnel should be trained in radioactive materials spill management and response. The Bureau of Radiological Health of the U.S. Department of Health and Human Services published a directory of personnel responsible for radiological health programs.[4] This resource lists persons in every state as well as in federal agencies who may be contacted in the event of a radioactive materials spill with potential impact on the public.

There are also publications describing the proper response by public safety officials such as police, sheriffs, ambulances and rescue squads, physicians and nurses.[5,6,7,8,9] Review of spill scenario will exemplify the principles of spill management.

RADIOACTIVE MATERIALS SPILL MANAGEMENT—CASE STUDY

In a well documented accident at the Department of Energy facilities, Hanford, Washington, a worker was injured and highly contaminated with ^{241}Am in 7M nitric acid. The prompt response by a highly trained health physics staff member was cited as a prime reason for control of the spill without sacrificing the required attention to the injured worker.[10] The health physics technician who responded initiated actions to care for the injured worker, transport him to medical care and to notify the proper supervision so that a full response to the incident could be initiated. All through the first few hours attention was given to containing the spread of contamination without compromising the care of the injured worker.

The Hanford case involves a radioactive materials spill in a nuclear facility with significant personnel injury. Fortunately, well-trained personnel responded properly and prevented the accident from becoming more serious.

In order to more clearly demonstrate the basic principles of radioactive materials spill management, a review of an accident scenario would be useful. To cover as many basic principles as possible, a transportation accident on a public access highway with personnel injury will be assumed. It will also be assumed that there has been a substantial spill of radioactive materials onto the roadway as a result of the accident. Many such scenarios have been studied in detail. In this discussion, accident events will be related to the basic principles of spill management given above.[11,12,13,14,15,16,17,18,19,20,21]

Initial Response
The first response to a transportation spill of radioactive materials with personnel injury would likely be by law enforcement officials who would be in charge of the initial response. Upon ascertaining that the accident involves radioac-

tive materials, the law enforcement officials should secure emergency medical aid for the injured and notify the appropriate local and/or state radiological health officials. He should also secure, to the extent possible, the area in the vicinity of the spill and hold any individuals who have been near the scene for identification and inspection by the radiological health specialists.

Upon arrival of emergency medical aid, the status of injured personnel must be determined. If the injuries are critical and immediate medical aid is required, the law enforcement and emergency medical personnel may be faced with the decision to provide aid with the possibility of radiation exposure. Providing aid to the injured may represent a radiation risk to emergency medical teams which may be high enough in rare cases to result in injury from radiation exposure. The most important guide to these first decisions is the shipping manifest which accompanies any shipment of radioactive materials. These documents should contain the generator of the radioactive materials, the radionuclide, quantity, physical and chemical form contained in the shipment and all information necessary to make an immediate evaluation of the radiological hazards. If these papers can be obtained and the information contained in them can be relayed to the radiological authorities, it may be possible to assess the risks of allowing medical attention to be given. Further, it may be possible to minimize the risks to medical personnel by following the advise given by radiological authorities.

If the shipping manifest is not obtainable, the decision of the emergency medical and law enforcement personnel to respond to the injured becomes more difficult. If they do provide care, they may be doing so with significant risk to themselves.

All persons involved in the response should be aware that time of exposure to the radioactive materials, distance from the radioactive materials and shielding are important principles in minimizing exposure. The less time spent near the spill, the lower the exposure; the greater the distance from the spill, the less the exposure and the more shielding materials between an individual and the radioactive material, the less the exposure. Risk to emergency medical personnel may be reduced by application of these principles. If medical aid requires transporting seriously injured individuals to hospital facilities immediately, every attempt should be made to minimize the spread of contamination to the emergency medical personnel, the ambulance, the route to the hospital and the hospital facilities. There is no good substitute for qualified health physics support, but if this support is not readily available, the state radiological health officials would be able to provide the proper guidance via telephone or radio communications on how to minimize the spread of contamination.

Radiological Response

Upon arrival of the appropriate radiological authorities or their designated representatives, the responsibility for management should transfer to them under the constraints of this accident scenario. At this point, we can assume that technically qualified personnel with the proper health physics equipment are not at the scene. All other personnel should follow the guidance and direction of these officials. It would be the responsibility of the radiological authorities to obtain information from law enforcement officials, emergency medical personnel and other persons involved in the discovery, reporting or early response to the accident. Based on this information, the radiological authorities will decide on the next steps to be taken to effectively contain and stabilize the spill. Clear and concise communication of events and actions is essential to proper management of the spill. To the extent possible, each individual involved should make notes regarding the specifics of the spill. These notes should include the names and addresses of all people directly or peripherally involved in the spill. The time various actions were taken and who took them should be recorded also. In general, any information which will be of value in evaluating the spill should be written down and these notes should be compiled by one individual during the incident. If it is possible, one person should be designated to record all information pertinent to the management of the spill.

The radiological authorities may, in the management of the spill, call upon law enforcement officials for logistical support, other government officials, the technical staff from the facilities from which the radioactive materials shipment originated or technical experts from other facilities. The goal of these efforts, after the safety of injured personnel is established, is to stabilize the spill. The main pathway of concern is airborne release of material due to fire, chemical reactions or suspension and transport due to winds. The airborne pathway offers the largest potential for the spread of contamination over the largest area in the shortest period of time. Action to mitigate airborne transport of material should be considered first during stabilization and containment of the spill. The physical form of the radioactive material will determine whether or not this release pathway is of concern. Air sampling for airborne radioactive materials should begin as soon as possible to either verify that no airborne release occurred or to quantify the magnitude of the release.

The next pathway of concern is the surface runoff pathway. The spill should be isolated from any free flowing streams, river or lakes in the vicinity of the spill. While this pathway of potential contamination spread is important, it is not as important as the airborne pathway. Airborne release presents immediate exposure to persons in the airborne plume of radioactive materials. Waterborne contamination will undergo dilution and sedimentation and generally this pathway does not pose an immediate hazard to the public health and safety. It may pose a significant clean-up problem.

In general, the spill of radioactive materials may be stabilized in a variety of ways depending upon its physical and chemical form and the specifics of the particular spill. Stabilization may be accomplished by covering the spill with clean dirt, absorbent materials or plastic sheeting. If the radioactive materials involved in a spill are highly radioactive, containment and stabilization should be done in a way to minimize exposure to personnel performing the work.

Spill Clean-Up

Once spill stabilization and containment have been achieved, arrangement to clean-up the spill must be made. The radioactive materials originally spilled, the non-contaminated materials used to contain and stabilize the spill and the residual contamination at or near the location of the spill will have to be packaged and transported to a secure location for interim storage until the eventual fate of materials can be decided. The goal during this phase of spill management is to return the spill site to its original condition prior to the accident. The urgency of the clean up depends on how vital is the area. Within the scenario used in this example, a public highway may need to be reopened or a detour could be set up. The specific conditions of the spill will dictate the urgency of clean up.

Post Clean-Up Radiological Assessment

After the clean-up of all bulk material involved in the spill, residual radioactive contamination may persist. Sampling of soil and other materials in the vicinity and environs of the spill for laboratory analyses, as well as direct readings radiation monitoring equipment, will reveal such residual contamination. In general, residual contamination should be removed so that any risks are minimal. Usually it is prudent to remove any contamination within practical limits. For example, if the road surface is contaminated, it may be advisable to presurface the road in the vicinity of the spill. The level to which an area is decontaminated may be determined by the state radiological authorities. The clean-up of this residual contamination need not be treated with the same urgency as the early response because the threat to public health and safety has been reduced to an acceptable level.

SUMMARY

The management and handling of a radioactive materials spill must be swift and effective to reduce or mitigate any adverse impacts on public health and safety. Spills within nuclear facilities generally pose less of a public health impact than spills in areas of public access. The essential elements of spill management include prior planning by agencies which may be required to respond to a spill. Any plan for the management of radioactive materials spills must be flexible enough to be applied in a variety of situations. The major elements of a radioactive materials spill plan, however, apply in every case. It is essential that communications be clear and effective, that the management of a spill be directed by a responsible party whose authority is recognized by everyone involved and that the actions, according to the principles discussed above, be taken to assure the safety of any injured personnel, containment and stabilization and clean up the spill and to verify through radiological surveys and sample analyses that the clean up is complete.

Finally, as with any accident, we learn from our mistakes. Any spill of radioactive materials, minor or major, should be assessed so that similar spills or accidents can be prevented.

REFERENCES

1. U.S. Nuclear Regulatory Commission, 1983. "Agency Procedure for the NRC Incident Response Plan," NUREG-0845.
2. U.S. Nuclear Regulatory Commission, 1983. "NRC Incident Response Plan," NUREG-0728, Revision I.
3. Kemeny, John G., 1979. "Report of the President's Commission on the Accident at Three Mile Island," U.S. Government Printing Office.
4. Bureau of Radiological Health, U.S. Dept. of Health and Human Services, 1981. "Directory of Personnel Responsible for Radiological Health Programs," HHS Publication, FDA 82-8027.
5. U.S. Atomic Energy Commission, 1972. "Emergency Handling of Radiation Accident Cases (Police)."
6. U.S. Atomic Energy Commission, 1972. "Emergency Handling of Radiation Accident Cases (Sheriffs)."
7. U.S. Atomic Energy Commission, 1972. "Emergency Handling of Radiation Accident Cases (Ambulance-Rescue Squads)."
8. U.S. Atomic Energy Commission, 1972. "Emergency Handling of Radiation Accident Cases (Physicians)."
9. U.S. Atomic Energy Commission, 1972. "Emergency Handling of Radiation Accident Cases (Nurses)."
10. McMurray, B.J., 1976. "Hanford Americium Exposure Incident: Accident Description," Health Physics, Vol. 45, No. 4.
11. Rhoads, R.E., 1977. "An Overview of Transportation in the Nuclear Fuel Cycle," BNWL-2066.
12. Elder, H.K., Project Coordinator, 1978. "An Assessment of the Risk of Transporting Spent Nuclear Fuel by Truck," PNL-2588.
13. Taylor, J.M., Daniel, S.L., 1982. "RADTRAN II: Revised Computer Code to Analyze Transportation of Radioactive Material," SAND80-1943.
14. DuCharme, Jr., A.R., Project Coordinator, 1978. "Transport of Radionuclides in Urban Environs: A Working Draft Assessment," SAND77-1927.
15. U.S. Atomic Energy Commission, 1974. "Everything You Always Wanted to Know About Shipping High-Level Nuclear Waste," WASH-1264.

16. Office of the Federal Register National Archives and Records Service, General Services Administration, Code of Federal Regulations, Title 49, Parts 100 to 199.
17. Greenberg, J., Brackenbush, L.W., Murphy, D.W., Burnett, R.A., Lewis, J.R., 1980. "Application of ALARA Principles to Shipment of Spent Nuclear Fuel," PNL 3261.
18. Franklin, A.L., 1980. "TRAX II: A Computer Program for Transportation Risk Assessment," PNL 3208.
19. Institute of Nuclear Materials Management, Inc. 1979. "Proceedings of the 19th Annual Meeting," Volume VII, Proceedings Issue Journal of the Institute of Nuclear Materials Management.
20. Davidson, C.A., Foley, J.J., 1982. "Transportation Technical Environmental Information Center Index," SAND82-1200.
21. Rao, R.K., Wilmot, E.L., Luna, R.E., 1982. "Non-Radiological Impacts of Transporting Radioactive Material.

Management of Radioactive Materials and Wastes: Issues and Progress. Edited by S. K. Majumdar and E. Willard Miller. © 1985, The Pennsylvania Academy of Science.

Chapter Twenty

PREPARATION OF HOSPITALS FOR HANDLING VICTIMS OF RADIATION ACCIDENTS

Michael A. Vince, M.Sc.
Medical and Health Physicist, and
Radiation Safety officer
St. Luke's Hospital
Bethlehem, Pennsylvania, 18015

The Joint Commission on Accreditation of Hospitals (JCAH) requires that responsible hospital personnel prepare, as part of their emergency policies and procedures manual, a well-defined plan, based on the capabilities and geographical location of the hospital, for "emergency management of individuals who have actual or suspected exposure to radiation or who are radioactively contaminated. Such action may include radioactivity monitoring and measurement; designation and any required preparation of space for evaluation of the patient including as required, discontinuation of the air circulation system to prevent the spread of contamination; decontamination of the patient through an appropriate cleansing mechanism; and containment, labeling, and disposition of contaminated materials[1]".

The likelihood is very remote that, at most, more than one or two radiation accident victims would ever appear, as a result of an accident, at the doors to the emergency care unit of the average community hospital. Except in bizarrely imaginative scenarios, it is very difficult to realistically conceive of an accident, of any sort, involving radiation or radioactive materials. In considering possible scenarios, a transportation accident is commonly thought to have a higher probability of occurring compared to other kinds of accidents which might be imagined. The U.S. Department of Transportation estimates that in 1981 there were approximately 100 billion individual shipments of hazardous material in this country. That includes materials that are flammable, explosive, toxic, and corrosive, as well as radioactive. Of the hazardous substances transported, 2,500,000 shipments (0.0025%) were radioactive materials.[2,3] Because of the large numbers of trucks that crowd the roads, it is tempting to suspect that the probability for accidents involving radioactive shipments might be higher on the highway. In fact, this is the case. The actual number of commercial transportation accidents involving the shipment of radioactive materials reported in 1981 was 8, all highway. In only two of these were there releases of radioactivity. There have *never* been any reported injuries or fatalities due to the release of radiation or radioactive materials resulting from a transportation accident, including rail, plane, ship or truck.[3]

The worst-case scenarios one might invent and which conceivably might produce a multitude of radiation victims, would be in the realm of a major nuclear reactor disaster, a military accident involving a nuclear warhead, an accidental release of radioactive materials to the environment from a radioactive waste disposal site, or the like. Such extraordinary situations are extremely improbable; but the consequences would be serious enough should one ever occur, that it is this kind of event for which there is needed advanced planning. Adequate preparation does not necessarily imply a high probability for occurrence but rather the attainment of the understanding and capabilities that would be needed should ever such an emergency arise.

The perceived risks involved with anything "nuclear" or "radioactive" are exaggerated to enormously unrealistic proportions in the public sector. Of course, images of mushroom clouds and the never-to-be-forgotten devastations of Hiroshima and Nagasaki cannot be dismissed offhandedly. Given these as legitimate public concerns to be avoided forever it is, nevertheless, patently unreasonable to juxtapose impression on the public consciousness of similar consequences from peacetime and humanitarian applications of nuclear energy and its beneficial by-products. Yet, this is what is being done repeatedly by exploitative and uninformed individuals and groups seeking financial advantages, publicity, political votes, misplaced public commendation, or other selfserving gains. No small force in perpetrating this wave of public "misinformation" has been the press, magazine and visual media journalists, often less informed and driven to sensationalism by motives similar to those from whom they derive the fallacious opinions which they present[4]. This criticism is not to suggest that the need is also exaggerated for establishing well-planned courses of action to serve the public welfare in the eventuality of a radiation accident.

However, the view contrary to those who would lead us to believe that nuclear catastrophe is imminent suggests that it would be unwise to heavily invest the medical and financial resources of every hospital in the country in preparation to handle victims of every conceivable kind of radiation accident[5]. Therefore, an hierarchy of preparation, in any particular geographic region, is advocated. Associated with location near industries using, manufacturing or disposing of radioactive materials is an increased potential for radiation accidents. Within regional hierarchies, those hospitals near nuclear facilities should maintain the highest levels of preparation and planning. The average community hospital will not be in this kind of situation and, therefore, will require a relatively simpler radiation emergency plan.

The establishment of such a basic plan will not represent a financial burden for the community hospital. Supplies for contamination control are readily available in emergency care units and surgical suites.[5] Hospitals which offer nuclear medicine services already have the components of a plan, the necessary equipment and monitoring instrumentation as a basic part of their normal safety and quality assurance routines. The available facilities and capabilities of each hospital will, in most cases, dictate the extent to which they prepare to respond to a radiation emergency.

With this perspective in mind, this chapter will be devoted to a generalized discussion of the inter- and intraorganizational structure of hospitals for handling radiation emergencies of the kind suggested above as well as the isolated remote minor accident involving radiation. The general elements of hospital planning for radiation accidents have been discussed by Saenger[6] and a detailed protocol for handling the radioactive patient was presented by Leonard and Ricks[7]. Minor additions and emphasis to parts of these earlier works will be summarized, reflecting experiences gained in receiving simulated radioactively contaminated victims in drills at St. Luke's Hospital of Bethlehem, Pennsylvania. Two accidents were simulated involving mock radioactive materials over a two year period. One such "accident" was staged at A-B-E Airport, Lehigh County, in 1981 and the other in the Saucon Valley in 1983. It should be mentioned that in neither case was the release of radioactive material possible, in reality, as portrayed. In planning mock radiation accident drills for emergency care units and support staff, one is best-advised not to pay too much attention to the logic of how the release occurred but rather that there are victims who must be treated, decontaminated and evaluated for the necessity of continued medical care. The first staged accident was minor. There was a trauma victim with simulated external radiation contamination in his wounds, a victim with ingested radioactivity and one with "radioactive contamination" in the eyes. There were also four persons with external contamination simulated on their hands and feet. The second staged accident was major in that there were 33 victims involved, a majority of whom had simulated trauma. Of these, one had external radioactive contamination in wounds and four persons were contaminated externally but had not suffered trauma. Although the methods of release of the radioactive contamination were highly imaginative, it was felt that the situations were representative of what might be expected in terms of receiving victims at a hospital.

REGIONAL MANAGEMENT OF CONTAMINATED VICTIMS

A regional approach in organizing hospitals for the management of victims of radiation accidents implies several levels of preparedness consistent with the capabilities of the participating hospitals[2]. The National Council on Radiation Protection and Measurements recommends that "every plant, laboratory and hospital that handles radioactive materials shall have a detailed emergency plan defining the lines of authority, responsibilities, and functions of the assigned qualified individuals[8]." Facilities which handle relatively large quantities of radioactive materials and wastes (hereafter referred to as nuclear facilities) should have on-site medical and health physics activities for the immediate evaluation and decontamination of emergency victims before they are transferred to hospitals for definitive care. In general, there should be expressed agreements and/or contracts with hospitals located in reasonable proximity to nuclear facilities to which patients who require medical attention may be sent. Hospitals which can enter such agreements should have, as a minimum, a nuclear medicine service which requires State or U.S. Nuclear Regulatory licenses to operate and, therefore, possess radiation detection equipment and have personnel with demonstrable proficiency in their proper usage. It is reasonable to expect hospitals which fall into this category to have a higher level of preparedness in terms of capabilities and equipment for managing radioactive or exposed patients.

A generalized flow pattern of radiation accident victims in a regionally organized group of hospitals is given in Figure 1. It shows two kinds of radiation accidents: one occurring within a facility which handles large quantities of nuclear materials, e.g., a nuclear power plant, which might involve victims from within the facility as well as from the surrounding environs and, the other, an accident occurring remotely from causes unrelated to the nuclear facility. Most conceivable accidents would produce few victims whose exposures would probably be minimal.

However, should the accident approach disaster proportions in which a large number of victims are involved with possible high levels of contamination or exposure, the responsibility of the receipt and care of these patients would be distributed among the hospitals in the region. Victims of accidents occurring within a nuclear facility would receive first aid and be decontaminated at the facility. Trauma victims requiring more care than could be administered within the nuclear facility would be decontaminated, if possible, and transferred to the nearby *primary hospital* with nuclear medicine facilities contracted to handle such victims. If the number of victims would exceed the capacity of the primary hospital, selected patients would be transported to more remote hospitals, as required, depending on the kind of care needed. Those patients with suspected or actual radiation exposure or internally incorporated radioactivity could be transported to distant hospitals having nuclear medicine services or to an Emergency Radiation Center. Radiation Emergency Centers (Appendix A.1) located throughout the United States are especially equipped with trained personnel, radiation detection instrumentation and specialized capabilities to provide radiochemical analysis, gamma spectrometry, chromosome analysis (Appendix A.3), white cell transfusions, bone marrow transplantation, radioactive sample counting, whole body counting and reverse isolation facilities[5]. Some centers have mobile units (Appendix A.2) with which

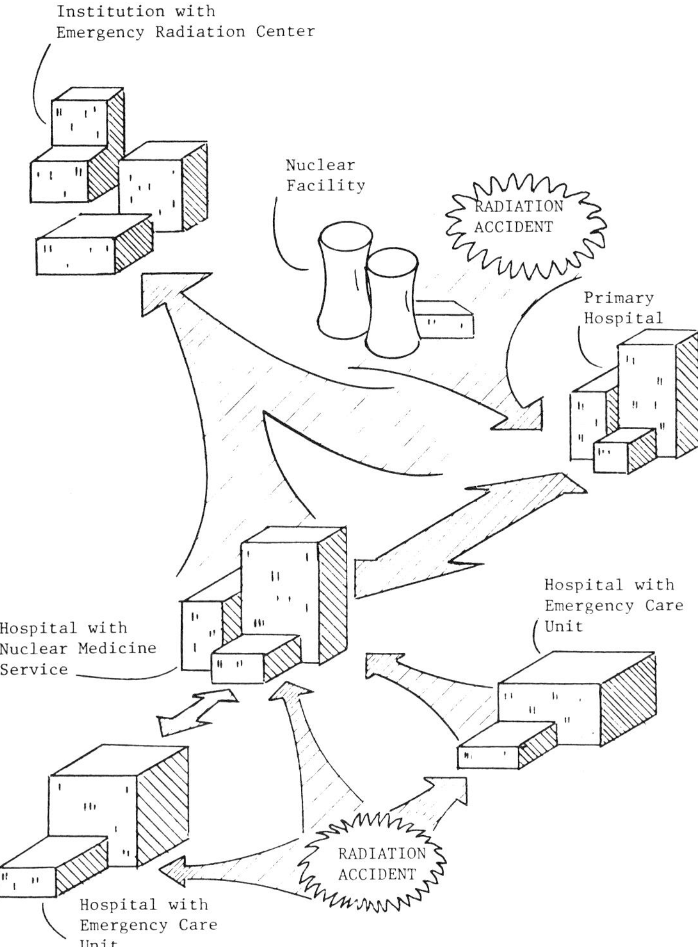

FIGURE 1. The regional management of radiation victims.

to lend assistance at the accident site or at the hospital requesting help. Hospitals should incorporate into their basic plan the names, addresses and telephone numbers of those facilities and agencies which would be able to lend support if needed. The availability of these services should be reconfirmed, through contact, annually.

After the initial evaluation of the patient's condition it may be prudent and necessary, despite the presence of trained personnel, to seek additional consultation. Calling in outside consultants may be indicated to verify the predicted consequences resulting from his internal contamination or external exposure and to plan the most appropriate course of action for the patient's extended care. A listing of available outside agencies and centers that could be utilized should be part of the Radiation Emergency Plan incorporated into the Hospital's Policy and Procedure Manual. It should include the telephone numbers and addresses of the appropriate Department of Energy Regional Coordinating Office - Interagency Radiological Assistance Plan or IRAP (Appendix A.4), the Inspection and Enforcement Regional Office of the U.S. Nuclear Regulatory Commission (Appendix A.5.), the State Bureau of Radiological Health, health or medical physicists from nearby institutions, and health physics personnel from nearby industrial or nuclear facilities.

The regional organization of hospitals to respond to radiation emergencies should be coordinated and implemented by state and local Emergency Medical Systems Councils or similar authorities having an overview of the needs of the region and the capabilities of its medical institutions.

COMMUNICATION

Primary hospitals, which have made agreements with nuclear facilities to be available as medical backup and to receive victims in the event of a large-scale emergency should be connected to the facility with direct communication links immuned to failure and overload. Community hospitals with nuclear medicine services should be contacted either by telephone or radio (if available) by the primary hospital or by emergency units (police, ambulance, fire and rescue squads) through a central emergency communications control center or directly. The exact nature of the communication linkup between emergency units and the hospitals in a region will depend on available equipment.

On notification that victims of a radiation accident will be arriving at an emergency department, as much advanced information as possible should be obtained by the person in charge. Of fundamental importance are the number of victims, the type and extent of their injuries, and the number known,or suspected, to be contaminated with radioactivity. Many emergency response units carry radiation monitoring instruments (survey meters or Geiger counters) and their personnel have been trained in their usage. If monitoring has been done at the accident site, victim status will probably be known and this information can be relayed to the hospital. If the accident occurs in a nuclear facility, the type of radioactive material and possibly the amount of radioactivity involved may also be known. Containers being transported in any type of vehicle (surface, air or sea) must be labeled with symbols and accompanied by documentation required by the Department of Transportation, giving the type and quantity of radioactive substance being carried[9].

If the accident involved only exposure to high level gamma rays or X rays and no potential contamination of victims by radioactive materials exists, that is essential information and should be communicated to the hospital as soon as it's known. This will influence the extent to which the emergency staff will prepare themselves and their department spaces for receiving the victims.

Communication of this nature implies a well-planned and understood protocol for emergency response units (police, ambulance, etc.), first in recognizing that radioactive materials are involved in the accident, and secondly, in handling the situation once that's known. The need for orientation programs and training of these units in this aspect of emergency services cannot be emphasized too strongly.

Because of the smallness of the probability that the average community hospital will ever see a radiation accident victim, placing the radiation emergency staff on call is unwarranted. Consequently, in the event of an actual radiation emergency, a hospital might be without trained personnel to handle the situation. Hospitals with nuclear medicine facilities and/or emergency care units should establish an Emergency Radiation Team, i.e., a staff of individuals trained in the handling and decontamination of radiation accident victims.

The basic radiation emergency plan of the hospital should include a tested telephone fan-out procedure for locating the personnel assigned to the Emergency Radiation Team in the event of an off-hour emergency. An estimate should be made of the minimum number of trained individuals needed to handle and supervise the course of action. These should include physicians, nurses, medical and health physicists, technologists, etc., depending on the designated responsibility of the institution in this kind of situation. The number of persons actually assigned to the Emergency Radiation Team should be, at least, four or five times that estimate. This will improve the chances of getting an adequate off-hour response should the occasion arise when it's needed. Small hospitals, not having nuclear medicine facilities, may or may not have radiation survey instruments and personnel knowledgeable in their use. They should have, as part of their basic radiation emergency plan, a list of telephone numbers of local individuals who may be called upon to aide in an emergency. Local Civil Defense Units are frequently the best resources.

VICTIM TYPES

There are three ways a victim may be affected in an accident involving a source(s) of ionizing radiation: by *irradiation,* by *external contamination,* and by *internal incorporation.*

When the victim has been *irradiated* by a concentrated flux of gamma or X rays, the radiation is effective only while he is actually in the radiation field. Out of the radiation flux, it has no effect on the victim's cells. There will be no residual radioactivity effecting the victim and radiation from the patient will not be detectable.

Tissue damage sustained during the exposure will not be seen immediately and the patient may remain asymptomatic if the exposure received was not very high. If the exposure was severe and acute, i.e., delivered in a relatively short period of time (a few hours or less), specific symptoms may be seen with the passage of time. The appearance of certain symptoms will also depend on whether the radiation absorbed dose was localized to one part of the body, e.g., the hand, or over the whole body. Familiarity with the "radiation syndrome," i.e., the appearance of dose-related symptoms with time, can be a useful gauge for physicians who may follow the patient over the full period of his recovery. The observance of these symptoms may be helpful in deciding the course of medical treatment for the patient and predicting the outcome. In most cases, the details of the "radiation syndrome" will have little bearing on the initial emergency care administered to the patient. Knowledge of the early symptoms due to acute exposure is helpful and should be recorded in the patient's chart if observed.

External contamination by a radioactive material in the form of a liquid, solid particle or dust can usually be eliminated quickly. It will be either attached to the patient's clothing, which can be removed, or on the patient's skin, which can be washed with mild soap and rinsed with water.

The third kind of situation, which represents a serious emergency in terms of the hazard to the patient, is the *internal incorporation* of the radioactive material either by inhalation, ingestion or entrance through an open wound. The radiation emitted from internally incorporated radionuclides can cause extensive cellular damage. Some elements, such as uranium, are also highly toxic, posing an additional threat. If possible, decorporation, i.e., removal of or blockage of the effects of internal radioactive contamination, should begin within one or two hours of the accident. It's possible that the emergency department personnel will not know which radionuclides are involved and identification may take days. Therefore, certain steps should be taken routinely to try to remove the most commonly encountered radionuclides[7]. Lincoln[13] has published detailed charts and references for this purpose which should be available to emergency department and emergency radiation team personnel. There are other guidelines available dealing with the appropriate techniques of handling internally incorporated radionuclides in the literature[8,13,14].

MINIMIZING THE RADIOACTIVE HAZARD

There are several precautions which should be taken to minimize the risk of radioactivity entering and contaminating the hospital. The easiest way to accomplish this is to leave the contamination at the accident site. Victims, whose medical conditions do not dictate otherwise, and if weather conditions permit it, should be stripped of their outer clothing, shoes and socks and wrapped in sheets or in blankets to contain any remaining radioactivity. These items should remain at the accident site in plastic bags or piled in a designated area until they can be properly monitored and disposed. Likewise, after the disposition of the victim is ascertained by the triage physician in the ambulance when it arrives and the patient is monitored for radioactivity by the Radiation Safety Officer, clothing and other items that may be contaminated should be removed from the victim and remain in the ambulance. Although these precautions are very effective, they do not guarantee that radioactivity will not enter the hospital with the patient.

Therefore, the entrance to the hospital through which the victims will enter should be evacuated of all persons except those few with specific duties related to receiving victims. If there is only one entrance to the emergency care unit, isolation may be impossible since it must remain available to receive victims of emergencies unrelated to the radiation accident. It may be possible to divide the single entrance into two entrance lanes to handle this situation. Traffic patterns can be controlled by roping off and labeling the route for the radiation accident victims with appropriate radiation warning signs and by utilizing security personnel for directing traffic. In some cases, it may be prudent to designate an area completely segregated from the emergency care unit, e.g., a morgue or a physical therapy unit, for receiving radiation accident victims, provided that it's at ground level and accessible. However, this is not recommended as a rule. The reason is that traumatized victims should not be separated from the equipment they may need to receive proper care.

The important idea to be maintained in planning is to isolate the entrance, decontamination and treatment areas to prevent the spread of radioactive contamination. As a corollary, methods of isolation and protection should be utilized which can be easily dismantled and will not hinder cleanup afterward. Excellent techniques for accomplishing this are discussed in detail by Leonard and Ricks.[7] The materials with which the designated areas are prepared and materials and garments that will be needed by the Radiation Emergency Team should be readily available and maintained in an adjacent or easily accessible storage area.

The decontamination area should be a separate room(s) large enough to hold more than one victim and the necessary medical personnel. Air from this area should not be ventilated to the rest of the hospital to prevent the spread of contamination. It may be necessary to shut off the entire ventilation system to accomplish this. The area should be immediately cleared of all unnecessary gear to maximize the available space. Items that cannot be removed should be draped with sheets to protect them from accidental contamination. Plastic or paper covering should be taped to the floor to prevent contamination and aid in cleanup. Part of the decontamination area, possibly an adjacent room, should be designated as a "clean" area and so marked to have a place to put decontaminated patients. Persons and equipment entering this area should be thoroughly monitored by an assigned individual with a survey instrument to assure that it remains free of radioactivity.

A description of a decontamination tray, made of galvanized iron or plastic, which is placed on the stretcher in which the patient lies, is described by Leonard and Ricks.[7] It's used to wash the contaminant away from the patient and to collect the water used. All hospitals do not have them and they are quite expensive to buy. An equally serviceable decontamination trough can be made with a sheet of flexible, waterproofed material draped over the raised sidebars of a trauma room stretcher which has a top that can be adjusted higher at one end than the other. The patient, lying in this makeshift trough can be washed with water which runs into a tub placed at the lower end. Several large plastic washtubs should be available for collecting contaminated water. Most external contamination is removable with small amounts of water so a few three gallon containers of water should be adequate.

Those members of the Emergency Medical Staff and of the Radiation Emergency Team who will physically participate in the decontamination procedures, including the Triage Physician, Triage Nurse, and the Radiation Safety Officer who enter the ambulance, shall wear protective outer clothing comprised of surgical gloves, gowns, masks, caps and *waterproof* shoe covers. It's helpful to have complete outfits assembled in packets, according to size, containing everything one needs. Boxes of small, medium and large size gloves should be available for those individuals attending the patient so that they may be changed frequently. Double gloving is recommended, i.e., an initial pair of gloves should be donned and taped to the wrists and the second pair, which are the ones exchanged as necessary, worn over them.

A film badge, with the team members name on it, should be attached to the street clothes or uniform under the protective clothing. This will be helpful in determining the total radiation exposure received during the entire operation. In addition, it's helpful for each member to wear a personal dosimeter attached to the outside gown at neck level which can be easily read and monitored by the individual or the Radiation Safety Officer as a safeguard against overexposure.

Finally, it is advised that all members of the team use the restroom facilities *before* the first victim arrives since they will not be allowed to leave the decontamination room once treatment and decontamination procedures begin.

HANDLING THE RADIOACTIVE PATIENT

The ambulance traffic arriving from the accident scene should be controlled by security personnel. Ambulances will be directed to the entrance of the area designated for receiving the radiation accident victims. They will be met by the triage team. The triage physicians's primary task is to determine if the victim has sustained trauma, not radiation-related, and the extent and seriousness of his injuries. In the meantime, the Radiation Safety Officer (R.S.O.) can determine if the victim has been radioactively contaminated with portable instrumentation. If contaminated clothes are still on the victim, they should be removed and remain in the ambulance unless the victim's need for immediate medical attention dictates otherwise. When a choice must be made by the emergency physician, the immediate care of serious medical problems takes *priority* over any attempts to decontaminate the patient. The radioactive patient shall be transferred from the ambulance to the stretcher and transported to the decontamination area by ambulance personnel.

There, after the patient is stabilized, decontamination may begin. The patient is carefully monitored with portable radiation detection instrumentation by the R.S.O. In the meantime, swab samples are taken from the patient's mouth, nose and ears and deposited in glass test tubes which are, then, labeled with the patient's identification and area from which the sample was taken. These are placed in a lead container and moved to the nuclear medicine department for analysis. A recording nurse should be on hand to manage the "bookkeeping" that's necessary. It's helpful to use a body diagram (prone and supine) of the patient on which contaminated areas can be indicated. Swab samples should also be taken of these areas.

Open wounds should be decontaminated first to reduce uptake of the radioactivity internally. Once the patient is stabilized, decontaminated and an assessment of his condition has been made, the patient may be moved to the "clean" area for further disposition. A complete description of the actions to be taken by the emergency staff and radiation emergency team should be clearly outlined in the Radiation Accident Protocol for the emergency department of the hospital. This should be simple and concise and should contain procedures for notifying emergency personnel, for obtaining on-site information, for preparing the designated radiation emergency are and the decontamination team, initial actions when the patient arrives, decontamination of the patient, removal of the patient from the decontamination area, procedures for releasing the decontamination team, and cleanup.[7]

CLEANUP

The importance of containment of radioactive contamination cannot be emphasized too strongly. This is a special consideration during cleanup. The Radiation Safety Officer and his assistants on the Radiation Emergency Team are responsible for surveying and decontaminating, if necessary, the ambulance, the ambulance attendants and the route from the ambulance entrance to the decontamination area. These procedures should be done expeditiously so the ambulance may be released to service as quickly as possible.

All materials, wastes, clothing and containers used during the incident must be monitored and disposed properly. Finally, each member of the decontamination team must remove their protective outer clothing, one piece at a time, beginning with the shoe covers and ending with the pair of inner gloves. Each article of clothing should be monitored with a survey meter as it is removed. Finally, the team member is monitored carefully to assure no contamination has been inadvertently transferred to his/her person. The last person to discard his protective clothing is the R.S.O. while being continuously monitored by an assistant.

DRILLS

Since radiation accidents are so infrequent, the hospital radiation accident protocol should be practiced. A drill simulating a radiation accident should be held at least once each year. This can be conveniently combined with hospital disaster drills. On a separate occasion, a telephone drill should be held to assure that the telephone fan-out procedure is effective. The handling of a radiation accident victim is a painstakingly time-consuming procedure. Except for handling life-threatening medical conditions it cannot be rushed. Each action must be carefully conceived and executed to assure containment of the radioactivity. An appreciation of the special precautions which must be taken can be gained only through well-conceived accident drills.

APPENDIX A.[5]

1. *Radiation Emergency Centers*

Hospital of the University of Pennsylvania
Philadelphia, Pennsylvania

University of Pittsburgh, Department of Radiation Health and Presbyterian University Hospital*
Pittsburgh, Pennsylvania

Oak Ridge Associated Universities Hospital*
Oak Ridge, Tennessee

Hospital of the Medical Research Center*
Brookhaven National Laboratory
Upton, Long Island, New York

Northwest Memorial Hospital
Chicago, Illinois

University of Cincinnati Medical Center
Cincinnati, Ohio

Los Alamos Medical Center
Los alamos, New Mexico

Kadlec Hospital
Richland, Washington

Donner Pavilion
Cowell Memorial Hospital
Berkeley, California
*These institutions also have whole body counting facilities.

Additional whole body counting facilities which are not Radiation Emergency Centers are located at

St. Luke's Hospital Center
Columbia University
New York, New York

M.D. Anderson Hospital
University of Texas Cancer Center Systems
Houston, Texas

National Naval Medical Center
Bethseda, Maryland

Argonne National Laboratory
Argonne, Illinois

Pacific Northwest Laboratories
Richland, Washington

Lawrence Livermore National Laboratory
Livermore, California

2. *Mobile Units*

Radiation Management Corporation
Hospital of the University of Pennsylvania
Philadelphia, Pennsylvania

Radiation Management Corporation
Northwest Memorial Hospital
Chicago, Illinois

Helgeson Nuclear Services, Inc.
Pleasonton, California

3. *Chromosome Analysis*

Department of Pathology
University of Pennsylvania School of Medicine
Philadelphia, Pennsylvania 19104

Laboratory of Human Genetics
Northwestern University Medical School
Chicago, Illinois 60611

Graduate School of Public Health
University of Pittsburgh
Pittsburgh, Pennsylvania 15213

Oak Ridge Associated Universities
P.O. Box 117
Oak Ridge, Tennessee 37830

Brookhaven National Laboratories
Upton, Long Island
New York, NY 11973

4. Department of Energy Coordinating Office
Interagency Radiological Assistance Plan (IRAP)

Regional Coordinating Office	*Post Office Address*	*Telephone for Assistance*
Brookhaven Area Office	Upton, Long Island, NY 11973	(516)345-2200
Oak Ridge Operations Office	P.O. Box E, Oak Ridge, TN 37380	(615)576-1005 or 576-7885
Savannah River Operations Office	P.O. Box A, Aiken, SC 29801	(803) 725-3333
Albuquerque Operations Office	P.O. Box 5400 Albuquerque, NM 87115	(505)264-4667
Chicago Operations Office	9800 S. Cass Avenue Argonne, IL 60439	(312)972-4800
Idaho Operations Office	550 Second Street Idaho Falls, ID 83401	(208)526-1515
San Francisco Operations Office	1333 Broadway Oakland, CA 94612	(415)273-4237
Richland Operations Office	P.O. Box 550 Richland, Washington 99352	(509)376-7381

5. United States Nuclear Regulatory Commission
Inspection and Enforcement Regional Offices

Region	*Address*	*Daytime*	Telephone Nights and *Holidays*
I Connecticut, Delaware, District of Columbia, Maine, Maryland, Massachusetts, New Hampshire, New Jersey, New York, Pennsylvania, Rhode Island, Vermont	Region I, USNRC Office of Inspection and Enforcement 631 Park Avenue King of Prussia, PA 19406	(215)337-5000	(215)337-5000
II Alabama, Florida, Georgia, Kentucky, Mississippi, North Carolina, Panama Canal Zone, Puerto Rico, South Carolina, Tennessee, Virginia, Virgin Islands, West Virginia	Region II, USNRC Office of Inspection and Enforcement 101 Marietta Street Suite 3100 Atlanta, Georgia 30303	(404)221-4503	(404)221-4503
III Illinois, Indiana, Iowa, Michigan, Minnesota, Missouri, Ohio, Wisconsin	Region III, USNRC Office of Inspection and Enforcement 799 Roosevelt Road Glenn Ellyn, IL 60137	(312)858-2660	(312)739-7711

IV			
Arkansas, Colorado, Idaho, Kansas, Louisiana, Montana, Nebraska, New Mexico, North Dakota, Oklahoma, South Dakota, Texas, Utah, Wyoming	Region IV, USNRC Office of Inspection and Enforcement 611 Ryan Plaza Drive, Suite 1000 Arlington, Texas 76012	(817)334-2841	(817)334-2841
V			
Alaska, Arizona, California, Hawaii, Nevada, Oregon, Washington, and U.S territories and possessions in the Pacific	Region V, USNRC Office of Inspection and Enforcement 1990 N. California Blvd. Suite 202 Walnut Creek, Calif. 94596	(415)486-3141	(415)273-4237

BIBLIOGRAPHY

1. Emergency Services, *In:Accreditation Manual for Hospitals*. 1983. Joint Commission on Accreditation of Hospitals, Chicago.
2. *A Review of U.S. Accident Experience Involving the Transportation of Radioactive Materials (RAM) 1971-1980.* 1983. Sandia National Laboratories, P.O. Box 5800, Albuquerque, NM 87185
3. Cooley, L. (1981). Transportation of radioactive materials in the United States, 1981. Presented at symposium on *The Preparation of Regional and Community Hospitals for Handling Victims of Radiation Accidents* held at Moravian College, Bethlehem, PA and cosponsored by the American Association of Physicists in Medicine and St. Luke's Hospital of Bethlehem. (Proceedings to be published.)
4. Taylor, L.S. 1983. What the public is told, and what it should know about radiation hazards. HPS Newsletter, Vol. XI, No. 7, Bethseda, MD.
5. Miller, K.L. and W.E. DeMuth. 1983. Handling radiation emergencies: no need to fear. J. Emerg. Nurs. 9:141-144.
6. Saenger, E.L. 1963. Hospital planning to combat radioactive contamination. JAMA 185:578-581.
7. Leonard, R.B. and R.C. Ricks, 1980. Emergency department radiation accident protocol. Ann. Emerg. Med. 9:462-470.
8. National Council on Radiation Protection and Measurements. 1980. Management of persons accidentally contaminated with radionuclides. Recommendations of NCRP, Report 65. Washington, D.C.
9. *A Review of the Department of Transportation (DOT) Regulations for Transportation of Radioactive Materials. 1977.* U.S. DOT, Materials Transportation Bureau, Washington, D.C.

10. Vince, M.A. (1981). Physics of a radiation accident. Presented at symposium on *The Preparation of Regional and Community Hospitals for Handling Victims of Radiation Accidents* held at Moravian College, Bethlehem, PA and cosponsored by the American Association of Physicists in Medicine and St. Luke's Hospital of Bethlehem. (Proceedings to be published).

11. Shapiro, J. 1981. *Radiation Protection, A Guide for Scientists and Physicians.* Harvard University Press, Cambridge, Massachusetts.

12. Lincoln, T.A. 1976. Importance of initial management of persons internally contaminated with radionuclides. *Am. Ind. Hyg. Assoc. J. 37:*16-21.

13. *Manual of Early Medical Treatment of Possible Radiation Injury.* 1978. Safety Series #47, International Atomic Energy Agency, Vienna.

14. Finkel, A.J. and E.Z. Hathaway. 1956. Medical care of wounds contaminated with radioactive materials. *JAMA 161*:121-126.

PART 5
Radiation Standards, Environment and Public Health

Because of the increasing possibility of radioactive contamination there has been a growing scientific literature on the subject. Few subjects have evoked more public attention than the risks of ionizing radiation on the individual. It is thus important that this serious topic be dealt with in a credible manner.

The initial chapter of this section develops a mathematical model for calculating individual doses associated with the release of radionuclides in the biosphere. The methodology focuses on the environmental pathways that are typically the most significant dose contributing routes. This is followed by a chapter that considers radiation protection standards and radiation risks.

Environmental monitoring of low-level radioactive materials is now a basic procedure in most countries where radioactive waste is produced. The International Committee on Radiological Protection has defined three broad objectives for environmental monitoring. These include, assessment of the actual or potential exposure of humans to radioactive materials or radiation in the environment, scientific investigation related to assessment of exposure, and improved communication to the general public.

While emphasis is sometimes placed on the problem of contamination of radioactive materials, the use of radioisotopes has provided great benefits to society. A wide variety of beneficial applications are now found in industry, agriculture, biomedical research, and medicine. For example, radioisotopes in industry provide quick answers to the flow of materials in enclosed systems, such as the flow of lubricants in engines, or in agriculture where radioisotopes have provided many answers regarding soils, plants, insects, microorganisms and animal nutrition. Radiation is playing an increasing role in diagnosing health problems. To illustrate, the use of very small amounts of radioactive carbon -14 in a clinical laboratory procedure provides a quick determination of bacterial infections in a patient.

The Three Mile Island nuclear accident provided a real-life situation to test how surrounding populations react to such a disaster. While the contamination of persons by radiation was extremely minimal, substantial anxiety was created and resultant apprehension persisted in the area. This chapter evaluates the psychological stresses and mental health problems that developed after the accident.

The final chapter considers some environmental and biological effects of radiation. In the 1950's and early 1960's the testing of nuclear bombs in the atmosphere created concern that the ozone of the stratosphere would be reduced. If this were to occur the number of incidents of cancer would increase significantly in the world. As a consequence atmospheric testing of nuclear bombs is now banned. The effects of radioactive wastes on the physical and biological environment is still in its initial stage. Much research is needed to determine the effects over a long time span.

Management of Radioactive Materials and Wastes: Issues and Progress. Edited by S. K. Majumdar and E. Willard Miller. © 1985, The Pennsylvania Academy of Science.

Chapter Twenty-One

METHODOLOGY FOR CALCULATING DOSES FROM RADIOACTIVITY IN THE ENVIRONMENT

James E. Fairobent[1], Edward F. Branagan, Jr.[2], Ph.D. and Frank J. Congel[2], Ph.D.
[1]Meteorology and Effluent Treatment Branch
[2]Radiological Assessment Branch
Division of Systems Integration
United States Nuclear Regulatory Commission
Washington, D.C. 20555

I. INTRODUCTION

The following sections briefly describe mathematical methods and procedures for calculating individual doses associated with the release of radionuclides to the biosphere. In particular, the models are derived and described to enable the reader to estimate individual doses from the primary environmental pathways as shown in Figure 1. The primary pathways (Figure 1, thick arrows) represent the most significant routes that radionuclides can follow in ultimately reaching individual human beings. Other pathways (Figure 1, thin arrows) would typically contribute less than 1% of the doses through the represented pathways. The dose equations are arranged so that the reader can enter necessary basic environmental data and perform the calculation. For flexibility, the equation derivation and parameters are briefly described to allow the reader to modify the equations when necessary.

II. GENERALIZED DOSE MODEL

All of the dose commitment models (hereinafter referred to as dose models) described in this chapter can be expressed as a linear multiplicative array as follows:

$$\text{Dose} = \frac{\text{Source}}{\text{Term}} \times \frac{\text{Dispersion}}{\text{Factor}} \times \frac{\text{Pathway}}{\text{Factor}} \times \frac{\text{Usage}}{\text{Factor}} \times \frac{\text{Exposure}}{\text{Duration}} \times \frac{\text{Dose}}{\text{Factor}} \tag{1}$$

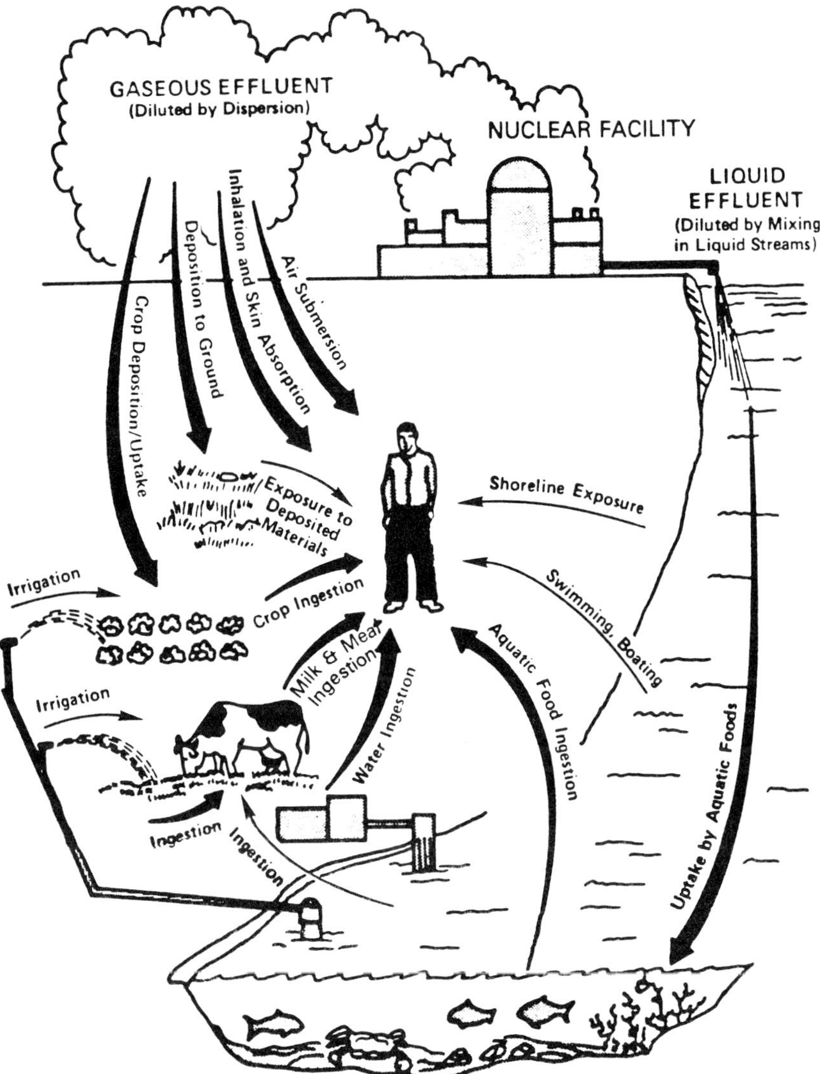

FIGURE 1. Potential pathways of exposure.

Each of the "blocks" in this array represents the results of a separate analysis. The desired units of the dose will determine the units of each one of the blocks. For this discussion, the units for dose will be given in millirems.

a—*Source Term*

The phrase, Source Term, is used to describe the quantity of radioactive material released to the environment. It is most commonly expressed in units of curies per unit time but is sometimes stated as the total activity released (curies). Because this chapter describes methods for calculating doses to individuals the amount of activity released is assumed to be given. In addition, the units for the source term will be fixed as curies per second (Ci/sec) averaged over the duration of the release.

b—*Dispersion Factor*

The phrase, Dispersion Factor, is used to relate the Source Term to the concentration in the environment at the location of interest. For atmospheric releases, it has units of Ci/m^3 per Ci/sec, or sec/m^3. Specific methods for estimating atmospheric dispersion are described in Section III.

c—*Pathway Factor*

The phrase, Pathway Factor, is used to describe the transfer of a radionuclide from one environmental compartment (e.g., the concentration of a radionuclide in air or water, or the deposition rate) to the last environmental compartment before it enters a human being. The pathway factor allows for the buildup and removal of radionuclides in the environment. The pathway factor is considered to be unity for three of the primary pathways—submersion, inhalation, and ingestion of drinking water. For food contaminated by airborne effluents or liquid effluents, the units for the pathway factor are pCi/kg of food per pCi/m^2-sec deposited on the ground, or pCi/kg of food per pCi/ℓ of water, respectively. Specific methods for calculating pathway factors are described in Section IV.

d—*Usage Factor*

The phrase, Usage Factor, is used to describe the quantities of air inhaled or food ingested at the location of an individual. Recommended values for the usage factor for a maximally exposed individual (that is, a hypothetical individual potentially subject to maximum exposure) are given in Table 1.

TABLE 1

Usage Factors for the Maximum Exposed Individual In Lieu of Site-Specific Data[a]

	Age Group			
Pathway	Infant	Child	Teen	Adult
Fruits, vegetables & grain and leafy vegetables (kg/yr)	—	550	670	580
Milk (ℓ/yr)	330	330	400	310
Meat & Poultry (kg/yr)	—	41	65	110
Fish (fresh or salt) (kg/yr)	—	6.9	16	21
Other seafood (kg/yr)	—	1.7	3.8	5
Drinking water (ℓ/yr)	330	510	510	730
Inhalation (m^3/yr)	1400	3700	8000	8000

[a] Values were taken from Table E-5 of USNRC (1977).

e—*Exposure Duration*

The phrase, Exposure Duration, is the total time over which exposure occurs, expressed as fraction of a year, and in the units of years.

f—*Dose Factor*

The phrase, Dose Factor, is used to describe the quantity of radiation absorbed by a tissue weighted by modifying factors to take into account the type of radiation (e.g., alpha, beta or gamma). Dose factors for exposure via the primary pathways—submersion, inhalation and ingestion—are contained in USNRC (1977). The units for dose factors are as follows: submersion, mrem/year per pCi/m^3; inhalation, mrem per pCi inhaled; and ingestion, mrem per pCi ingested.

III. ATMOSPHERIC DISPERSION

The most common model used to describe atmospheric dispersion is the Gaussian plume model, which assumes that effluent material is normally distributed around the plume centerline. The distribution of material within the plume can then be represented by characteristic standard deviations (σ's) in 3-dimensions (referred to as the x-, y-, and z-axes) around the plume centerline. Although the Gaussian model is widely used because of the relative ease of calculation, the model is based on fundamental concepts of atmospheric turbulent diffusion, and provides results which are in reasonable agreement with experimental diffusion data.

Considering a release from a continuous point source, which is transported in the along-wind direction (x-axis) at a mean wind speed, $\bar{u}$, and integrated over the time interval of release, the concentration of material at a particular location at the ground along the plume centerline (downwind distance, x) can be represented by the expression

$$\bar{\chi}(x,0,0) = Q(\pi\sigma_y\sigma_z\bar{u})^{-1} \exp \left(\frac{H^2}{-2\sigma_z^2} \right) \tag{2}$$

where, $\quad \bar{\chi} = $ average concentration (Ci/m^3)
$\quad\quad Q = $ source strength per unit time (Ci/sec)
$\quad\quad \sigma_y = $ standard deviation of plume concentration in the horizontal (y-axis)(m)
$\quad\quad \sigma_z = $ standard deviation of plume concentration in the vertical (z-axis)(m)
$\quad\quad \bar{u} = $ representative mean wind speed (m/sec)
$\quad\quad H = $ the height of the plume centerline, usually determined from the sum of the physical stack height and plume rise minus any increase in terrain between the release point and receptor (m)

This equation assumes that diffusion in the along-wind direction is small compared to transport and can be neglected, and that total reflection of the plume takes place at the earth's surface. For convenience, both sides of this equation are divided by Q to derive a relative concentration parameter, $\bar{\chi}/Q$ (with units of Ci/m^3 per Ci/sec or sec/m^3) which can be determined solely from available meteorological information.

Releases from free-standing stacks which are at least 2.5 times the heights of adjacent or nearby structures are usually assumed to be elevated. Releases from building penetrations or vents are usually assumed to be at ground-level. For a continuous point source release at ground level, the relative concentration of material at a particular location at the ground along the plume centerline (downwind distance, x) can be represented by

$$\bar{\chi}/Q(x,0,0) = (\pi\sigma_y\sigma_z\bar{u})^{-1} \tag{3}$$

The selection of appropriate σ_y and σ_z values and a representative mean wind-speed require some interpretation of available meteorological data. Figures 2 and 3 present σ_y and σ_z values as functions of atmospheric stability class and downwind data. Atmospheric stability is not measured directly, but inferred from other measurements and observations. Two of the most common indicators of atmospheric stability class are vertical temperature gradient (delta-T) and the standard deviation of horizontal wind direction (sigma-theta). Tables 2 and 3 present stability classifications based on these indicators. The delta-T method is most useful for estimating atmospheric stability class during low wind speed, stable conditions. Delta-T is a poor indicator of unstable conditions, and should not be considered the best stability indicator for evaluating diffusion from elevated release points. Sigma-theta is a calculated parameter which is dependent on the response characteristics of the wind vane. Sigma-theta is only useful for wind speeds greater than 1.5 m/sec. When delta-T measurements and sigma-theta calculations are not available, atmospheric stability class may be inferred objectively by considering cloud cover and height, solar angle (a function of

TABLE 2

*Classification of Atmospheric Stability by Vertical Temperature Difference**

Stability Classification	Pasquill Categories	Temperature Change with Height ($= C/100$ m)
Extremely unstable	A	$\Delta T/\Delta z \leq -1.9$
Moderately unstable	B	$-1.9 < \Delta T/\Delta z \leq -1.7$
Slightly unstable	C	$-1.7 < \Delta T/\Delta z \leq -1.5$
Neutral	D	$-1.5 < \Delta T/\Delta z \leq -0.5$
Slightly stable	E	$-0.5 < \Delta T/\Delta z \leq 1.5$
Moderately stable	F	$1.5 < \Delta T/\Delta z \leq 4.0$
Extremely stable	G	$4.0 < \Delta T/\Delta z \leq$

*USNRC (1972)

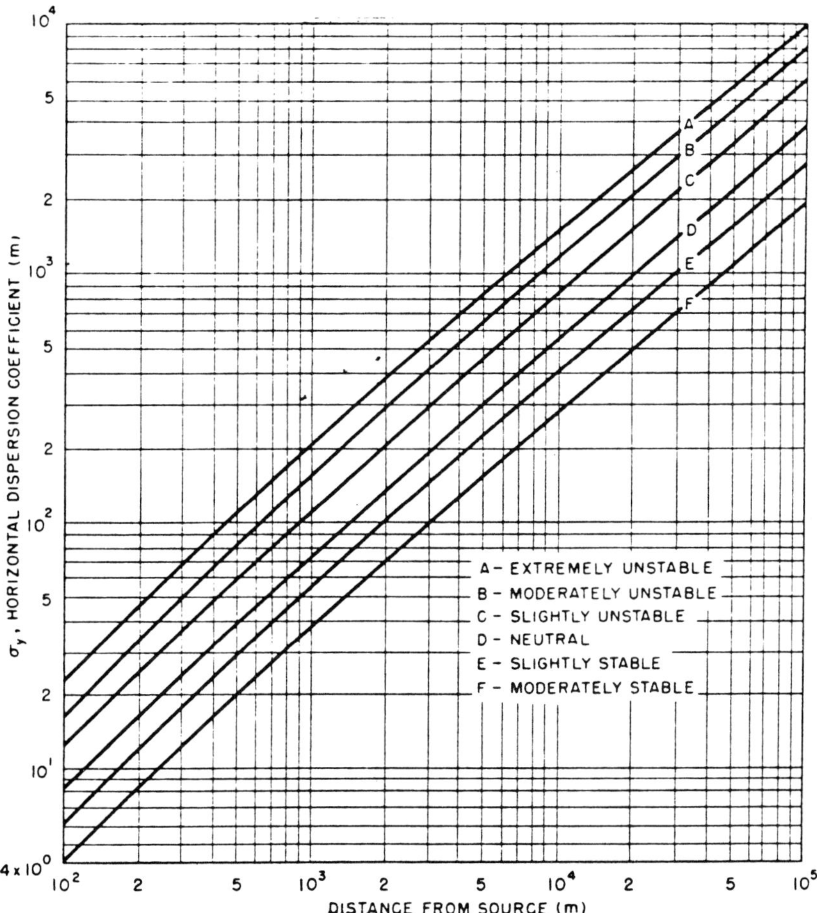

FIGURE 2. Lateral diffusion without meander and building wake effects, σ_y, vs. downwind distance from source for Pasquill's turbulence types (atmospheric stability) (Slade (1968))

$$\sigma_y(G) = \frac{2_\sigma}{3}y \ (F)$$

time, date, and location), and wind speed. Table 4 presents stability classifications based on these conditions.

Wind speeds should be representative of the height of release. Generally, for releases considered at ground-level, a measurement at the 10m level will suffice for diffusion estimates. For elevated releases, wind speed measurements near the release height are most appropriate. The atmosphere is seldom truly motionless, although wind speed may be lower than that necessary to initiate instrument response. When wind speed is too low to be measured (referred to as a "calm" condition), an arbitrary value such as 0.25 m/sec may be used in Equations 2 and 3.

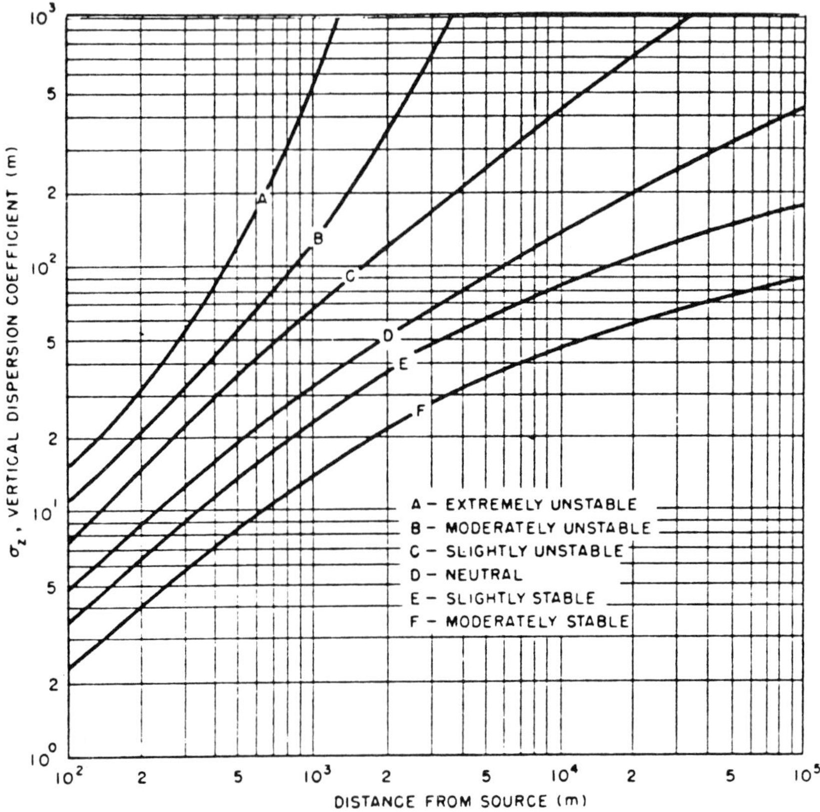

FIGURE 3. Vertical diffusion without meander and building wake effects, σ_z, vs. downwind distance from source for Pasquill's turbulence types (atmospheric stability) (Slade (1968))

$$\sigma_z(G) = \frac{3_\sigma}{5}z \; (F)$$

Meteorological parameters are constantly varying with time, and instantaneous or near-instantaneous measurements may provide non-representative or misleading information. Meteorological measurements used in short-term diffusion estimates should be averaged for at least 15 minutes and not longer than one hour.

The Gaussian plume model described above is independent of wind direction. The plume is assumed to be transported in a straight-line in the downwind direction. (Wind direction is reported as the direction *from* which the wind is blowing, and the downwind direction will be 180° opposite of the wind direction.) As wind direction changes, plume location will change and consideration will have to be made for these changes in determining integrated doses. Wind direction may also be significantly influenced by prominent terrain

TABLE 3

Classification of Atmospheric Stability by Standard Deviation of Horizontal Wind Direction (σ_θ)*

Stability Classification	Pasquill Categories	σ_θ** (degrees)
Extremely unstable	A	$\sigma_\theta \geq 22.5$
Moderately unstable	B	$22.5 > \sigma_\theta \geq 17.5$
Slightly unstable	C	$17.5 > \sigma_\theta \geq 12.5$
Neutral	D	$12.5 > \sigma_\theta \geq 7.5$
Slightly stable	E	$7.5 > \sigma_\theta \geq 3.8$
Moderately stable	F	$3.8 > \sigma_\theta \geq 2.1$
Extremely stable	G	$2.1 > \sigma_\theta$

*Use of abc to represent atmospheric stability when wind speeds are less than 1.5 m/s should be substantiated. If abc is to be used as an indicator of vertical diffusion (atmospheric stability), adjustments to the sampling interval may be needed to eliminate wind fluctuations in the horizontal which do not occur in the vertical, especially during nighttime conditions. USNRC (1972).

**Determined for a 15 minute to one hour period for horizontal diffusion.

TABLE 4

*Meteorological Conditions Defining Pasquill Turbulence Types**

A: Extremely unstable conditions   D: Neutral conditions**
B: Moderately unstable conditions   E: Slightly stable conditions
C: Slightly unstable conditions   F: Moderately stable conditions

Surface wind speed, m/sec	Daytime insolation			Nighttime conditions***	
	Strong	Moderate	Slight	Thin overcast or > 3/8 cloudiness	≤ 3/8 cloudiness
<2	A	A—B	B		
2—3	A—B	B	C	E	F
3—5	B	B—C	C	D	E
5—6	C	C—D	D	D	D
<6	C	D	D	D	D

*From F.A. Gifford, Turbulent Diffusion Typing Schemes: A Review, *Nucl. Saf.*, 17(1): 71(1976)

** Applicable to heavy overcast day or night.

***The degree of cloudiness is defined as that fraction of the sky above the local apparent horizon that is covered by clouds.

features such as river valleys, and wind direction may also vary in coastal zones under the influence of sea- and lakebreeze circulations.

IV. PATHWAYS OF EXPOSURE

A. *Airborne Effluents*

1. Plume Exposure

The simplest application of the generalized dose model is for exposure to dispersed noble gases. Exposure results from the penetration of the body and/or skin by gamma rays and beta particles produced by the radioactive gases. For this situation, the "pathway factor" is unity and the generalized equation becomes:

$$D = \frac{10^{12} \text{ pCi}}{\text{Ci}} \times Q \times \chi/Q \times f \times R_s \qquad (4)$$

where
D = total body dose, mrem
Q = source strength, Ci/sec
χ/Q = atmospheric dispersion factor, sec/m³
f = fraction of year exposure takes place, year
R_s = noble gas submersion dose factor from Table 5, $\dfrac{\text{mrem/yr}}{\text{pCi/m}^3}$

TABLE 5

Dose Factors for Exposure to a Semi-Infinite Cloud of Noble Gases

Nuclide	Beta Dose to Skin, $\dfrac{\text{mrem-m}^3}{\text{pCi-yr}}$	Gamma Dose to Whole Body, $\dfrac{\text{mrem-m}^3}{\text{pCi-yr}}$
Kr-83m	—	7.56E-08
Kr-85m	1.46E-03	1.17E-03
Kr-85	1.34E-03	1.61E-05
Kr-87	9.73E-03	5.92E-03
Kr-88	2.37E-03	1.47E-02
Kr-89	1.01E-02	1.66E-02
Kr-90	7.29E-03	1.56E-02
Xe-131m	4.76E-04	9.15E-05
Xe-133m	9.94E-04	2.51E-04
Xe-133	3.06E-04	2.94E-04
Xe-135m	7.11E-04	3.12E-03
Xe-135	1.86E-03	1.81E-03
Xe-137	1.22E-02	1.42E-03
Xe-138	4.13E-03	8.83E-03
Ar-41	2.69E-03	8.84E-03

*From Appendix B of USNRC (1977).

2. Inhalation

Calculation of doses due to inhalation of radioactive materials is performed using the following equation:

$$D_o^A = \frac{10^{12} \text{ pCi}}{\text{Ci}} \times Q \times \chi/Q \times R_o^A \times f \tag{5}$$

Where Q, χ/Q, and f are defined above and

D_o^A = organ O dose for recptor age A, mrem

R_o^A / organ O inhalation dose factor for receptor age A, mrem/yr per pCi/m^3

Note that the values of R_o^A selected for solving specific problems must correspond to the characteristics of the individuals of interest. The values of R_o^A in Table 6 are equal to the product of the pathway factor, usage factor and the dose factor. Values of the individual factors can be found in USNRC (1977).

TABLE 6

Pathway Dose Factors for Inhalation, mrem/yr pCi/m^3

Nuclide	Critical Organ	Infant	Child	Teen	Adult
H-3	T. Body*	6.5E − 04	1.1E − 03	1.3E − 03	1.3E − 03
C-14	Bone	2.6E − 02	3.6E − 02	2.6E − 02	1.8E − 02
P-32	Bone	2.0E + 00	2.6E + 00	1.9E + 00	1.3E + 00
Cr-51	GI-LLI	3.6E − 04	1.1E − 03	3.0E − 03	3.3E − 03
Mn-54	Lung	1.0E + 00	1.6E + 00	2.0E + 00	1.4E + 00
Fe-55	Lung	8.7E − 02	1.1E − 01	1.2E − 01	7.2E − 02
Fe-59	Lung	1.0E + 00	1.3E + 00	1.5E + 00	1.0E + 00
Co-58	Lung	7.8E − 01	1.1E + 00	1.3E + 00	9.3E − 01
Co-60	Lung	4.5E + 00	7.1E + 00	8.7E + 00	6.0E + 00
Ni-63	Lung	2.1E − 01	2.8E − 01	3.1E − 01	1.8E − 01
Zn-65	Lung	6.5E − 01	1.0E + 00	1.2E + 00	8.6E − 01
Sr-89	Lung	2.0E + 00	2.2E + 00	2.4E + 00	1.4E + 00
Sr-90	Bone	4.1E + 01	1.0E + 02	1.1E + 02	9.9E + 01
Y-91	Lung	2.5E + 00	2.6E + 00	2.9E + 00	1.7E + 00
Mo-99	Lung**	1.4E − 01	1.4E − 01	2.7E − 01	2.5E − 01
I-131	Thyroid	1.5E + 01	1.6E + 01	1.5E + 01	1.2E + 01
I-133	Thyroid	3.6E + 00	3.8E + 00	2.9E + 00	2.2E + 00
Cs-134	Liver	7.0E − 01	1.0E + 00	1.1E + 00	8.5E − 01
Cs-137	Liver	6.1E − 01	8.3E − 01	8.5E − 01	6.2E − 01

*Total Body

**For Mo-99, the gastrointestional tract-lower large intestine (GI-LLI) is the critical organ for the teen and adult.

TABLE 7

*Pathway Dose Factors for Infants, mrem/yr per pCi/m²-sec**

Nuclide	Critical Organ	Milk
H-3	T. Body	2.4E − 03
C-14	Bone	3.2E + 00
P-32	Bone	1.6E + 05
Cr-51	GI-LLI	4.2E + 00
Mn-54	Liver	3.1E + 01
Fe-55	Bone	1.1E + 02
Fe-59	Liver	3.4E + 02
Co-58	GI-LLI	5.0E + 01
Co-60	GI-LLI	1.7E + 02
Ni-63	Bone	3.0E + 04
Zn-65	Liver	1.7E + 04
Sr-89	Bone	1.1E + 04
Sr-90	Bone	1.0E + 05
Y-91	GI-LLI	4.4E + 00
Mo-99	Kidney	3.1E + 02
I-131	Thyroid	1.0E + 06
I-133	Thyroid	9.6E + 03
Cs-134	Liver	5.4E + 04
Cs-137	Liver	4.9E + 04

*The pathway dose factors for H^3 and C^{14} are in units of mrem/yr per pCi/m³.

3. Ingestion

The dose from ingesting food "j" is calculated with the following equation:

$$D_o^{A,j} = 10^{12} \, \frac{pCi}{Ci} \times Q \times \chi/Q \times d \times f \times R_o^{A,j} \tag{6}$$

where: Q, χ/Q, and f are defined above and

    d = deposition velocity (a typical annual average deposition velocity is 0.01 m/sec), m/sec

    $R_o^{A,j}$ = organ O ingestion dose factor for receptor age A for food j, mrem/yr per pCi/m²-sec[1]

Pathway dose factors for the critical organ for each nuclide are listed in Tables 7 to 10 for infants, children, teenagers, and adults, respectively. The pathway

---

[1] For H^3 and C^14 the pathway dose factors are in units of mrem/yr per pCi/m³, and the deposition velocity does not enter into the dose calculation.

TABLE 8

*Pathway Dose Factors for Children, mrem/yr per pCi/m²-sec**

Nuclide	Critical Organ	Vegetation	Milk	Meat	All
H-3	T.Body	4.0E − 03	1.6E − 03	2.3E − 04	5.9E − 03
C-14	Bone	3.5E + 00	1.6E + 00	5.3E − 01	5.6E + 00
P-32	Bone	3.7E + 03	7.6E + 04	7.3E + 03	8.7E + 04
Cr-51	GI-LLI	6.1E + 00	4.8E + 00	4.2E − 01	1.1E + 01
Mn-54	Liver	6.5E + 02	1.7E + 01	6.4E + 00	6.7E + 02
Fe-55	Bone	7.6E + 02	8.7E + 01	3.6E + 02	1.2E + 03
Fe-59	Liver	6.3E + 02	1.7E + 02	5.2E + 02	1.3E + 03
Co-58	GI-LLI	3.7E + 02	5.9E + 01	8.0E + 01	5.1E + 02
Co-60	GI-LLI	2.1E + 03	1.9E + 02	3.0E + 02	2.6E + 03
Ni-63	Bone	4.6E + 04	2.5E + 04	2.5E + 04	9.6E + 04
Zn-65	Liver	2.7E + 03	9.9E + 03	9.0E + 02	1.4E + 04
Sr-89	Bone	3.5E + 04	5.6E + 03	4.1E + 02	4.1E + 04
Sr-90	Bone	1.4E + 06	9.3E + 04	8.6E + 03	1.5E + 06
Y-91	GI-LLI	2.4E + 03	4.4E + 00	2.0E + 02	2.6E + 03
Mo-99	Kidney	1.7E + 01	1.7E + 02	2.4E − 01	1.9E + 02
I-131	Thyroid	4.8E + 04	4.3E + 05	5.4E + 03	4.8E + 05
I-133	Thyroid	8.1E + 02	3.9E + 03	1.3E − 04	4.7E + 03
Cs-134	Liver	2.6E + 04	2.9E + 04	1.2E + 03	5.6E + 04
Cs-137	Liver	2.4E + 04	2.5E + 04	1.0E + 03	5.0E + 04

*The pathway dose factors for H³ and C¹⁴ are in units of mrem/yr per pCi/m³.

dose factors were computed with the "PARTS" computer code which is described in Boegli et al. (1978). Note that the transfer factors in Tables 7 to 10 for milk and meat assume that the animal feeds entirely from pasture grass, rather than from stored feeds. If the animal obtained a considerable fraction of its annual diet from stored feeds (e.g., with a storage time of 90 days), then the calculated transfer factor would be less than the values in Tables 7-10; the exact value would depend on the decay constant of the nuclide, the approximate storage time of the feed, and the fraction of the animal's diet obtained from stored feeds (see USNRC (1977)).

B. *Exposure Via Liquid Effluents*

Ingestion of food is the major pathway of concern in evaluating exposure via liquid effluents. Doses from exposure of individuals from other pathways such as swimming and boating are typically very small compared with doses from ingestion of food. For the cases in which doses from ingesting irrigated

TABLE 9

*Pathway Dose Factors for Teenagers, mrem/yr per pCi/m²-sec**

Nuclide	Critical Organ	Vegetation	Milk	Meat	All
H-3	T. Body	2.6E − 03	9.9E − 04	1.9E − 04	3.8E − 03
C-14	Bone	1.4E + 00	6.7E − 01	2.8E − 01	2.4E + 00
P-32	Bone	1.8E + 03	3.1E + 04	3.8E + 03	3.7E + 04
Cr-51	GI-LLI	1.0E + 01	7.5E + 00	8.4E − 01	1.8E + 01
Mn-54	GI-LLI	9.2E + 02	2.3E + 01	1.1E + 01	9.5E + 02
Fe-55	Bone	3.1E + 02	3.5E + 01	1.9E + 02	5.4E + 02
Fe-59	GI-LLI	9.7E + 02	2.5E + 02	1.0E + 03	2.2E + 03
Co-58	GI-LLI	5.9E + 02	9.1E + 01	1.6E + 02	8.4E + 02
Co-60	GI-LLI	3.2E + 03	2.9E + 02	6.0E + 02	4.1E + 03
Ni-63	Bone	1.9E + 04	1.0E + 04	1.3E + 04	4.2E + 04
Zn-65	Liver	1.9E + 03	6.6E + 03	7.8E + 02	9.3E + 03
Sr-89	Bone	1.5E + 04	2.3E + 03	2.2E + 02	1.8E + 04
Sr-90	Bone	8.3E + 05	5.5E + 04	6.7E + 03	8.9E + 05
Y-91	GI-LLI	3.1E + 03	5.4E + 00	3.3E + 02	3.4E + 03
Mo-99	Kidney	1.3E + 01	1.0E + 02	1.9E − 01	1.1E + 02
I-131	Thyroid	3.1E + 04	2.2E + 05	3.6E + 03	2.6E + 05
I-133	Thyroid	4.6E + 02	1.7E + 03	7.2E − 05	2.2E + 03
Cs-134	Liver	1.6E + 04	1.8E + 04	9.7E + 02	3.5E + 04
Cs-137	Liver	1.4E + 04	1.4E + 04	7.8E + 02	2.9E + 04

*The pathway dose factors for H^3 and C^{14} are in units of mrem/yr per pCi/m³.

food may be of concern, the reader is referred to Appendix A of USNRC (1977). The dose from ingesting water and seafood is calculated with the following equation:

$$D_o^{A,j} = 10^3 \ mC_i/C_i \times Q \times F_i \times F_a \times f \times R_o^{A,j} \tag{7}$$

Where   Q is defined above
   F_i = the reciprocal of the initial dilution stream, sec/ℓ
   F_a = any additional dilution provided by the receiving water body, unitless
   f = duration of release, hours
   $R_o^{A,j}$ = organ 0 ingestion dose factor for receptor age A for food j, mrem/hr per mCi/ℓ.

A pathway dose factor for the critical organ for each nuclide is listed in Table 11 for the adult age group, which is the critical age group due to the larger usage

TABLE 10

*Pathway Dose Factors for Adults, mrem/yr per pCi/m²-sec**

Nuclide	Critical Organ	Vegetation	Milk	Meat	All
H-3	T. Body	2.3E − 03	7.6E − 04	3.2E − 04	3.4E − 03
C-14	Bone	8.7E − 01	3.6E − 01	3.3E − 01	1.6E + 00
P-32	Bone	1.5E + 03	1.7E + 04	4.6E + 03	2.3E + 04
Cr-51	GI-LLI	1.2E + 01	6.4E + 00	1.6E + 00	2.0E + 01
Mn-54	GI-LLI	9.4E + 02	2.1E + 01	2.2E + 01	9.8E + 02
Fe-55	Bone	2.0E + 02	2.0E + 01	2.3E + 02	4.5E + 02
Fe-59	GI-LLI	9.7E + 02	2.0E + 02	1.8E + 03	3.0E + 03
Co-57	GI-LLI**	1.8E + 02	2.4E + 01	9.4E + 01	3.0E + 02
Co-58	GI-LLI	6.1E + 02	7.9E + 01	3.1E + 02	1.0E + 03
Co-60	GI-LLI	3.1E + 03	2.4E + 02	1.1E + 03	4.4E + 03
Ni-63	Bone	1.2E + 04	5.7E + 03	1.6E + 04	3.4E + 04
Zn-65	Liver	1.3E + 03	3.9E + 03	1.0E + 03	6.2E + 03
Sr-89	Bone	9.8E + 03	1.2E + 03	2.6E + 02	1.1E + 04
Sr-90	Bone	6.7E + 05	3.9E + 04	1.0E + 04	7.2E + 05
Y-91	GI-LLI	2.7E + 03	4.0E + 00	5.2E + 02	3.2E + 03
Mo-99	Kidney	1.4E + 01	5.6E + 01	2.3E − 01	7.0E + 01
I-125	Thyroid***	2.5E + 04	9.0E + 04	3.2E + 03	1.2E + 05
I-131	Thyroid	3.8E + 04	1.4E + 05	4.9E + 03	1.8E + 05
I-133	Thyroid	5.3E + 02	9.9E + 02	9.3E − 05	1.5E + 03
Cs-134	Liver	1.1E + 04	1.1E + 04	1.2E + 03	2.3E + 04
Cs-137	Liver	9.1E + 03	8.1E + 03	9.6E + 02	1.8E + 04

*The pathway dose factors for H^3 and C^{14} are in units of mrem/yr per pCi/m³.
** The dose conversion factor is from Dunning et al. (1981).
***The dose conversion factor is from ICRP (1978).

factors for adults than for the other age groups. The pathway dose factors were calculated with the LADTAP II computer code (Simpson et al., 1980).

V. SUMMARY

This chapter has described mathematical models for calculating doses to individuals associated with the release of radionuclides to the biosphere. Because this chapter was intended to meet the needs of general circumstances, the discussion has focused on the environmental pathways that are typically the most significant dose contributing pathways. The reader is referred to USNRC (1977), and Till et al. (1983) for more detailed information.

TABLE 11

Pathway Dose Factors for Exposure to Liquid Effluents, mrem/hr per mCi/liter of Water

Nuclide	Critical Organ	Drinking Water	Freshwater Fish	Seafood Ingestion Saltwater		All	
				Fish	Invertebrates	Freshwater	Saltwater
H-3	T. Body	8.7	2.3E − 01	2.3E − 01	5.6E − 02	9.0E + 00	2.8E − 01
C-14	Bone	2.4E + 02	3.1E + 04	1.2E + 04	2.3E + 03	3.2E + 04	1.4E + 04
Na-24	T. Body	1.4E + 02	4.1E + 02	2.7E − 01	1.8E − 01	5.5E + 02	4.6E − 01
P-32	Bone	1.6E + 04	4.6E + 07	1.3E + 07	3.3E + 06	4.6E + 07	1.7E + 07
Cr-51	GI-LLI	5.6E + 01	3.2E + 02	6.4E + 02	7.6E + 02	3.8E + 02	1.4E + 03
Mn-54	GI-LLI	1.2E + 03	1.3E + 04	1.8E + 04	3.2E + 03	1.4E + 04	2.2E + 04
Fe-55	Bone	2.3E + 02	6.6E + 02	2.0E + 04	3.1E + 04	8.9E + 02	5.1E + 04
Fe-59	GI-LLI	2.8E + 03	8.1E + 03	2.4E + 05	3.9E + 05	1.1E + 04	6.3E + 05
Co-57	GI-LLI*	3.8E + 02	5.5E + 02	1.1E + 03	2.6E + 03	9.2E + 02	3.7E + 03
Co-58	GI-LLI	1.3E + 03	1.8E + 03	3.6E + 03	8.6E + 03	3.1E + 03	1.2E + 04
Co-60	GI-LLI	3.4E + 03	4.8E + 03	9.6E + 03	2.3E + 04	8.2E + 03	3.2E + 04
Ni-63	Bone	1.1E + 04	3.1E + 04	3.1E + 04	1.8E + 04	4.2E + 04	5.0E + 04
Ni-65	GI-LLI	1.4E + 02	4.2E + 02	4.2E + 02	2.5E + 02	5.6E + 02	6.6E + 02
Zn-65	Liver	1.3E + 03	7.4E + 04	7.4E + 04	4.4E + 05	7.5E + 04	5.1E + 05
Sr-89	Bone	2.6E + 04	2.2E + 04	1.5E + 03	3.5E + 03	4.8E + 04	5.0E + 03
Sr-90	Bone	3.2E + 05	2.8E + 05	1.8E + 04	4.4E + 04	6.0E + 05	6.2E + 04
Y-91	GI-LLI	6.5E + 03	4.6E + 03	4.6E + 03	4.4E + 04	1.1E + 04	4.9E + 04
Mo-99	Kidney	8.1E + 02	2.3E + 02	2.3E + 02	5.6E + 01	1.0E + 03	2.9E + 02
Tc-99m	GI-LLI	3.4E + 01	1.5E + 01	9.9E + 00	1.2E + 01	4.9E + 01	2.2E + 01
I-125	Thyroid**	1.0E + 05	4.5E + 04	3.0E + 04	3.6E + 04	1.5E + 05	6.6E + 04
I-131	Thyroid	1.6E + 05	7.0E + 04	4.7E + 04	5.6E + 04	2.3E + 05	1.0E + 05
I-133	Thyroid	3.0E + 04	1.3E + 04	8.7E + 03	1.0E + 04	4.3E + 04	1.9E + 04
Cs-134	Liver	1.2E + 04	7.1E + 05	1.4E + 04	2.1E + 03	7.2E + 05	1.6E + 04
Cs-137	Liver	9.1E + 03	5.2E + 05	1.0E + 04	1.6E + 03	5.3E + 05	1.2E + 04

*Dose conversion factor was taken from Dunning et al (1981).

**Dose conversion factor was taken from ICRP (1978).

REFERENCES

Boegli, J.S., R.R. Bellany, W.L. Britz, R.L. Waterfield, W.C. Burke, and F.J. Congel (1978), "Preparation of Radiological Effluent Technical Specifications for Nuclear Power Plants," NUREG-0133.

Dunning, D.E., Jr., G.G. Kilough, S.R. Bernard, J.C.Pleasant, and P.J. Walsh (1981), "Estimates of Internal Dose Equivalent to 22 Target Organs for Radionuclides Occurring in Routine Releases from Nuclear Fuel-Cycle Facilities," NUREG/CR-0150, Volume 3.

International Commission on Radiological Protection (1978), "Limits for Intakes of Radionuclides by Workers," ICRP Publication 30, Supplement to Part 1.

Simpson, D.B., and B.L. McGill (1980), "User's Manual for LADTAP II-A Computer Program for Calculating Radiation Exposure to Man from Routine Release of Nuclear Reactor Liquid Effluents." NUREG/CR-1276.

Slade, D.H. (1968), *Meteorology and Atomic Energy*

Till, J.E., and H.R. Meyer, ed (1983), "Radiological Assessment: A Textbook on Environmental Dose Analysis," NUREG/CR-3332, ORNL-5968.

U.S. Nuclear Regulatory Commission (1972), Regulatory Guide 1.23, "Onsite Meteorological Programs."

U.S. Nuclear Regulatory Commission (1977), Regulatory Guide 1.109, Rev. 1, "Calculation of Annual Doses to Man From Routine Releases of Reactor Effluents for the Purpose of Evaluating Compliance with 10 CFR Part 50 Appendix I."

Management of Radioactive Materials and Wastes: Issues and Progress. Edited by S. K. Majumdar and E. Willard Miller. © 1985, The Pennsylvania Academy of Science.

Chapter Twenty-Two

RADIATION PROTECTION STANDARDS AND RADIATION RISKS

A.T. Sabo

Director, NES Licensing, Safeguards and Safety
Westinghouse Electric Corporation
Box 355
Pittsburgh, Pennsylvania 15230

In this decade, few subjects have evoked as much public attention as ionizing radiation and its risks to individuals. It is important that this matter be dealt with in a credible manner. Radiation is everywhere. It has been part of our lives from the moment of our existence. Radiation is, thus, part of our natural environment on earth. For an individual, however, radiation exposures and subsequent health affects can only be linked statistically. A radiation produced cancer is indistinguishable from cancer caused, for example, by workplace chemicals. Consequently, many studies attribute cause and effect relationships to only radiation while ignoring the cause and effect relationships from other materials.

Each person on earth receives certain amounts of background radiation (Table 1). If an average exposure dose of 125 mrem/yr is selected for western Pennsylvania residents, a calculation, using an average life span of 72 years, reveals a lifetime exposure of 9000 mrem (9rem) from background radiation. Individuals in various sections of the United States would receive different levels of exposure. Individuals living in Denver, Colorado, for example, receive on an average 200 mrem per year, or a lifetime exposure from background radiation of 14,400 mrem (14.4 rem).

Background radiation provides about 67.6 percent of the total radiation received by an individual in a lifetime (Figure 1). The second largest single source of radiation received in a lifetime is from medical sources of which chest and dental X-rays are significant. Medical irradiation totals about 30.7 percent. These two sources total 98.3 percent. Minor sources include fallout, occupational exposure, releases from the nuclear industry and miscellaneous sources.

When considering radiation there are four energy sources -alpha, beta, gamma and neutrons. The penetration power of each type of radiation varies greatly (Figure 2). Alpha radiation is stopped by a sheet of paper while beta radiation will be stopped by a few tenths of an inch of aluminum. Gamma rays will

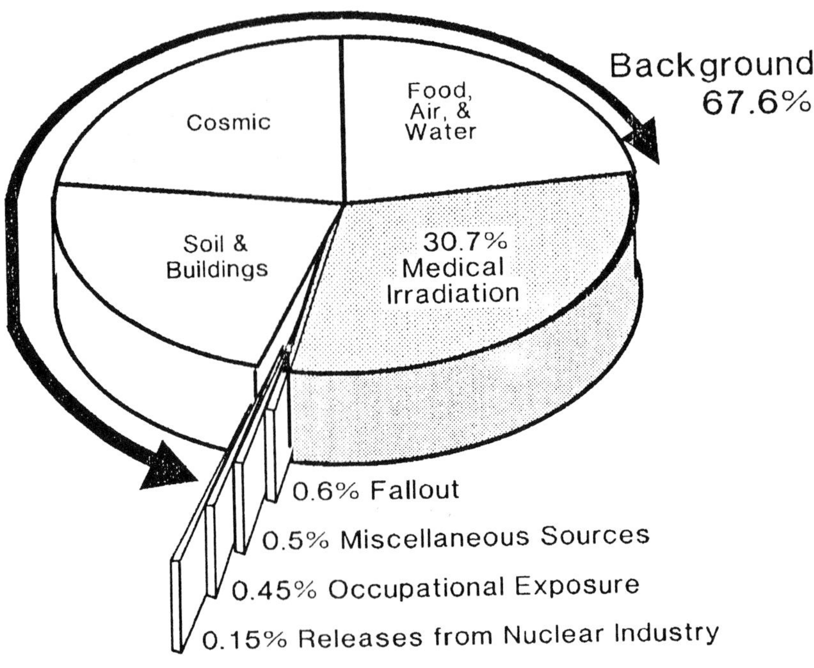

FIGURE 1. Average individual radiation exposure.

PENETRATING POWER OF VARIOUS TYPES OF RADIATION

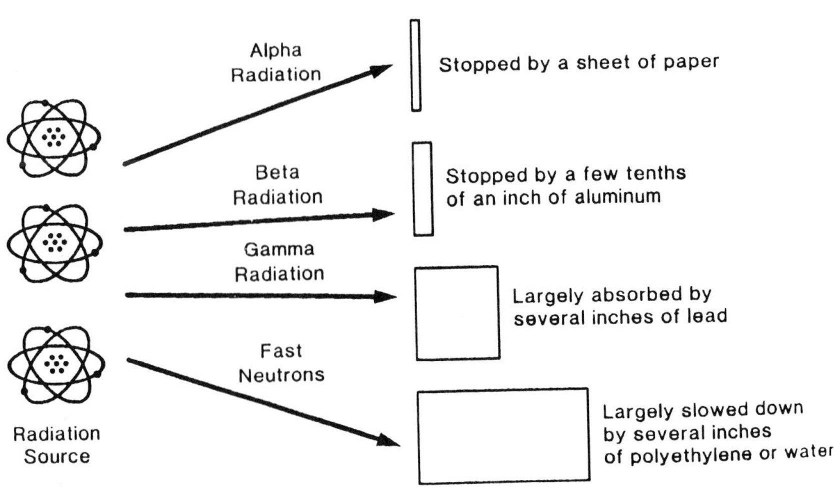

FIGURE 2. Penetrating power of various types of radiation.

be largely absorbed by several inches of lead while fast neutrons are largely slowed down by several inches of polyethylene or water.

RADIATION STANDARDS

There are a number of governmental groups, both national and international, that are developing, issuing information and implementing programs in the radiation area. At the international level, there are two groups of importance. The first is the International Commission on Radiological Protection (ICRP). The ICRP has issued two recent reports that impact on United States radiation protection practices. The first report is "Basic Safety Standards for Radiation Protection," commonly referred to as the ICRP-26 report. The second ICRP report is "Limits for Intakes of Radionuclides by Workers," commonly referred to as the ICRP-30 report. ICRP reports are issued identifying fundamental principles upon which appropriate radiation protection measures can be based while leaving to the national protection bodies the responsibility of formulating specific codes of practice which are subsequently promulgated into rules and regulations. In brief, ICRP reports give recommendations and are intended to be used as advisory guides.

The second international group that has an important role in radiation practices is the United Nations Scientific Committee on the Effects of Atomic Radiation (UNSCEAR). This group collects, analyzes, and disseminates information in the radiation area by issuing scientific reports. Because it is a branch of the United Nations, it is widely used by countries when discussing radiation matters.

Within the United States, the National Academy of Sciences (NAS) is responsible for maintaining and evaluating scientific information on a broad national basis. The Academy functions through special groups of scientists who are assigned to study all information in a certain area and to issue reports to the academy on their findings. One of the groups NAS uses is the National Council on Radiation Protection and Measurements (NCRP).

The National Council on Radiation Protection and Measurement is a nonprofit corporation chartered by Congress in 1964 to collect, analyze, develop, and disseminate in the public interest, information and recommendations on radiation. The NCRP reports are widely used as reference reports by the U.S. nuclear industry and NRC licensees. Copies of all NCRP reports are available for purchase directly from the NCRP.

The group charged with determining and establishing U.S. radiation rules and regulations is the Environmental protection Agency (EPA). EPA is the basic U.S. group responsible for collecting, analyzing, and establishing regulatory radiation practices and exposure limits to be implemented within the United States.

TABLE 1

Annual Average Radiation Dosage

Sources	Annual Dosage (mrem/yr)
Cosmic radiation at sea level	40
Elevation above sea level (e.g. Pittsburg, PA)	12
Ground (U.S. average)	15
Building construction materials	57
Water, food, air (U.S. average)	30
Air travel (6000 miles/yr.)	4
Black and white television (3 hours/day)	<0.1
Color television (3 hours/day)	0.1 - 6
Chest X-ray (1 per year)	50
Dental X-ray (1 per year)	20
U.S. average dose	150 - 200
Typical nuclear power plant at boundary	5 - 10*
	*NRC Design Objectives

One of several other governmental groups charged by Congress to implement and enforce radiation practices in the U.S. is the Nuclear Regulatory Commission (NRC). The NRC legally establishes rules and regulations which must be followed by all U.S. NRC licensees. In certain specific instances, the NRC will formally sign an agreement with a state agency permitting the state to assist in this responsibility in a limited manner. The NRC and Agreement State Agencies issue specific licenses, which permits nuclear plants to possess, use and transport radioactive materials according to the rules specified in Title 10 of the Code of Federal Regulations.

Another American group is the Committee on the Biological Effects on Ionizing Radiation, known as the BEIR III Committee. This committee, in 1980, completed an indepth scientific study of the risks of radiation. The BEIR III Committee and the ICRP Report 26 both recommend that radiation exposures to individuals be evaluated using "RISK" criteria.

RADIATION RISK

Within recent times the concept has developed that radiation protection practices should be based on equating radiation exposures in terms of risks instead of dose. Radiation risk is best defined as the probability of harm from radiation exposure. The term risk coefficient means the risk per unit dose equivalent.

The Nuclear Regulatory Commission has established limits of radiation (Figure 3). Generally stated the limits are 5 Rem per year after age eighteen. Higher yearly doses can be experienced provided the individual's average lifetime exposure remains at 5 Rem/yr and no more than 3 Rem per three month period.

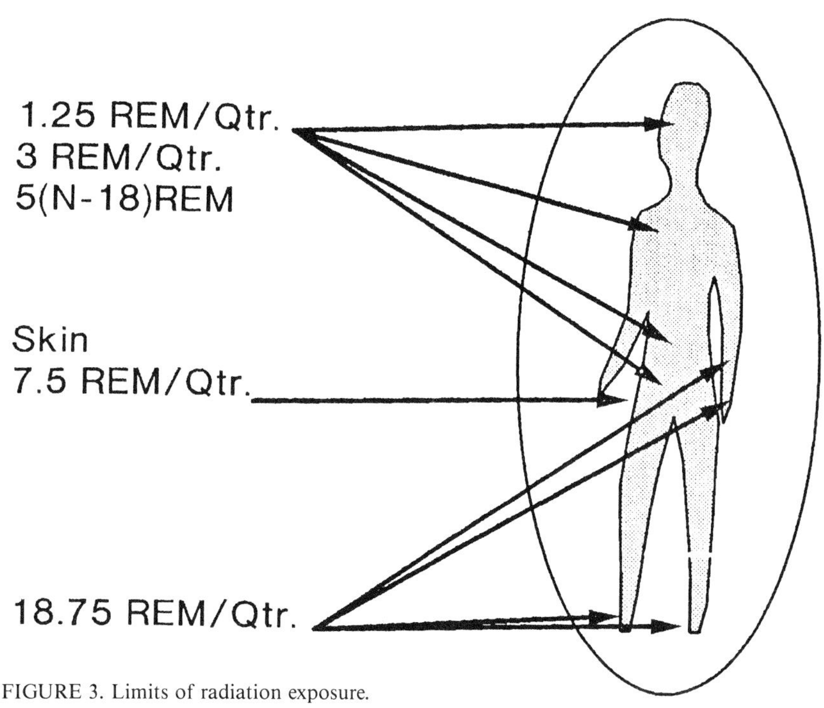

FIGURE 3. Limits of radiation exposure.

TABLE 2

Estimates of Excess Cancer Incidence from Exposure to Low-Level Radiation

Source	Number of Additional Cancers Estimated to Occur in 1 Million People After Exposure of Each to 1 Rem of Radiation
BEIR, 1980	160-450
ICRP, 1977	200
UNSCEAR, 1977	150-350

All scientific groups have agreed that the current average limit of 5 Rem/yr will not result in an individual being subjected to unacceptable risks. Nevertheless, it is cautioned that radiation exposures should be maintained as low as reasonably achievable and that all exposures be evaluated as necessary. Significant advancements to control radiation risks, involving robotics and advanced tooling, are now becoming a regular part of work at nuclear plants. This has resulted in reducing radiation exposures on specific jobs thus reducing the total exposure required to complete various work functions.

As Figure 4 indicates radiation has various combination of effects on living cells. Initially a living cell may be killed or damaged by radiation or radiation

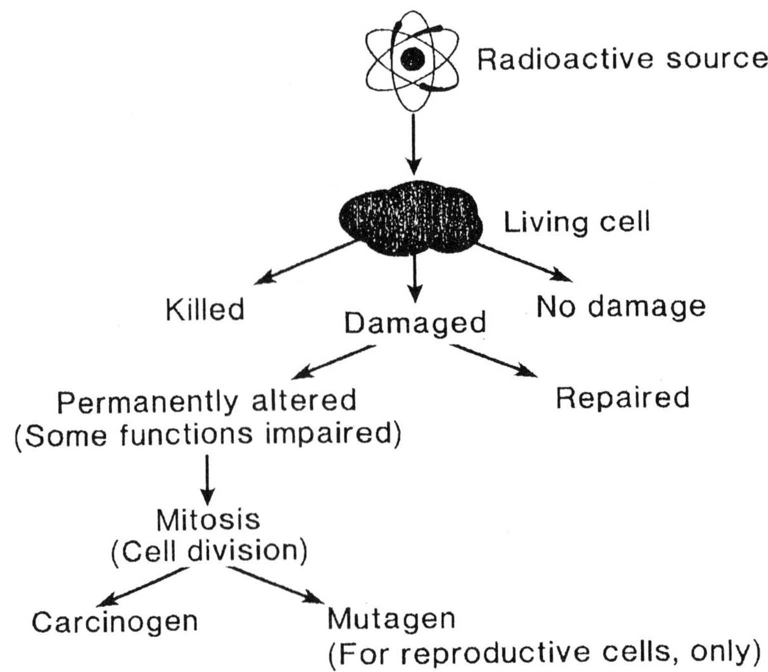

FIGURE 4. Effects of radiation on living cells.

TABLE 3

Estimated Loss of Life Expectancy from Health Risks

Health Risks	Estimates of Days of Life Expectancy Lost, Average
Smoking 20 cigarettes/day	2370 (6.5 years)
Overweight (by 20%)	985 (2.7 years)
All accidents combined	435 (1.2 years)
Auto accidents	200
Alcohol consumption (U.S. average)	130
Home accidents	95
Drowning	41
Natural background radiation, calculated	8
Medical diagnostic x-rays (U.S. average), calculated	6
All catastrophes (earthquake, etc.)	3.5
1 rem occupational radiation dose, calculated (industry average for the higher-dose job categories is 0.65 rem/yr.)	1
1 rem/yr. for 30 years, calculated	30

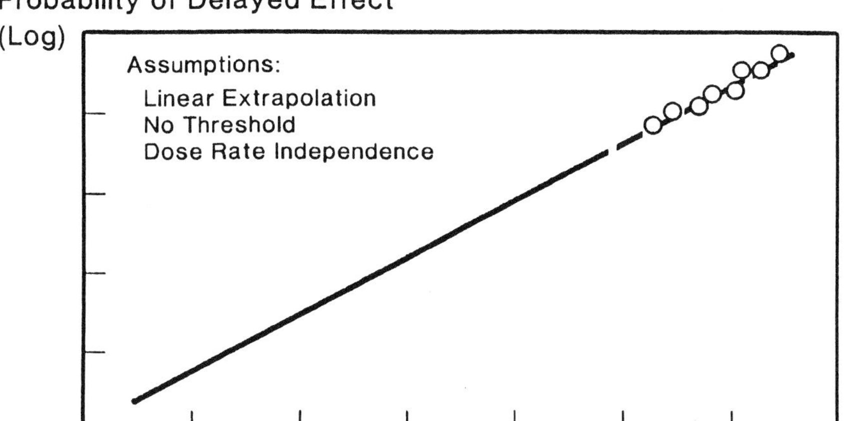

FIGURE 5. Linear dose response model.

may have no effect on it. If the living cell is damaged it may be repaired or permanently altered. Finally, the altered living cell will divide as a carcinogen or mutagen.

A number of models have evolved to reveal the dose response per unit of exposure. A basic one is commonly referred to as the Linear Dose Response Model (Figure 5). It indicates a straight line response regardless of the amount of the radiation dose. It has been found, however, that no accurate measurements exist to confirm that in the low dose area (less than 100 Rem) the response is linear.

Because the linear dose model has not been accepted by all scientists, two other models have been developed. There are the linear quadratic and threshold models (Figure 6). However, there is still discussion as to which model represents best the actual dose—response predictions. The NRC has issued instructions to NRC licensees concerning biological risks from occupational radiation exposure in NRC Regulatory Guide 8.29, "Instruction Concerning Risks from Occupational Radiation Exposure." In this guide the NRC uses the Linear Dose Response model for its predictions.

Data indicates the risk of cancer from radiation is exceptionally low. Current data from the American Cancer Society indicates that in the 20 to 65 year age groups (basic working years), if 10,000 workers are given a one Rem exposure, three additional cancer cases might occur (Table 2). In contrast, it is estimated of 10,000 worker, 25 percent, or 2,500 workers, will develop cancer from all possible causes, such as smoking, alcohol, food, drugs, air pollutants, and natural background radiation (Table 3). Thus a response to one Rem would

Effects (Cancer Risks)

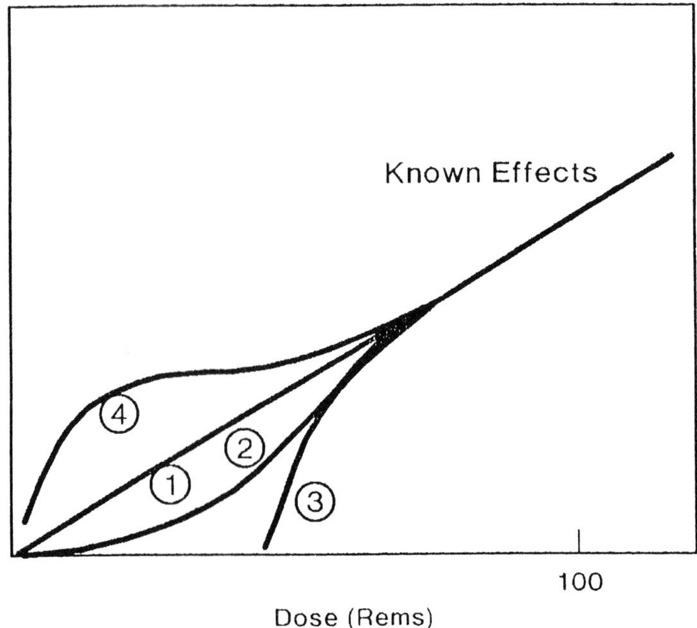


```
Curve 1 - Linear model
Curve 2 - Linear-quadratic model (Low risk for small doses)
Curve 3 - Threshold model
Curve 4 - Linear quadratic model (Higher risk for small doses)
```

FIGURE 6. Effects of radiation varies with doses at low levels.

increase the normal incidence from 2,500 cases to 2,503 cases, or an increase of about 3 hundredths of one percent.

CONCLUSIONS

Although the dangers of radiation are well-known, there are also many risk-free benefits from radiation, that are now commonly used to advance our knowledge and way of life. The use of activation analysis in which materials are bombarded with neutrons is now used to identify and measure various elements. Archeologists use radiation to estimate the period and probable place where an object was made. It is also used to diagnose problems in forensic medicine, crime detection and mineral exploration. In biochemistry scientists tag and follow molecules in complex interactions. Activation analysis is also used in sterilization and plant breeding. These are only a few of the beneficial uses of radiational processes that are routinely used in our current technological oriented society.

PERTINENT LITERATURE

Eric H. Hall. Radiation & Life. 1974, Pergamon Press.

Management of Radioactive Materials and Wastes: Issues and Progress. Edited by S. K. Majumdar and E. Willard Miller. © 1985, The Pennsylvania Academy of Science.

Chapter Twenty-Three

ENVIRONMENTAL MONITORING OF LOW-LEVEL RADIOACTIVE MATERIALS

William A. Jester,[1] Ph.D. and Charley Yu,[2] Ph.D.
[1]Associate Professor of Nuclear Engineering
The Pennsylvania State University
Breazeak Nuclear Reactor
University Park PA. 16802
and
[2]Assistant Environmental Scientist
Arsonne National Laboratory
9700 S. Cass Ave.
Arsonne, IL 60439

Some of the current rationale behind the environmental monitoring of low-level radioactive materials are as follows:

Committee 4 of the International Commission on Radiological Protection (ICRP)[1] defined three broad objectives for environmental monitoring: 1) assessment of the actual or potential exposure of humans to radioactive materials or radiation present in their environment or the estimation of the probable upper limits of such exposure; 2) scientific investigation, sometimes related to the assessment of exposures, sometimes to other objectives; 3) improved public relations. Various regulations have been written requiring environmental monitoring to ensure that the public is not being exposed to excessive amounts of radiation from natural sources or from human activities. An example of the monitoring of natural sources of radiation is a requirement of the Environmental Protection Agency's (EPA) National Interim Primary Drinking Water Regulations[2] whereby U.S. water supply companies must have their drinking water monitored at least once every four years for radionuclides, primarily the naturally occurring radium-226.

A much more extensive and frequent environmental radiation monitoring program is required by the Nuclear Regulatory Commission (NRC) around the various nuclear facilities of the nation[3,4]. The objectives of environmental monitoring of a nuclear facility, such as a nuclear power plant, radioactive waste disposal site, or spent fuel reprocessing plant, can be divided into three phases[5], i.e., preoperational, operational, and postoperational phases.

The objectives of the preoperational phase of the environmental monitoring are: (1) to determine the levels and variations of the levels of radiation and radioactive materials in the neighborhood of the facility before it goes into operation, (2) to investigate critical pathway to humans, and (3) to evaluate sampling and analytical methods. The data collected in this phase constitutes a base line to which levels and variations measured after the facility is in operation may be compared.

The objectives of the operational phase of environmental monitoring are (1) to provide data for comparison with preoperational data, (2) to show compliance with applicable radiation standards, and (3) to detect any environmental changes produced by the operation of the facility.

The objectives of the postoperational phase of the environmental monitoring are: (1) to verify confinement capability of any stored or resident radionuclides; and (2) to determine impact of decommissioning activities on local environment.

Thus, as seen above, the most important objective of any environmental monitoring program for the three phases of nuclear facility operations is to determine population dose resulting from facility operations, as it indicates the health impact on humans.

SOURCES OF RADIOACTIVE MATERIALS

Throughout history, mankind has been exposed to radiation from the environment. Prior to the beginning of this century, environmental radioactivity was strictly a natural phenomenon, which resulted from three sources: cosmic rays, highly energetic radiation bombarding the earth from outer space; cosmogenic radionuclides, produced by interaction with cosmic rays; and primodial radionuclides, which existed at the time of formation of the earth, along with the primodial radionuclide-decay products.

With the discovery of how to produce X-rays and the discovery of the existence of naturally-occurring radioactive materials, there was a rapid growth in the use of X-rays and naturally-occurring radionuclides, especially in medicine during the first half of this century. After World War II, these natural radiation sources were augmented by radiation-producing devices such as particle accelerators, nuclear reactors, nuclear weapons, and consumer products. In addition, there were available a large number of accelerator and reactor-produced radionuclides for use in medicine, industrial applications, and research.

A distinctive characteristic of most natural radioactivity is that it involves the whole population of the world. In addition, biological life has been exposed to an ever-decreasing level of background radiation throughout history. Thus the intensity of natural radiation may be used as a reference level for comparison with the more localized man-made sources of radioactivity.

A. Natural Sources of Radiation

Cosmic Rays

Cosmic radiation is radiation of extraterrestrial origin which continually bombards the earth's atmosphere. Cosmic radiation comes from our sun, and from other sources both inside and outside our galaxy. Primary cosmic rays are those which have not yet interacted with matter in the earth's atmosphere, lithosphere, or hydrosphere. They consist of very high energy particles[6,7] about 70% protons; 20% alpha particles; 0.7% nuclei of lithium, beryllium, and boron; 1.7% nuclei of carbon, nitrogen, and oxygen; and 0.6% nuclei of atomic mass greater than that of neon.

Secondary cosmic rays are those which are produced by interactions of the primary rays and atmospheric nuclei and consist largely of mesons, electrons, protons, neutrons, and gamma rays. At sea level, nearly all the observed cosmic radiation consists of secondaries, with about 80% being mesons, and 20% electrons.

The cosmic radiation intensity increases with altitude; in Denver, Colorado, for example, it is approximately double what it is in New York City. The cosmic ray intensity is also a function of latitude, being about 15% lower near the equator than at the poles. In the United States, the cosmic ray dose rate changes from a low of 350 μSv* per year in Florida to over 1000 μSv per year in Wyoming. The values in Table 1 are typical whole body doses in the United States[7,8,9] showing an average annual cosmic dose of 450 μSv per year.

Cosmogenic Radionuclides

Approximately 20 radionuclides are known to be continuously produced in the atmosphere by cosmic ray interaction with air. The natural production of cosmogenic radionuclides in the atmosphere shows latitudinal and elevational patterns similar to those of cosmic ray intensities[6,7]. About 70% of the cosmogenic radionuclides arise in the stratosphere, while about 30% are formed in the troposphere.

With the exception of tritium and carbon-14, the cosmogenic radionuclides contribute very little to the environmental radiation levels. Both carbon-14 and tritium are weak beta emitters and thus are not a hazard until they enter the body incorporated in our food, air, and water. They then become part of our body tissue. The estimated annual dose from these two radiation sources is about 10 μSv per year, with the bulk of this coming from carbon-14.

Primodial Radionuclides

Primodial radionuclides are those radionuclides which existed at the time of formation of the earth. These radionuclides can be divided into two groups: those whose half lives are longer than the age of the earth and those that are

---

*Note that 1 Sv (Sievert) = 100 rem, 1 μSv = 1x10^{-6} Sv..

TABLE 1

U.S. Average Annual Radiation Doses from Natural and Man-Made Sources

Natural Radiation	Dose Rates* μSv/year	% of Total
External Irradiation		
Cosmic rays	450	25
Terrestrial Irradiation	300	17
Soil, Rocks, Building materials		
Internal Irradiation	250	14
Taken into the body from		
food, water, and air		
Subtotal Natural	1000	56
Man-Made Radiation		
Medical Uses		
X-rays & radio-pharmaceuticals	750	41
Nuclear power	3	0.2
Fallout from past weapons testing	40	2
Consumer products	0.3	—
Occupational exposure	10	0.6
Subtotal Man-Made	803	44
GRAND TOTAL	1803	100

*These numbers will vary somewhat in different publications.

radioactive products of a few of these long-lived radionuclides. The first group contains about 20 known radionuclides with half-lives ranging from uranium-235, with a half-life of 7×10^8 years to dysprosium-156, with a half-life of about 10^{18} years. The second group contains 40 radionuclides all of which are decay products of either uranium-235, uranium-238, or thorium-232. Each of these three radionuclides undergoes successive alpha and beta decay events until it reaches one of the stable isotopes of lead.

The most important radionuclide in the first group is potassium-40, which is a significant external and internal background radiation source. Potassium is one of the major elements in the earth's crust, and 0.012% of all potassium atoms are potassium-40. Thus this radionuclide is more abundant than many elements on the earth's surface, including copper, cobalt, zinc, and lead. As a gamma-ray emitter, this radionuclide in earth materials contributes to a significant portion of the radiation dose from terrestrial irradiation (Table 1). Potassium is also an essential element in the body, comprising about 0.2% of the body mass. Thus this radionuclide on the average contributes over half the average internal irradiation dose to the body through the emission of gamma and beta rays.

The widespread distribution of uranium and thorium throughout the earth's crust also leads to a significant contribution to the external irradiation dose. This occurs through the decay of their gamma-emitting daughters, especially for lead-212, a daughter product of thorium-232 decay, and lead-214 and bismuth-214, daughters of uranium-238 decay. In some regions of the world, the presence of either uranium or thorium in the soil leads to doses as high as 0.12 Sv per year.

The most serious components of internal irradiation are the radium and radon decay products of uranium and thorium. Radium, especially radium-226 from uranium-238 decay, enters the body through food and drinking water and is deposited in the bone. It is estimated that this contributes about 10 μSv per year to the average radiation dose. Radon is a gas and thus is in the air all about us, especially radon-222 from the decay chain of uranium-238. The presence of radon gas and its subsequent decay products in the lungs can lead to an average lung dose of about 300 μSv per year, primarily from alpha radiation. Since this is not a whole body dose, this dose is not listed in Table 1. It is now believed that the presence of radon gases in the air is a major cause of lung cancer. With the trend toward cutting down air leakage from homes and buildings for energy conservation, indoor environmental radon concentrations are growing, thereby increasing this hazard.

B. Man-Made Sources of Radiation

Medical Uses

The medical uses of radiation include the use of diagnostic x-rays, nuclear medicine, radiation therapy, CT scans, and other medical techniques whereby patients are exposed either to a radiation field or to radionuclides which are introduced into their bodies.

In recent years, despite the ever-increasing medical use, there has actually been a decline in the average dose to the public from this source. This has been accomplished through the use of image intensifying screens for x-ray work, the more careful use of radiation shielding in the use of x-rays and in radiation therapy, and the availability of a greater variety of lower dose-producing radionuclide-labeled compounds for nuclear medical applications.

Nuclear Fuel Cycle

A potential large scale source of environmental radioactivity is present in the nuclear fuel cycle, but as seen in Table 1, under current conditions, this constitutes only a very small source of environmental exposure to humans. For example, the maximum dose to a person who should remain continuously at the site boundary of a typical 1000 megawatt nuclear power station would be less than 20 μSv per year.

The world installed nuclear generating capacity in 1982 was about 400 GW(e), which is about 2.6 times that produced in 1981. Projection of the world nuclear generating capacity in the year 2000 is about 1300 GW(e)[8]. Along with the generation of electric power, radioactive materials are also generated in the nuclear fuel cycle. The nuclear fuel cycle includes the mining and milling of uranium ores, conversion to various chemical forms, enrichment of the isotopic content

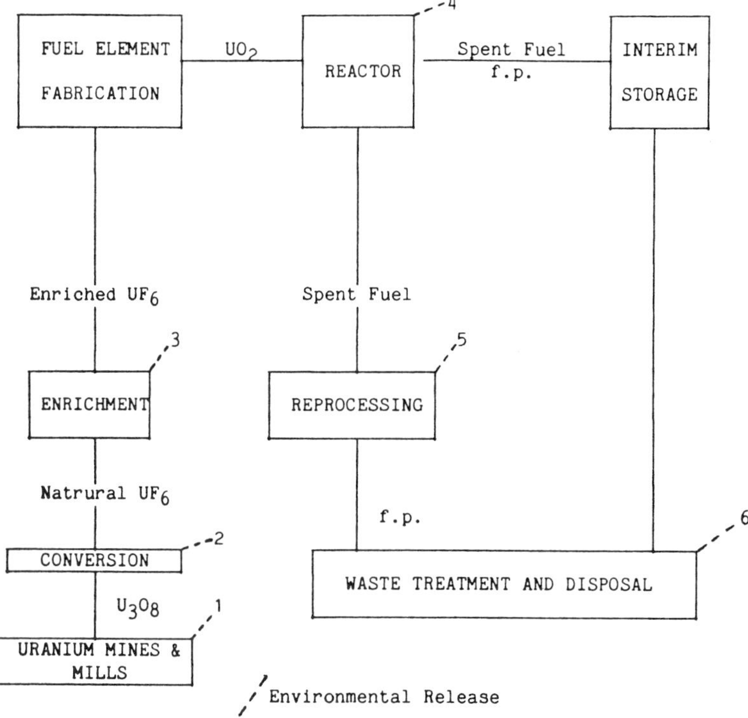

1. U, Th, Ra, Rn, Processing Chemicals
2. UF_6
3. UF_6
4. ^{3}H, ^{37}Ar, ^{85}Kr, ^{131}I. ^{133}Xe and other volatile activation and fission products
5. ^{3}H, ^{85}Kr, ^{129}I, ^{131}I and transuranics
6. long-lived activation and fission products and transuranics

FIGURE 1. Basic elements in the nuclear fuel cycle, showing the kinds of materials that can possibly be released

of ^{235}U, fabrication of fuel elements, production of power in nuclear reactors, reprocessing of irradiated fuel and recycling of fissile and fertile nuclides recovered, transportation of materials between the various installations, and, finally, disposal of radioactive waste. These steps are shown in Fig. 1. This figure shows the kinds of materials, both radioactive and nonradioactive, that can be released to the environment.[7]

The reactor is the dominant feature of the nuclear fuel cycle. Almost all the artificial radionuclides associated with the nuclear fuel cycle are present in the irradiated nuclear fuel. Some neutron activation of structure and cladding materials does take place, and the naturally-occurring radionuclides are present at the uranium mining and milling stage. At each step of the fuel cycle, releases of small quantities of radioactive materials to the environment may

TABLE 2

Six Critical Radionuclides from the Nuclear Fuel Cycle

Name	Symbol(s)	Half-Life	Principal Radiation Emitted	Possible Means of Production	Possible Environmental Forms	Critical Organ	Biological Half-Life
Tritium	^{3}H or T	12.3 y	β^- (weak)	Fission ^{10}B(n,α)	$H_2(g)$, H_2O H^+, OH^-, $CH_4(g)$	Whole Body	10 d
Carbon-14	^{14}C	5.7 x 10^3 y	β^- (weak)	^{14}N(n,p)	CO_2^-/HCO_3^- CO_2, $CH_4(g)$	Body Fat	10 d
Cobalt-60	^{60}Co	5.3 y	β^- (weak)	^{59}Co(n,γ)	Co^{2+}, $Co(OH)_2(s)$	Whole Body	4.5 d
Strontium-90	^{90}Sr	27.7 y	β^-	Fission	Sr^{2+}, $SrCO_3(s)$	Bone	50 y
Iodine-131	^{131}I	8.04 d	β^-,γ	Fission	I^-, IO_3^-, $I_2(g)$ $Ch_3I(g)$	Thyroid	10 d
Cesium-137	^{137}Cs	30 y	β^-,γ	Fission	Cs^+	Bone	140 d

β^- — beta particle; n — neutron; α — alpha particle; p — proton; γ — gamma photon; g — gas; s — solid; Half-life — time for half the radionuclides to decay away; Biological half-life — time for half the atoms to leave the critical organ.

occur. If the radionuclides released have long half-lives and rapid environmental mobility, such as krypton-85 does, they will become globally distributed. The current worldwide krypton-85 concentration is between 20 and 40 pCi/m^3. Essentially all of this has come from nuclear weapons testing and nuclear power plants.

In a nuclear reactor core, uranium-235, uranium-238, and/or plutonium-239 are fissioned, producing more than 80 direct radioactive fission products and over 200 secondary fission products from the decay of the direct fission products. An intense neutron field is also produced in the core region resulting in radionuclides in the core components and reactor coolant through neutron activation. Most of these fission products have very short half-lives and decay away to stable nuclides within the reactor fuel.

From an environmental concern, the most important fission products are tritium, strontium-90, iodine-131, and cesium-137 (see Table 2). Tritium has been previously discussed. Strontium-90 and its short-lived daughter, yttrium-90, are beta emitters, and if strontium-90 enters the body it can become part of the bone. Iodine-131 is both a strong beta and gamma emitter which has the potential of leaving a reactor as a gas. The chemistry of iodine is such that the radioiodine can be taken up by dairy cattle and concentrated in the milk. Once the radioiodine enters the body it is then concentrated in the thyroid. Thus large scale release of iodine-131 could pose a serious health problem, especially to small children who might drink the contaminated milk.

Cesium-137 is both a strong beta and gamma emitter so that it can pose both an external and an internal hazard. Like strontium-90, this radionuclide is also a bone seeker.

The most significant neutron activation products are tritium, carbon-14, and cobalt-60. Tritium is produced by the neutron irradiation of boron used in reactor control. Carbon-14 is produced through the irradiation of oxygen and nitrogen in the reactor coolant. Both tritium and carbon-14 have been previously discussed. Cobalt-60 results from the neutron activation of cobalt in steel reactor components. This radionuclide is a strong gamma and beta emitter. It can readily be complexed by organic constituents in ground surface waters and thus can move relatively rapidly through the environment. Therefore it has the potential of being both an internal and an external radiation hazard.

Nuclear Explosions

Nuclear explosions have been conducted since 1945. Intensive atmospheric nuclear test programs took place during 1954-1958 and 1961-1962. Continued individual tests have occurred since 1964. Recent deposition of fallout radioactivity has been largely due to the high yield test (4 Mt) which occurred in November 1976. The latest atmospheric nuclear test was of intermediate yield (0.6 Mt) and occurred in October 1980. Both of these tests were conducted by the Peoples' Republic of China.

Man-made radioactive material from nuclear tests was the cause of widespread contamination of the environment. Much of this material was initially injected into the upper atmosphere, then dissipated on earth as fallout. The type and composition of the nuclear device markedly affects the kinds of radionuclides produced, while the size and location of the detonation determine the quantity of radioactivity which is released to the atmosphere. Table 3 lists the approximate yields of several important radionuclides produced in nuclear explosions as fission products and air and soil neutron activation products. As can be seen from this table, many of the radionuclides of concern are the same as those of concern from the nuclear fuel cycle.

The use of nuclear weapons can add significantly to radioactivity in the environment. For example, as a result of atmospheric testing, the amount of carbon-14 in the atmosphere in 1962 was 50 percent higher than it was prior to the start of nuclear weapons testing. A study of a group of New Yorkers showed that their average cesium-137 content was 1.4 Bq per kilogram of body weight in 1961 and 4.1 Bq per kilogram in 1963. With the test ban between the USA and the USSR, there was a steady reduction in cesium-137 levels so that by 1969 the levels had dropped to about 0.74 Bq per kilogram of body weight.

Consumer Product Exposure

The estimate of the average population exposure from consumer products varies widely up to several μSv per year. Radiation dose comes from such sources as television and other types of electronic equipment, luminous dials, smoke detectors, and microwave ovens. The government and industry are continually working to reduce public exposure from such sources.

Occupational Exposure

Occupational exposure is exposure obtained by individuals as a result of their occupation. It includes workers in the nuclear power field as well as medical personnel, industrial workers, and researchers who, in the course of their work, are exposed to radiation fields and/or radioactive materials. It also includes professionals such as airline personnel who are exposed to higher levels of cosmic radiation as they fly through the upper atmosphere. A 6000 mile trip on a jet liner will give an individual about a 40 μSv dose.

METHODS OF MONITORING FOR LOW LEVEL RADIOACTIVE MATERIALS IN THE ENVIRONMENT

A. Methods of Sample Treatment

The radioactive sample to be analyzed must initially be selected in such a manner that it accurately represents the material of interest. Then the sample

TABLE 3

Approximate Yields of the Principal Radionuclides per Megaton of Explosion

Sources Radionuclides	Half-life		Yield (PBq/Mt)*
Fission Products			
^{89}Sr	52.7	d	590
^{90}Sr	27.7	y	3.9
^{95}Zr	64	d	920
^{103}Ru	40	d	1500
^{106}Ru	368	d	78
^{131}I	8.04	d	4200
^{136}Cs	13.7	d	32
^{137}Cs	30.0	y	5.9
^{140}Ba	12.8	d	4700
^{141}Ce	32.5	d	1600
^{144}Ce	284	d	190
Activation Products (air)			
^{3}H	12.3	y	3.7×10^{-5}
^{14}C	5730	y	1.26
^{39}Ar	269	y	2.2×10^{-3}
Activation Productsd (soil)			
^{24}Na	15	h	1×10^{7}
^{32}p	14.3	d	7×10^{3}
^{42}K	12	h	1.1×10^{6}
^{45}Ca	165	d	1.7×10^{3}
^{56}Mn	2.6	h	1.3×10^{7}
^{55}Fe	2.9	y	6.3×10^{3}
^{59}Fe	46	d	8.1×10^{1}

*NOTE: 1 PBq (peta becquerel) = 10^{15} disintegrations/sec

treatment procedures may be applied to increase the concentration of the radionuclide of interest. Sample treatment procedures include:[10] (1) bulk segregation by physical methods (e.g., evaporation, filtration), chemical methods (e.g., ashing), and biological processes; (2) chemical purification; and (3) isotope enrichment or selection.

Bulk Segregation

Removal of the major part of the sample matrix is often accomplished by a relatively simple bulk concentration process. This frequently serves as a first step for more complex chemical processes. These processes can be categorized as physical, chemical, and biological processes. Among the physical concentration techniques, evaporation and volatilization are probably the most common. Probably the most widespread separation technique employed is the evaporation of water samples followed by the measuring of the gross alpha and beta activity of the non-volatile residue. Filtration, particularly in the case of air samples, is also widely used to collect airborne particles. The more commonly used chemical concentration techniques include ion-exchange, combustion, low-temperature ashing, coprecipitation, and adsorption. Ion exchange is the main technique for chemically concentrating ^{131}I from bulk milk samples.

Chemical Purification

Chemical separation can be used to isolate groups of elements and single elements. Chemical purification techniques include precipitation and coprecipitation; volatilization; solvent extraction; diffusion; ion-exchange and gas-chromatography; and selective electrodeposition or reduction. Precipitation and coprecipitation are basic steps in the analysis of strontium-89 and strontium-90 in water and biological samples.

Isotope Enrichment

Sample treatment involving isotopic separation according to mass number (A) is complementary to chemical separation according to atomic number (Z). Isotope enrichment techniques include mass spectrometry, distillation, electrolysis, thermal diffusion, and gas-chromatography. A commonly used technique for increasing the detection of tritium in water is the electrolysis of the water sample resulting in a residual water sample significantly enriched as to its tritium content.

B. Lower Limits of Detection

The lower limit of detection (LLD) is the minimum activity of a radionuclide that must be incorporated in a sample to give evidence of its existence. This LLD is a characteristic of the detection system. Since counting time cannot be made indefinitely long, the counting time and the background activity determine the detection limit.

If the confidence and error probabilities in the determination of an activity-limit are chosen to be 0.95 and 0.05, respectively, it can be shown that the LLD at 95% confidence level is given by:[11,12]

$$\text{LLD} = \frac{2k\sqrt{2}\sqrt{B}}{60E \cdot T \cdot Q} = \frac{4.65\sqrt{B}}{60E \cdot T \cdot Q} \quad (\text{Bq/quantity})$$

where

    k = 1.645; the value of the standardized normal deviate that is exceeded with probability 0.05

    B = the background count

    E = counting efficiency in counts per disintegration

    T = counting time in minutes

    Q = quantity of sample (volume or mass)

Table 4 lists typical sources of LLD for environmental samples.

TABLE 4

Typical Lower Limits of Detection for Environmental Sample Analysis

Isotope	Sources	LLD		
		Air (mBq/m^3)	Water (Bq/ℓ)	Soil (mBq/g)
^{3}H	CR, FP, AP	70	4	—
^{14}C	CR, AP, ^{14}N(n,p) ^{17}O(n,α)	7.0	1	—
^{32}P	CR, AP	—	0.04	—
^{40}K	PR	4.0	4	2
^{54}Mn	AP	—	0.4	2
^{55}Fe	AP	0.4	0.4	2
^{65}Zn	AP	—	0.7	4
^{85}Kr	FP	40	—	—
^{90}Sr	FP	0.04	0.04	2
^{95}Zr	FP	—	2	—
^{106}Ru	FP	—	2	—
^{131}I	FP	0.2	1×10^{-2}	—
^{137}Cs	FP	0.4	0.4	2
^{140}Ba	FP	0.4	0.4	—
^{140}La	FP	0.4	0.4	—
^{222}Rn	PR	40	—	—
^{226}Ra	PR	—	0.07	20(1)*
^{232}Th	PR	—	—	1.5
^{235}U	PR	—	—	1.5
^{239}Pu	AP	—	—	1.5
^{241}Am	AP	—	—	1.5
Gross Alpha		—	0.04	200
Gross Beta		0.1	0.04	—

*By ^{222}Rn emanation techniques

Symbols:

CR—cosmogenic radionuclides	AP—activation products
FP—fission products	PR—primodial radionuclides

C. Measurement of Gamma-Ray and X-Ray-Emitting Nuclides

The most valuable technique of measuring low-level radioactive nuclides is gamma-ray spectrometry. The measurement system consists of one or more scintillation or semiconductor detectors coupled to a multichannel pulse-height analyzer. This system provides for rapid simultaneous measurement of many gamma emitting radionuclides in the same sample. The choice of an analyzing system for gamma-ray emitters is based on considerations such as energy resolution, intrinsic detector efficiency, sample size, time available for measurement, and the level of activity in the sample.

X-ray emitters are analyzed using thin semiconductor detectors or gas proportional counters connected to a multichannel pulse-height analyzer.

Scintillation Detectors

The most commonly used scintillator in low-level assay is NaI(Tl). It has been produced in single crystals of up to 0.75 m in diameter[13] and considerable thickness (0.25 m). Such large detector volumes, combined with the relatively high density (3.67×10^3 kg/m^3) and high atomic number, produce gamma-ray detectors with very high efficiency. The light-conversion efficiency of Na(Tl) is the highest among all the inorganic scintillators.

The efficiency of the counting system must be calibrated by using radioactivity standards in the same geometry as the sample. When the energy of the sample is below 250 KeV, the calibration must be carried out near the peak of interest because of the nonlinear response of NaI(Tl). The counting efficiency may be increased by increasing either the sample size or the size of the detector. For small samples, the efficiency is greatly increased by the use of well crystals that give almost 4π counting geometry.

A NaI(Tl) detector has poor resolution compared to a semiconductor detector. Thus peak area analysis and spectrum stripping techniques are often required in order to interpret quantitatively the scintillation spectrometer data, especially with sources emitting several gamma rays.

Semiconductor Detectors

The most powerful measurement technique presently used in low-level assay is gamma-ray spectrometry using semiconductor detectors. The most successful semiconductor detectors are made of silicon and germanium. The advantages of semiconductor detectors, compared to other types of radiation counters, are high energy resolution; linear response over a wide energy range; fast pulse risetime; and ability to operate in a vacuum. Most of the current systems utilize Ge(Li) or IGe detectors and have counting efficiencies between 5 and 40% relative to 7.6 x 7.6-cm NaI(Tl) scintillation detectors for 1.33 MeV gamma rays. They have energy resolution (FWHM) of about 2.0 keV, which is more than an order of magnitude better than NaI(Tl). Thin planar detectors, either Si(Li), Ge(Li) or IGe, are used in x-ray spectroscopy. Typical resolutions for this type of detector are about 200 eV at 5.9 keV.

The semiconductor detector is operated at liquid nitrogen temperatures. This is necessary to minimize the electronic noise and thermal excitation of carriers.

Gas-Proportional Counters

Gas-proportional counters are used for environmental samples containing radionuclides which are primarily x-ray emitters such as ^{37}Ar, ^{55}Fe, ^{125}I, and ^{129}I. Gas-proportional counting gives high resolution but has low efficiencies except in the x-ray range. Background can be reduced to very low limits. X-ray proportional counters are usually cylindrical, having very thin beryllium windows and a gas of high Z (noble gas) at a pressure of about 1 atm. The energy resolution (FWHM) is about 1-2 keV at 20 keV. Thus, gas-proportional counters are superior to scintillation counters in this energy range.

Of the three types of x-ray detectors mentioned—scintillation, semiconductor, and gas-proportional detectors, the Si(Li) semiconductor detector has the best energy resolution for x-rays. K_α and K_β lines can be resolved by the Si(Li) detector[13].

D. *Measurement of Beta-Particle-Emitting Nuclides*

When nuclides emit no gamma and x-rays, or emit them at a very low probability per decay, then the methods mentioned in the previous section are not appropriate. Such nuclides as ^3H, ^{14}C, ^{32}P, ^{35}S, ^{45}Ca, ^{85}Kr, ^{89}r, ^{90}Sr, and ^{90}Y are assayed by detection of their beta particles. Some properties of these radionuclides are listed in Table 5. Note that ^3H has a low energy beta of 0.0186 MeV. Thus tritium is the only nuclide in Table 5 that cannot be measured with a Geiger Muller probe with window thickness typically about 2.5 mg/cm^2.

Since the beta-particle energy spectrum is continuous, quantitative identification of a sample containing several beta-particle-emitters by spectrum analysis needs radiochemical separation of nuclides. An example is when ^{89}Sr and ^{90}Sr are contained in the sample. The ^{90}Sr is determined by chemically separating out its beta emitting daughter ^{90}Y and determining its activity[14,15]. The ^{89}Sr is then determined by subtracting the ^{90}Sr activity from the total strontium activity. Due to the short range of beta particles in solids and liquids, absorption and scattering by sample matrix must be determined for quantitative measurements.

The common methods used for assay of beta-particle emitters are gas-multiplication and liquid-scintillation.

Gas-Multiplication Counters

A typical gas-multiplication counting system for measuring low-level beta-particle emitting nuclides consists of either a GM or a proportional counter, a heavy shield for environmental gamma rays, and the source. Usually a guard detector operated in anticoincidence with the main detector is used to exclude counts from penetrating cosmic-ray particles. The gas-multiplication detectors may be divided into two types, i.e., solid source and gaseous source detectors. Counting systems for solid sources have been extensively developed commercially. They have automatic sample changing and data readout capability. The counting efficiency of the system should be calibrated, and the experimental conditions of calibration and counting unknown sources must be maintained as constant as possible.

Gas-multiplication detectors with internal sources are widely used for ^3H and ^{14}C sources, and they are sometimes used for ^{35}S and ^{85}Kr. In this system, the gas to be analyzed is mixed with a non-radioactive counting gas. The advantages of this system are high sensitivity and stability, and the disadvantages are the time and care required for source and sample preparation.

TABLE 5

Nuclear Properties of Beta-Particle-Emitting Nuclides

Isotope	Half-life	Maximum Beta Energy MeV (%)	Range of Beta (mg/cm²)	Gamma Energy MeV (%)
^{3}H	12.3 y	0.0186 (100)	0.58	—
^{14}C	5730 y	0.155 (100)	28	—
^{32}P	14.3 d	1.708 (100)	790	—
^{35}S	88 d	0.1674 (100)	32	—
^{45}Ca	165 d	0.254 (100)	60	0.0125(1.4x10⁻⁵)
^{85}Kr	10.76 y	0.06 (99.6)	6.5	0.514 (0.4)
		0.15 (0.4)	27	—
^{89}Sr	52.7 d	1.463 (100)	650	—
^{90}Sr	27.7 y	0.546 (100)	176	—
^{90}Y	64 h	2.27 (100)	1090	1.75 (0.02)

Liquid Scintillation Counters

Liquid scintillation counters are widely used for the assay of low levels of beta-particle emitting radionuclides. While all of the radionuclides listed in Table 5 can be assayed by the method of liquid-scintillation counting, it is especially useful for the low energy beta emitters such as ^{3}H, ^{14}C, and ^{35}S. Liquid scintillation counting systems consist of a scintillator cocktail containing the internal dispersed beta sample, two photomultipliers operated in coincidence, a temperature-control system, electronic circuitry to process the pulses, and a heavy radiation shield to reduce the background from environmental gamma rays. The source to be analyzed may be introduced into the scintillator in various forms. ^{14}C can be assayed as ^{14}CO$_2$ dissolved in an organic base, or ^{14}CO$_2$ can be converted to BaCO$_3$ and counted as a suspension. Tritium-containing samples can also be dispersed or dissolved directly in a scintillator solution.

E. Measurement of Alpha-Particle-Emitting Nuclides

Environmental alpha-particle emitters are primarily certain members of the uranium-235 -238, thorium decay series, and certain transuranic elements. The range of alpha particles in liquid and solid samples are very short, usually less than 0.01 cm. This dictates that the amount of matter between source and detector be minimized. Because of the high LET values and high energy (4-9 MeV) of alpha particles, they produce detector pulses much larger than those produced by most background sources. Thus the counting system can be designed to have very small background counting rates. Methods for measuring low-level alpha-particle radiation include scintillation counting, gas-filled detector, and nuclear-track detector.

Scintillation Counting

One of the earliest scintillators used for measuring alpha particles was crystals of ZnS(Ag). These detectors are capable of providing a high degree of discrimination between alpha particles and low LET radiation such as beta and gamma radiation. Other scintillators that are useful for the detection of alpha particles are the CsI(T1) and CaF$_2$(Eu). Liquid-scintillation counting systems can also be used for the assay of some samples, but the background count rates may be so high as to preclude their application to very low-level sources.

Generally, the alpha-particle emitting source is sealed in a thin capsule, one side of which is transparent and coated on its inner surface with scintillator crystals. An example of its use is the analysis of ^{226}Ra. Samples containing ^{226}Ra isotopes can be assayed if the sample is sealed so that there is no appreciable loss of ^{222}Rn. The ^{226}Ra contribution to the total alpha particle can be estimated by observing the ingrowth of ^{222}Rn activity.

Gas-Filled Detectors

Current and pulse mode ionization chambers can be used to measure the gaseous daughters ^{222}Rn and ^{226}Ra. A gas-filled proportional detector can be used to detect alpha particles from solid sources. This source may be inside the detector, or outside next to a thin window. Such systems can be designed to discriminate between alpha and beta particles from the same sample. A widely used example of this type of system is the low background gas flow proportional counters used to simultaneously determine the gross alpha and beta activity of samples such as evaporated water samples or air particulate filter samples.

Nuclear-Track Detectors

The nuclear track detector is based on the damage created in a solid along the path of a high LET particle such as an alpha particle. The damage along the path, latent track, may become visible under an ordinary optical microscope after etching with suitable chemicals under appropriate conditions. The counting of the tracks can be either by direct observation or by using a spark counter. The detector materials usually used are mica, quartz, cellulose nitrate, and soda-lime glass. The sensitivity of nuclear track detectors is high; they can be used to assay extremely low activity levels (~ 37 μBq). The disadvantages of nuclear track detectors are long exposure times, extremely tedious measurements, and poor resolution (~ 250 keV).

DISCUSSION

In no field of environmental monitoring do sensitivities exceed those of radiation monitoring. This is because in radiation monitoring one is detecting the decay of individual atoms. In this field, the limits of detection are steadily being lowered as new and improved methods of radionuclide concentration and detection are being developed.

It can be argued that even existing techniques are too sensitive, in that they detect the presence of radionuclides at concentrations many orders of magnitude below the levels at which harmful biological effects have been observed, and that such information can then be used to frighten the public.

A major problem in most of these techniques for environmental monitoring is that it usually takes between days and weeks before the monitoring results are available to persons responsible for interpreting whether or not a potential hazardous situation is beginning to develop. Thus, there is a need for the development of more sensitive real time environmental monitors capable of both qualitative and quantitative analysis of automatically collected radioactive environmental samples.

REFERENCES

1. International Commission on Radiological Protection, 1965, *Principles of Environmental Monitoring Related to the Handling of Radioactive Materials.* ICRP Publication No. 7, Pergamon Press, New York.
2. Environmental Protection Agency, 1976. *National Interim Primary Drinking Water Regulations.* EPA-570/9-76-003.
3. U.S. Nuclear Regulatory Commission. *Standard for Protection Against Radiation*, 10 CFR 20.
4. U.S. Nuclear Regulatory Commission, 1982. *Licensing Requirements for Land Disposal of Radioactive Waste,* 10 CFR 61.
5. Denham, D.H., P.A. Eddy, K.A. Hawley, R.E. Jaquish, and J.P. Corley, 1981. *Technology, Safety and Costs of Decommissioning a Reference Low-Level Waste Burial Ground,* NUREG/CR-0570 Addendum, Pacific Northwest Laboratory.
6. Eisenbud, M., 1973. Environmental Radioactivity, 2nd ed., Academic Press, New York.
7. Whicker, F.W. and V. Schultz, 1982. Radioecology: Nuclear Energy and the Environment, Vol. 1, CRC Press, Inc., Boca Raton, Florida.
8. United Nations Scientific Committee on the Effects of Atomic Radiation 1982. *Ionization Radiation: Sources and Biological Effects,* 1982 Report to the General Assembly, with Annexes, United Nations, New York.
9. Choppin, G.R. and J. Rydberg, 1980. Nuclear Chemistry Theory and Applications, Pergamon Press, New York, p. 342.
10. International Commission on Radiation Units and Measurements, 1972. *Measurement of Low-Level Radioactivity,* ICRU Report 22, Washington.

11. National Council on Radiation Protection and Measurements, 1978. *A Handbook of Radioactivity Measurements Procedures.* NCRP Report No. 58, Washington.
12. Medrano, G., 1979. Design and Implementation of an Integrated Radiochemical Laboratory for Radiological Environmental Analysis. *In: Advances in Radiation Protection Monitoring,* IAEA, Vienna.
13. Tsoulfanidis, N., 1983. Measurement and Detection of Radiation, McGraw-Hill Book Company, New York.
14. Dutton, J.W.R., N.T. Mitchell, E. Reynolds, and L. Woolner, 1974. Analytical Systems Applied to Monitoring the Aquatic Environment in the Control of Radioactive Waste Disposal, *In: Environmental Surveillance Around Nuclear Installations, vol. I,* IAEA, Vienna.
15. National Council on Radiation Protection and Measurements, 1976. *Environmental Radiation Measurements,* NCRP Report No. 50, Washington.

Management of Radioactive Materials and Wastes: Issues and Progress. Edited by S. K. Majumdar and E. Willard Miller. © 1985, The Pennsylvania Academy of Science.

Chapter Twenty-Four

DERIVED BENEFITS WHICH LEAD TO RADIOACTIVE WASTE

Kenneth L. Miller, M.S., C.H.P.
Associate Professor of Radiology
Director, Division of Health Physics
The Milton S. Hershey Medical Center
The Pennsylvania State University
Hershey, PA. 17033.

The vast majority of scientific knowledge has been obtained over the past one hundred years. Radioisotopes have aided significantly in our search for knowledge and they continue to provide an easily obtainable and economic source of answers to many questions. One of the first experimental uses of radioisotopes[1] was performed by G. de Hevesy and F. Paneth in 1913. They used a naturally occurring radioisotope of lead in an experiment to determine the solubility of lead sulfate and lead chromate in water. This tracer study gave them the most accurate determination to date. In 1923 de Hevesy used this same radioisotope of lead (radium-D) in the first application of radioisotopes in biological research[2] when he studied the movement of lead in bean seedlings. Man-made sources of radioactive iodine were first used in the treatment of thyroid disease in 1941[3]. For certain types of thyroid cancer, radioiodine is still the most successful treatment. Over the past forty years this radioisotopes has saved thousands of lives.

Abundant supplies of useful radioisotopes were not readily available until after the second World War. Nuclear Reactors provide the means for economically producing numerous radioisotopes. The following is but a sampling of the many ways in which these radiosotopes provided us with benefits:

INDUSTRIAL

Radioisotopes in industry provide quick answers to the flow of materials in enclosed systems, such as the flow of lubricants in engines. The rate of surface wear can be readily determined by making one surface radioactive. After a short operating time, the amount of radioactive material in the oils or on other surfaces in contact with the radioactive surface can be easily measured. In similar fashion various mechanical factors such as friction and wear of pistons, rings, valves, bearings, etc. can be tested in engines while they are in operation. When added to automobile tires or road surfaces radiosotopes provide clues to better tread design and improved compositions for highway surface. Radioisotopes of iron and other metals have been used in determining effectiveness of mixing in metallurgical processes such as the formation of alloys. Radioactive phosphorus has provided information on the formation of grain boundaries in iron while tritiated water has helped to trace the migration of water through roofing materials.

In the petroleum industry radiation sources have been used to determine hydrodynamic properties of oil bearing soils and rock formations through the measurement of variations in scattered radiation as the source is lowered through the various types of rock formation in test wells.

Oil and water flow patterns in subterranean geologic formations have also been determined using radioisotopes injected into wells. Trace amounts of radioisotopes can be used to indicate the boundaries between different sources of oil being pumped through transcontinental pipe lines or in pipelines being used by several different companies. Detection of the radioisotope permits appropriate and accurate separation of the oil at the distribution point. In pipe lines radioisotopes can quickly pinpoint areas of blockage or breaks such as the break illustrated in Figure 1.

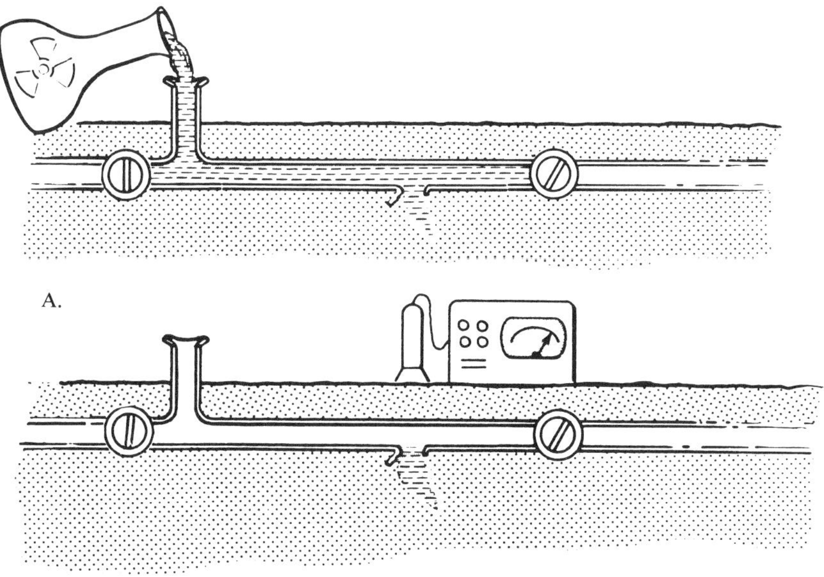

FIGURE 1. After radioactivity is added to the underground pipe (A) and flushed away, the break is easily found with a suitable radiation detector (B).

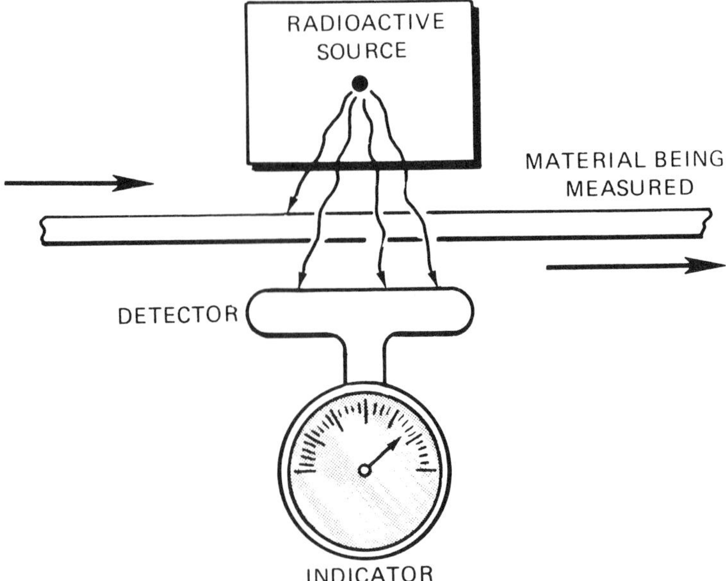

FIGURE 2. Gamma Ray Transmission.

FIGURE 3. Gamma rays from source A pass through the tank and are detected with the radiation detector, B.

Radioisotopes incorporated into various gauges play an important part in quality control in various industrial processes. Gamma rays which are transmitted (Figure 2) through products such as metals, plastics, paper and glass can be easily detected. The detected radiation intensity varies quickly as the thickness of material changes. These changes in detected radiation levels are used to provide instantaneous feedback to rollers to control thickness or sprayers to control surface coatings. Similarly, gamma ray transmission through processing tanks and lines is used to control flow of ingredients for uniform mixing and density control of liquids, powders, granules and slurries. Figure 3 shows a device used to halt the flow of ingredients once they reach a predetermined level in the tank or to provide a direct measurement of the density of the mixture inside.

Beta rays from similar sources are used to measure density and thickness when used in the manner illustrated in Figure 4. The number of beta rays which are reflected or backscattered provide a direct measurement of weight, thickness or density of the material being measured. Such devices are known as backscatter gauges, and have been used extensively to determine and control the thickness of coatings such as the amount of adhesive on tape; the amount of tin coating on steel used for the making of tin cans and the thickness of precious metal platings such as platinum and gold. Similar devices are also used to determine corrosion inside boilers or pipes and on ship hulls.

Radioisotope gauges have become very popular in industry because they are economical, require little power and maintenance and can be used in remote processing areas normally inaccessible to personnel. By rigidly controlling the amount of material in the finished product, these gauges help to decrease waste and save on raw materials. The radiation from these gauges can be shielded so they present no hazard to the user and the radiations will neither harm nor make radioactive the materials being examined.

In industry and construction there are many areas where it is impossible to

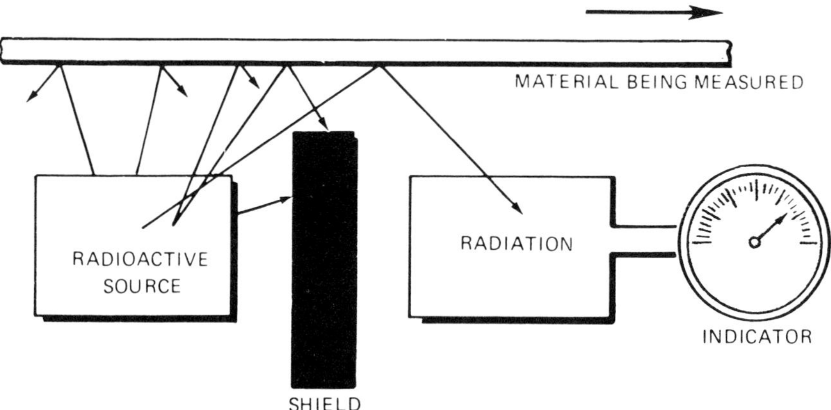

FIGURE 4. Beta ray backscatter gauge.

take or to use an x-ray machine. Under these circumstances the gamma rays from radioactive cobalt or iridium have proven very useful in obtaining radiographic images of metal welds or determining voids or cracks in materials. Cobalt-60 gamma rays have even been used to obtain x-rays of the Liberty Bell[4] and to determine its condition prior to moving it to a new pavilion on Independence Mall for better viewing during the Bicentennial. Devices containing sources of radioiodine-125 have been developed as a substitute for medical x-rays for use in the field where x-ray machines cannot go or sources of electricity are not available; similar devices have recently been approved[5] for use in hospitals.

Radiation's ability ot ionize air has led to the incorporation of sources in consumer products such as smoke detectors, static eliminators and dust prevention devices. Radioisotopes are incorporated in luminescent materials for use on watch and instrument dials and to provide illumination for aircraft exit signs, runway markers and navigation beacons. Although radium was originally used for this purpose, it has been replaced by less hazardous radioisotopes of H-3 (tritium), krypton-85 and promethium-147.

Radioisotopes are used in chemical processing to determine residence times of chemicals in solutions, flow rates, efficiency of separation in distillations and the fate of catalysts. Cobalt-60 gamma rays are used to initiate various chemical reactions such as that between ethylene and bromine to produce ethylbromide, a chemical used in drug manufacture; or for polymerization of ethylene to produce materials for packaging, laboratory tubing and various other uses. Likewise, radiation is used to induce cross-linking (chemical bonding) in polymers, a process that gives added strength to polyethylenes. Irradiated stretched tubing can be heat shrinked to provide adherent insulation on pipes, or close fitting plastic bags on frozen turkeys. Radiation is also used in the manufacture of detergents which are readily biodegradable. Irradiation of wood which has had its pores filled with monomethylmethacrylate causes polymerization of the monomer and its grafting to the cellulose of the wood. The resultant is a toughened wood product with a finish essentially distributed throughout the wood, making it ideally suitable for uses such as furniture veneers, wear resistant parquet flooring and billiard balls.

Radiation sources have been incorporated into analytical devices which are used for chemical analysis. One such device, which contained the radioisotope Curium-242, was loaded aboard Surveyor 5 and provided information on the composition of the moon. Still other sources have been incorporated into power generating devices which are used to provide power for satellites. Radioisotope power generators have also been evaluated for use in cardiac pacemakers.

ENVIRONMENTAL

Radioisotopes in procedures, devices and gauges have played an important role in environmental studies involving air and water pollution. Such devices can help determine deep water currents in lakes and oceans or sand and silt drift and accumulation in waterways. Carbon-14 as $^{14}CO_2$ has greatly increased our understanding of photosynthesis in both land plants and marine algae. Tracer studies which indicate water movement have provided better answers to snow melt and watershed management. Other studies have provided answers to the effect of environment on various substances.

AGRICULTURAL

In agriculture radioisotope techniques have yielded many answers regarding soils, plants, insects, microorganisms and animal nutrition. Some of the first agricultural studies involved the use of radioactive phosphorus-32 in analysis of plant nutrition. These studies provided answers to the proper and most productive use of fertilizers. Other studies have indicated the best techniques for plant disease, weed and insect control. At one time, the screwworm fly[6] caused millions of dollars in livestock losses yearly. The fly would lay its eggs in open wounds of livestock and the navels of newborns and the resulting maggots would burrow into the animal causing eventual death. The release of billions of radiation induced sterile males of this insect led to its successful eradication in the late 50's. Since that time, this same procedure has been successfully used to control other insect populations. Radiation from sealed sources has proven applicable in a variety of sterilizing procedures ranging from the elimination of insects from grain before storage to the destruction of anthrax bacterium in Australian goat hair.

In the processing of agricultural products, transmission gauges have proven useful in quality control. Such devices can readily distinguish thin shelled from thick shelled eggs thereby minimizing the risk of cracking in shipping. These gauges are also used to insure that each carton of milk has been filled to the proper height.

Radioisotopes have played an important role in nutritional studies involving vitamins, hormones and enzymes and their requirements for proper growth and well being of farm animals. Additional studies have produced answers to questions regarding the mechanisms of cholesterol formation and improved methods of artificial insemination. Sperm from prime breeder bulls can now be collected and preserved for use long after the death of the bull.

One of the most valuable usages of radiation in agriculture may prove to be in food preservation. Radiation exposure can extend the shelf life of highly perishable foods such as fruit, vegetable, poultry and seafood from several days to several weeks. Irradiation is as effective as chemical preservatives for cured meats and may present a lesser hazard than the chemicals. Radiation preserved meats have been served on a number of space missions, including the space shuttle.

Irradiation of seeds has been shown to decrease germination time and results in more rapid growth of plants which are more disease resistant, and they produce greater yields which are higher in nutrients. This same process can also be used to destroy insects in grains, such as wheat, rice and corn, greatly extending their storage times.

The various usages for food preservation and growth stimulation have not been approved for routine use in the United States because of fear that radiation might produce hazardous chemicals in the foods thus irradiated. Over thirty years of research around the world has not shown this to be the case. In 1981 the World Health Organization stated that there is no evidence for adverse effects from foods irradiated at low dose levels (up to one million rads). This has prompted the Food and Drug Administration to announce it is considering a new policy to facilitate the use of radiation in food preservation.

BIOMEDICAL RESEARCH

Academic and biomedical research at colleges and universities plays a key role in the development and testing of new products, equipment, techniques and procedures. Radioisotopes have played a ubiquitous role ranging from determination of the turnover of body water in athletes to the diffusion rates of various materials into semiconductor grade silicon. In the biological sciences alone, nearly every aspect of research involves radioisotope utilization in everything from the formation of macromolecules to the innermost workings of cells, from studies of the immune response systems to unlocking the key to the cause of cancer. Radioisotopes have provided quick and efficient answers to the kinetics of drugs and how and why they are effective in the body. The fate and action of new drugs can be determined quite easily when the drug is labeled with trace amounts of radioactive atoms. The uniqueness of radioactive atoms in this type of research is due to the fact that radioactive atoms are chemically identical to non-radioactive atoms of the same element. Therefore, radioactive atoms can be substituted for non-radioactive atoms in compounds and molecules without altering their functioning in any way. The radioactive compounds or molecules can be used to readily determine the specific action or involvements of the compound or molecule under question.

When researchers wanted to learn the effect of the mechanical heart on the various components of the blood of calves in which the mechanical heart was being studied, radioisotopes were used. A sample of the calf's blood was obtained and separated into its various components. The white and red blood cells and platelets were tagged with different radioisotopes and then reinjected into the calf. Evaluation of the radioactivity in subsequent blood samples, tissue samples and the mechanical heart itself provided easily obtainable information which would not have been readily available with any other technique. The effectiveness of chemotherapeutic agents in the treatment of breast and other cancers has been determined using the same agent tagged with trace amounts of radioisotopes.

The flow patterns of blood to the far reaches of the brain, other body organs and tumors are being determined using radioactive carbonized microspheres which are injected into the blood stream. These microspheres are so small that they only become trapped in the capillary beds. Analysis of the radioactivity in the capillary beds yields an exact mapping of the blood flow patterns.

The impact of such research from a societal standpoint, albeit large, cannot be determined. Indicators of improved life style, cures for disease, lessening of pain, and prolongation of life are difficult to obtain. Certain diseases have been completely eliminated and hopefully we will never again see such scourges as the Black Plague of the 14th Century which claimed the lives of 25 million or the influenza pandemic of 1918 which claimed 20 million. From a purely economical standpoint it has been estimated[7] for 1983 that in Pennsylvania alone, the dollar value of academic research contracts, involving radioisotope projects, is close to two hundred million dollars. Additionally, these research contracts are providing direct employment for approximately 5,300 individuals.

MEDICAL

Radiation plays an important role in sterilizing medical supplies and devices which cannot be subjected to heat or chemical sterilization, e.g. plastics (used in heart valves and catheters), pharmaceuticals, surgical dressings, prostheses, syringes, surgical instruments, sutures, ointments and solutions.

The use of very small amounts of radioactive carbon-14 in a clinical laboratory procedure provides a quick determination of bacterial infections in patients. Radioactive sources in detector cells of gas chromotography units and other types of analytical instruments can determine the presence of minute amounts of environmental pollutants such as pesticides, kepones and PCB's.

There are well over 100 clinical laboratory tests, known as radioimmunoassay (RIA)[7,8], which yield precise information on physiological and biological processes. These procedures involve antigens tagged with trace amounts of radioisotopes, principally I-125 and H-3. The tagged antigens compete with unlabeled antigens found in a sample of the patient's blood for specific binding to antibodies, allowing for the formation of an antigen-antibody complex. The high sensitivity of radioisotope counting techniques combined with the exact specificity of antibody recognition allows for the detection of minute amounts of antigen in the patient's blood.

RIA has been applied as a measuring tool to yield information relative to physiology, disease and hormone levels, and it is also used in the measurement of specific drug levels. In 1981[9] there were nearly 3½ million RIA procedures

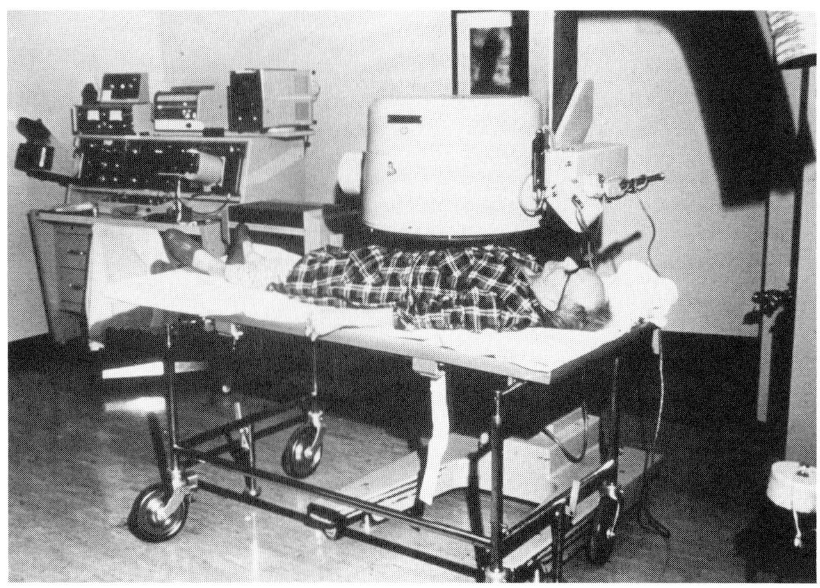

FIGURE 5. Nuclear medicine scan patient positioned under a gamma camera for a liver scan.

performed in Pennsylvania hospitals.

The imaging ability of diagnostic x-rays is due primarily to density differences which exists between bone and soft tissue or soft tissue and air. Diagnostic x-rays do not provide good images of those organs or regions of the body which are surrounded by material of similar density. Therefore, the soft tissue organs of the body such as the heart, thyroid gland, liver, spleen, brain, etc. do not image well with x-rays. Certain procedures have been developed which can introduce denser materials into some of these organs such as is done with cardiac catheterizations, barium swallows, barium enemas and angiography procedures (the injection of contrast materials into the blood vessels). However, these procedures do not answer all the questions, do not permit imaging of all the organs and provide very little information on physiological processes. Many of these deficiencies are overcome by nuclear medicine imaging procedures.

Many radioactive compounds have been developed which are physiologically specific for various organs of the body. When injected into the blood stream, these compounds are readily removed from the blood and concentrated in the organ of interest. Following injection of the radioactive compounds, specialized radiation detectors (known as gamma cameras) (Figure 5) are positioned over the organ of interest. These gamma cameras detect the gamma rays coming from the compounds and determine their exact location within the organ. The resultant images provide a fairly precise picture of the distribution of the material throughout the organ and indicate the shape, size and physiological functioning of it. Several examples which might serve to clarify these procedures are as follows:

THYROID IMAGING

When radioactive iodine-123 (half-life of 13 hours) is administered orally, it is quickly removed from the stomach by the blood system. An appreciable percentage of this iodine is then removed from the blood by the thyroid gland where it becomes organified[10] and incorporated into thyroid hormone as it is stored. The gamma rays coming from the iodine-123 can be easily imaged to give a picture of its distribution within the gland. The answers available[11] from the imaging include: (a) determination of thyroid size, (b) indications of any thyroid enlargement, (c) evidence for thyroid carcinoma, (d) detection of nodules, (e) possible cause of hyper or hypothyroidism, (f) location of aberrant or ectopic thyroid tissue, (g) indications of unilateral subacute or acute thyroiditis and (h) a means of performing postoperative evaluations of the gland.

Figure 6 indicates the difference between the images of a normal and a diseased thyroid gland.

LIVER IMAGING

Intraveneously administered molecules of technetium-99m (6 hours half-life) tagged sulfur colloid is readily phagocytized and removed from the blood by the reticuloendothelial cells of the liver. The images shown in Figure 7 were made through the detection of gamma rays coming from Tc-99m S.C. in the livers of two different patients. Such images provided indications[12] of (a) liver size, shape and position, (b) indications of abdominal masses, (c) detection of abscesses, hematomas, tumors and cysts, (d) location of lesions for biopsy, (e) pre-and post-treatment of liver metastases and (f) evaluation of diseases such as cirrhosis and hepatitis.

There are dozens of similar types of nuclear medicine imaging procedures routinely performed to evaluate the spleen, the kidneys, brain, bone and other body organs.

In addition to 650,000 nuclear medicine procedures[9], there are approximately 77,000 radiation therapy treatments performed each year in Pennsylvania. Although many of these treatment procedures are performed using radiation from accelerators (machines similar to high energy x-ray machines), a large number are performed using the high energy gamma rays from radioisotopes such as cobalt-60 (commonly referred to as cobalt treatments or cobalt-teletherapy). Additional treatments utilize the radiation coming from sealed sources[13] such as cesium-137, radium-226, gold-197, or iridium-192. These sealed

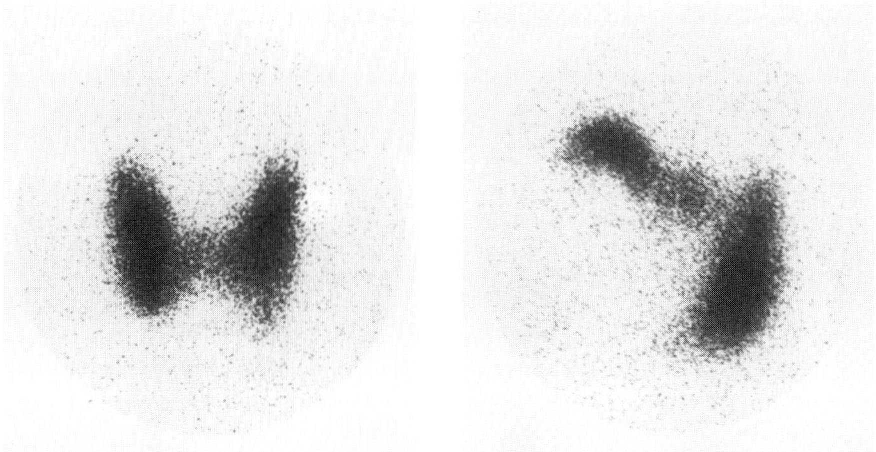

FIGURE 6. Normal thyroid scan on left, diseased thyroid scan on right.

FIGURE 7. Normal liver scan on left, diseased liver scan on right.

sources (the radioactive material is sealed in metal capsules so the radiation, but not the radioactive material, can get out) are fastened in the body using special holders for periods ranging from one to several days. After the prescribed treatment time, the sources are removed and returned to a shielded vault or safe and the patient can be discharged without representing a hazard to anyone. Occasionally low energy radiation sources such as iodine-125 seeds are permanently implanted in the body. Since the seeds are permanently fixed in the tissue and the radiation is essentially all absorbed by the patient's body, the patient can usually be released from the hospital after a short observation period.

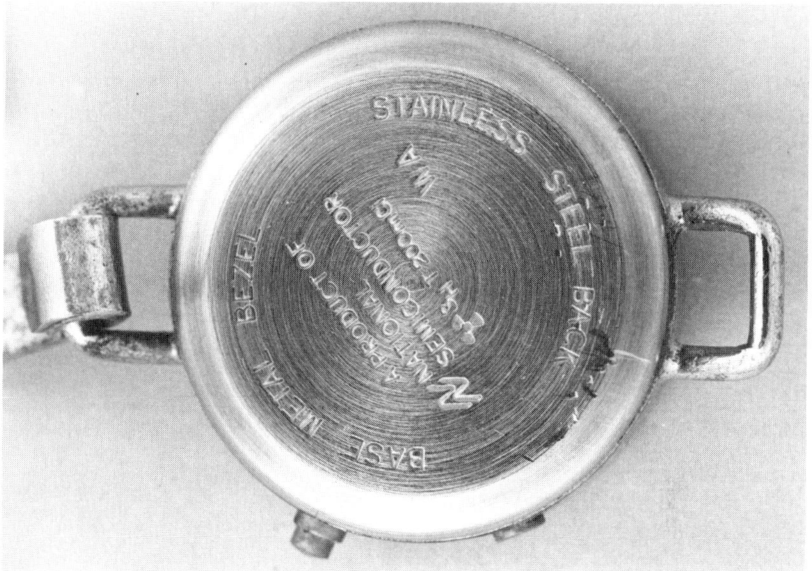

FIGURE 8. Watch containing 200 millicuries of Tritium (H-3).

ACTIVATION ANALYSIS

Non-radioactive atoms can be transformed[14] (activated) into radioactive atoms if they are bombarded with neutrons, charged particles or high-energy photons. Subsequent measurement of the radiation emitted by the activated atoms provides a means of analyzing for trace elements in the sample. When neutrons are the activation source the process is known as neutron activation analysis. Neutron activation analysis has found widespread application in forensic science. Sampling kits have been developed for use by police departments in determining if a suspect fired a gun. This procedure involves activation analysis of gun powder residue on a wipe sample of the suspect's hand. Analysis of air samples have been performed to determine if a victim were actually shot in the room in which he was found. Still other tests have been performed to determine the source of poisons such as arsenic.

At the Pennsylvania State University[15] neutron activation analysis has recently been applied in the study of pollution histories of geological regions. Samples taken from a 1250 year old California Redwood yielded specific information on the history of pollutants in the area. Other studies[16] have indicated fluctuations in diet during periods of social change and political consolidation through activation analysis of prehistoric bone samples. Studies are also being conducted to determine the flow rates of non-digestible particles through the digestive system of dairy cows.

Radioisotope usage has assumed an integral role in our scientific, technical, medical and industrial world. Radioisotopes are used because they represent the easiest, most economical and oftentimes the only way of getting the job done. Unfortunately, the use of radioisotopes can lead to an unwanted and troublesome byproduct, radioactive waste. It is imperative that such radioactive waste be processed and disposed of in a way that renders it innocuous. Failure to provide adequate and safe radioactive waste disposal capabilities would be comparable to "cutting off our noses to spite your faces". Typically, a truckload of university generated radioactive waste contains about the same amount of radioactivity as the watch pictured in Figure 8.

REFERENCES

1. Choppin, G.R. and Rydberg, J.: *Nuclear Chemistry Theory and Applications,* Pergamon Press, New York, 1980, 404-406.
2. Radioisotopes in Industry, Philip S. Baker, Domenic A. Fuccillo, Jr., Martin W. Gerrard and Robert H. Lafferty, Jr., an "Understanding the Atom" booklet prepared for the U.S. Atomic Energy Commission, Washington, D.C., 1965.
3. Beierwaltes, W.H.; Johnson, P.C., and Solari, A.J.: *Clinical Use of Radioisotopes,* W.B. Saunders Company. Philadelphia, 1957. 99-130.
4. Medical Radiography and Photography, published by Eastman Kodak Company, Rochester, N.Y. Vol. 52, No. 2, 1976,
5. Federal Register, Vol. 48, No. 125, Tuesday, June 28, 1983, 29677-29678.
6. Atoms in Agriculture, Thomas S. Osborne, an "Understanding the Atom" booklet prepared for the U.S. Atomic Energy Commission, Washington, D.C., 1968.
7. Skelly, D.S.: "Basic Principles of Radioimmunoassays" in*Nuclear Medicine in Vitro, 2nd Edition,* B. Rothfeld, Ed., J.B. Lippincott Company, Philadelphia, 1983, 45-100.
8. Narra, R. and Miller, W.: "Radioimmunoassay" in *Textbook of Nuclear Medicine Technology, 2nd Edition,* P.J. Early, M.A. Razzak, D.B. Sodee, Eds., C.V. Mosby Company, Saint Louis, MO, 1975, 295-318.
9. Low Level Radioactive Waste Disposal, a fact book prepared by the Pennsylvania Radwaste Working Group, September, 1983.

10. Baum, S.; Vincent, N.R.; Lyons, K.P.; Wu, S.Y.; Gurkin, S.C.: *Atlas of Nuclear Medicine Imaging* published by Appleton-Century-Crofts, New York, 1981, 123-157.
11. Okerlund, M.D.; Hoffer, P.B.; Matin, P.: "Thyroid Imaging" in *Clinical Nuclear Medicine,* P. Matin, Ed., Medical Examination Publishing Co., Inc., New York, 1981, 258-278.
12. Stadalnik, R.C., "Hepatic Scintigraphy" in *Clinical Nuclear Medicine,* P. Matin, Ed., Medical Examination Publishing Co., Inc. New York, 1981, 59-76.
13. Stovall, M.: "Brachytherapy Dosimetry" in *Handbook of Medical Physics, Volume I,* R.G. Waggener, J.G. Keriakes, R.J. Shalik, Eds., CRC Press, Inc., Boca Raton, 1982, 197-216.
14. Hendee, W.R.: *Radioactive Isotopes in Biological Research,* John Wiley and Sons, New York, 1973, 229-236.
15. 26th Annual Progress Report of the Pennsylvania State University Breazeale Nuclear Reactor, July 1981, PSBR 315-498194, 30-41.
16. 25th Annual Progress Report of the Pennsylvania State University Breazeale Nuclear Reactor, July 1980, PSBR 315-498093, pg. 53.

Management of Radioactive Materials and Wastes: Issues and Progress. Edited by S. K. Majumdar and E. Willard Miller. © 1985, The Pennsylvania Academy of Science.

Chapter Twenty-Five

THREE MILE ISLAND NUCLEAR ACCIDENT AND ITS EFFECT ON THE SURROUNDING POPULATION

George K. Tokuhata, Dr. P.H., Ph. D.

Director
Division of Epidemiology Research
Pennsylvania Department of Health
P.O. Box 90
Harrisburg, PA 17108
and
Professor of Epidemiology and Biostatistics
Graduate School of Public Health
The University of Pittsburgh

On the morning of March 28, 1979, a series of "unlikely events" at the Three Mile Island (TMI) nuclear plant led to a loss-of-coolant accident which became the most serious accident yet to occur in commercial nuclear power generation[1]. For several hours after the reactor first tripped, the reactor core was allowed to overheat. Up to 10 million curies of radioactivity have been estimated to have escaped into the atmosphere during a tense week of worldwide concern over the fate of the nuclear plant and its surrounding population.[2-3]

The maximum possible dose to a hypothetical person standing unprotected anywhere along the border of the plant site for the duration of the accident was estimated as no more than 100 millirems[4], the approximate equivalent of one year natural background radiation in the area. The average likely dose to persons living within 5 miles of the plant was estimated at 9 millirems[5]. At these low doses of radiation, no major health effects on the exposed population can be expected. The long-term health effects from the TMI radiation exposure to the more than 2,164,000 persons living within 50 miles of the plant at that time was projected as one excess cancer death over the lifetimes of these residents. The total number of excess health effects from TMI radiation, including all cases of cancer (fatal an nonfatal) and genetic ill health to all future generations, was estimated as two[4].

Despite these radiation estimates and learned opinions of several technical groups, including those from government, industry, national laboratories and universities, substantial amount of anxiety was created and resultant apprehension remained in the area. The public questioned the validity of the estimated radiation dose to local residents and also the health risk from that dose. The apprehension was due, in part, to the fact that radiation is invisible and its effects potentially pernicious. It was felt that even nonlethal doses are capable of causing immediate detrimental effects, especially on the unborn and the very young, as well as latent cancers and other chronic conditions. Many local residents actually believed that they received very high doses of radiation and some of them in fact developed a "radiation syndrome," a form of iatrogenic disease.

Health authorities in both the Commonwealth of Pennsylvania and the Federal government agreed that, because of the confusion and uncertainty surrounding the TMI accident from the beginning and because the nuclear accident was the first of its kind, the exposed population should be followed and studied for many years in order to monitor any possible changes in health status.[6-7] Also, because of the high levels of psychological distress experienced by the local residents during the crisis period and the likelihood of distress continuing over the many years needed to clean up the damaged reactor, psychological health and its sequelae were perceived as important outcomes to monitor independently of the issue of radiation exposure.

Psychological Stress and Health

Stress is an organismic state that can contribute, under the proper circumstances, to changes in body function, which, if intense or chronic, may lead to disease. In other words, stress can trigger a multiplicity of organismic reactions, some of which may contribute to illness, while others may result in normal adaptive responses.

Psychophysiological studies[8] indicate that life situations which threaten the security of the individual would evoke attempts at adaptive behavior and also evoke significant alterations in the function of most bodily tissues, organs and systems. These physiological changes, in turn, will lead to a lowering of the body's resistance to disease. It is assumed that certain events require more intense and prolonged coping efforts than do others. The greater the strains on the coping mechanisms, the more likely that an inadequate or inappropriate response will be utilized, thus eliciting idiosyncratic or pathological physiological reactions.

It is also important to recognize that an understanding of a life event's impact must take into account the physical susceptibility of the individual, the meaning of an event, the person's ability to cope with a variety of stresses and the individual's social support network. With the exception of extreme and sudden life-threatening situations, no raw stimulus is a universal stressor. The true consequences of stress arise from the manner in which the organism responds to the presumed danger. It is the way in which the organism handles perceived stressors—the defenses it mobilizes and the alarm reactions ignited—that constitutes the true nature of the stress.[9]

The psychosomatic approach,[10] on the other hand, identifies certain personality type and life history that would make them more vulnerable to certain diseases. Whenever a stimulus is perceived to threaten a fundamental human need, the stress response also will be inititated. Imagination can produce its own stressors and prompt a neuroendocrine-autonomic response that itself poses a real threat to the organism.

Stress can cause disease by lowering or exaggerating the immune response,[11] creating endocrine problems through either hypoactivity or hyperactivity,[12] altering the balance of autonomic control, resulting in changes in the cardiovascular, respiratory, secretory, and visceral system,[13] altering sleep patterns, with attendant impact on protein metabolism, hormone secretion and other vegetative functions,[14] and by affecting the functions of the brain itself, which can have profound impact upon health through a variety of mechanisms, including changes in eating and health habits, such as exercise, drug, alcohol or cigarette consumption.[15] Numerous studies have shown that the pituitary-adrenal axis may be activated or inhibited by fear, anger, rage, pain or adverse environmental conditions.

The stress associated with the TMI nuclear accident cannot be considered as a single unique experience because the prolonged recovery period following the accident gives rise to numerous additional stressors. However, it is unlikely that any given psychological stressor will be etiologically specific for any given disease. The important point is that a range of health outcomes, both mental and physical, need to be assessed in studies of stress or disaster since certain individuals may be more susceptible to health sequelae than others.

There are a number of studies in humans which have found an association between prenatal anxiety/stress and gestational, perinatal, and developmental pathology including complications of pregnancy[16-17] and infant growth and development.[18] These findings suggest a number of practical and scientific questions to be addressed within the context of the TMI Health Effects Research Program. The *first* is whether or not the local population, including pregnant women, as a whole experienced any detectable stress effects. The *second* question concerns factors which render individual pregnant women, particularly vulnerable to stress effects. As reviewed earlier, stress may be associated with morbidity only in the absence of supportive interpersonal relations.

While the specific mechanism of stress induced morbidity is not yet fully understood, there may be several different explanations with respect to pregnancy outcome; e.g., stress-anxiety induced changes (1) in maternal behavior, such as increased smoking, drinking or medication while pregnant, (b) in obstetric practice, such as increased prescription of analgesics and psychotropic drugs or use of special diagnostic procedures, (c) in maternal-infant bonding and child-rearing practices, and (d) in the hypothalamic-adrenocortical mechanism.[19]

Subsequent to the TMI nuclear accident, the Pennsylvania Department of Health developed a comprehensive plan for a variety of epidemiologic and sociological studies designed to assess the impact, both immediate and long-term, of the accident upon the local population. Some of the short-term studies are still in progress while long-term followup studies are being planned. Investigators of other organizations and institutions have also conducted short-term studies. In this paper, findings from certain major studies are briefly summarized in three categories: (a) psycho-behavioral studies, (b) physical health studies, and (c) long-term epidemiologic surveillance.

Psycho-Behavioral Studies

Although no immediate radiation health effects were recognized during the nuclear accident, and probably no delayed or late radiation health effects are to be expected, what emerged from this experience was that the major health effect of the accident appears to have been on the mental health of the people living in the region of TMI and of the workers at the nuclear power plant.

There was immediate mental distress produced by the accident among certain groups of the general population living within 20 miles of Three Mile island.[20] The highest levels of distress were found among adults living within 5 miles of TMI, or those with preschool children; and among teenagers living within 5 miles of TMI, those with preschool siblings, or whose families left the area. Workers at the TMI nuclear plant experienced more distress than workers at the Peach Bottom nuclear plant in Pennsylvania which was studied for comparison purposes. The level of distress was higher among the nonsupervisory employees and stress continued in the months following the accident.

Health-related behavioral studies conducted by the Pennsylvania Department of Health in collaboration with the Hershey Medical Center[21] indicated that persons who are younger, more educated, married and female were especially distressed during the crisis. The greater responsiveness of younger, married persons was probably due to their concerns about the effects of radiation on their present and future children and, since radiation effects often have a long latency, concerns about their own future health. However, these demographic variables did not relate to changes in the level of distress over time. People who actively coped had high distress during the crisis and tended to maintain that distress over time. Persons with poor mental or physical health tended to have high distress scores and to maintain their distress over time.

The number of persons with severe distress dropped shortly after the crisis, but between 10% and 20% of local residents residing close to TMI remained distressed nine months after the crisis. Persons residing close to TMI used more alcohol, tobacco, sleeping pills and tranquilizers during the two week period immediately following the crisis than before, but the use of these substances which were mediated through coping with the crisis situation did not persist beyong that time.

The October 1980 survey conducted by the Pennsylvania Department of Health in collaboration with the Hershey Medical Center indicated that the level of anxiety and stress declined more among residents within 5 miles of TMI than among those living more than 40 miles away. Thus, 18 months after the accident, the previously significant differences in stress-related symptoms, both behavioral and somatic which existed between the close and the far groups were no longer present. However, differences still persisted through October, 1980 as far as perceived threat of TMI and attribution of the recognized symptoms to TMI were concerned.

An in-depth epidemiologic study of psychological impact in a more *psychiatric context* was conducted by Bromet at the Western Psychiatric Institute.[22] Her study covered three selected "high risk" groups in the TMI area, namely, (a) TMI employees, (b) mothers with preschool children, and (c) mental health clinic patients. People residing around the undamaged nuclear plant at Shippingport in western Pennsylvania were used as controls for comparison. One year after the accident, the condition of psychiatric outpatients near TMI did not differ significantly from that of counterpart in the control group. She also found that TMI workers experienced only slightly higher rates of clinical depression and anxiety as compared with Shippingport workers. But, mothers of preschool children living within 5 miles of TMI suffered far more anxiety and depression than did mothers living near Shippingport. Bromet also found that mothers who evacuated in the height of the accident had more distress one year later than mothers who did not evacuate. Mothers living within 5 miles of TMI reported more distress symptoms than mothers living farther away from the plant. It was concluded that manifestations of clinical levels of mental health effects occurred primarily during the 2-month period after the accident, but sub-clinical levels of symptomatology were elevated as late as one year following the accident. There was evidence that social support bore an important relationship to these symptoms. Bromet's findings support a view that the burden of the stress was determined more by the actual experience, such as actual living in the vicinity of TMI, rather than by the perception of the stressful situation.

Related to the psychological stress caused by the TMI accident was *crisis evacuation* during the accident by local residents. Although the level of radiation exposure was minimal, a substantial number of residents in the vicinity of the TMI plant left the area primarily because of their perception of imminent danger associated with radiation. The Governor of Pennsylvania advised pregnant women and small children to evacuate. Within hours of the Governor's advisory and with mounting media coverage of the accident, which was often confusing, mass evacuation occurred. Some 64% of the population in the 5-mile area left their homes some time during the nuclear crisis. It is important to document individual evacuation as it can be related to estimating radiation exposure and the future health effects studies.

A total cross-sectional population census conducted by the State Health Department supported by the Federal Centers for Disease Control and Bureau of the Census shortly after the accident within five miles of the plant revealed that evacuation behavior was related to several demographic variables. Specifically, more younger people evacuated and for longer periods than older people. More females evacuated than males. The more educated and white collar workers evacuated somewhat more than the less educated and blue collar workers. The strongest predictor of evacuation was the presence of one or more preschool children in the household. Distance of residence from the damaged plant was inversely correlated with the decision to evacuate. There were no major differences in the pattern of evacuation between medical personnel and other residents in the same community, i.e., nurses and young women behaved similarly while physicians and middle-age men were alike in their evacuation behavior.

Radiation Exposure and Health Risks

Nuclear accidents, such as the 1979 episode at TMI, are potentially harmful to health if the amount of ionizing radiation absorbed by humans is substantially high. However, whether health is affected by exposure at the low levels characteristic of natural background radiation is a matter of conjecture. Observations at higher radiation intensities have implied, but are difficult to measure, that the risk of certain health effects may be increased even at the lowest dose levels. These effects may include any one or combination of the following: (a) damage to genes and chromosomes, or *mutagenic* effects, (b) damage to the growth and development of the embryo and fetus, or *teratogenic* effects, and (c) damage to cells that increases the risk of their forming cancer, or *carcinogenic* effects[23-24].

However, since health effects of radiation at the levels of natural background cannot be distinguished individually from similar effects produced by other causes, the effects of low-level radiation are estimated only by extraporation from observations at higher radiation doses and dose rates, based on tentative assumptions about the relevant dose-effect relationships. In the present state of our knowledge, such estimates must be regarded as highly uncertain at best[25-26].

The accidental radiation received by people residing in the vicinity of Three Mile Island (TMI) came almost entirely from xenon-133 (half-life, 5.3 days), xenon-135 (half-life, 9.2 hours), and traces of radioactive iodine (principally iodine-131, half-life, 9.0 days), which escaped intermittently from the plant as gases[27-28]. These radioactive gases followed prevailing winds and increased the level of ionizing radiation along their path. However, the increase was short-lived because xenon dispersed rapidly and because radioactive iodine was present only in barely detectable amounts. No release of long-lived fission products, such as strontium-90, cesium-137, and plutonium-239, was detected.

Based on the available measurements, it is estimated that the maximum cumulative whole-body gamma radiation dose to anyone off site was less than 100 mrem, that the average cumulative dose to those within 10 miles of the plant was approximately 8 mrem, and that the average cumulative dose to those within 50 miles of the plant was about 1.5 mrem. Because these estimates make no allowances for shielding, they are generally considered to represent over-estimates[27-28]. Additional exposure of the population came from the *beta radiation dose to the skin* and from the *inhalation dose to the lung.* It is estimated that the total dose to the skin could have been much larger than the whole-body gamma dose by a factor of 3 to 4 if the protective effects of shelter and clothing are neglected.[27] The inhalation dose is estimated to have constituted no more than 3% to 7% of the dose to the whole body.

The risk of cancer is generally assumed to be increased by low-level radiation, but it is clear from observations at intermediate-to-high dose levels that the risk may vary depending on the type of cancer in question, age at the time of irradiation, the quality of radiation, and other factors. According to a linear, nonthreshold extrapolation model, with no allowance for biological repair at low doses and low dose rates, cancer risks are regarded by many experts as being likely to overestimate the risks of low-level radiation. For this reason, some experts prefer a linear-quadratic model, which yields risk estimates that tend to be 25%-50% smaller[29-30]. If these risk coefficients are applied to the population of about 2.2 million people residing within 50 miles of Three Mile Island, they predict a lifetime risk of less than one extra fatal cancer and less than one extra nonfatal cancer.

It is generally assumed that irradiation can cause genetic damage in human germ cells that is transmissible to future generations in the form of various inherited diseases. It has been estimated that the incidence of genetic abnormalities in humans would be doubled by a dose of 20 rem -200 rem[25-26] and, that the number of descendants of the population within 50 miles of TMI who are likely to be affected by genetic disorders resulting from the TMI accident would be approximately one.

The risks of teratogenic effects of radiation on the human embryo and fetus are more difficult to estimate, owing to the paucity of relevant data. The evidence at hand implies, however, that the risks of such effects are smaller per unit dose than are the risks of carcinogenic and mutagenic effects[25-26]. On this basis, it may be inferred that such effects are unlikely to result from the TMI accident in view of the small magnitude of the radiation dose.

Physical Health Studies

Although increased risks of cancer, birth defects, and genetic abnormalities are potential long-term consequences of low-level irradiation, few if any such effects of the TMI accident are likely to be observed, because the collective dose of radiation received by the population within a 50-mile radius of the plant was so small.

In order to evaluate the potential effect of radiation and/or acute stress upon reproductive process, an epiodemiologic study was conducted to determine whether the incidence of *spontaneous abortion* was greater than expected near the Three Mile Island nuclear plant during the months following the March 28, 1979 accident. All persons including those who were pregnant living within five miles of TMI were registered shortly after the accident, and information on pregnancy at the time of the accident was collected. After one year, all pregnancy cases were followed up and outcomes ascertained. Using the life table method, it was found that, given pregnancies after four completed weeks of gestation counting from the first day of the last menstrual period, the estimated incidence of spontaneous abortion (miscarriage before completion of 16 weeks of gestation) was 15.1 percent for women pregnant at the time of the TMI accident. Combining spontaneous abortions and stillbirths (delivery of a dead fetus after 16 weeks of gestation), the estimated incidence was 16.1 percent for pregnancies after four completed weeks of gestation. Both incidences are comparable to baseline studies of fetal loss, indicating that the effects of the TMI accident upon spontaneous abortion was negligible, if any.

The crisis at Three Mile Island presented a natural experiment in disaster response, although this disaster was substantively different from any before it. Not only was this the first to involve a nuclear plant, but no one was bodily hurt, no property outside the nuclear facility was physically damaged and, it is generally believed, no excess deaths or illness will be detected as a result of the accident. Nevertheless, a disaster situation was experienced psychologically and emotionally by the nearby population.

A study was conducted by the Pennsylvania Department of Health to determine the effect of the 1979 nuclear accident at Three Mile Island on *residential mobility* and subsequent population composition. The entire population living within five miles of TMI was registered shortly after the accident and traced one year later to identify movers. The results of this analysis showed that the rate at which people moved remained the same the year after the accident as before, and that approximately 15% of those who moved (changed address) gave TMI as the main reason for their decision to move. The study also found that those moving because of TMI had attributes highly associated with mobility in general. When those attributes were controlled in analysis, attitudes about TMI were virtually the same among movers and nonmovers. On the other hand, demographic characteristics of new people moving into the area were not different from those who had moved out. However, attitudes about TMI were significantly more positive among the newly moved-in people than among the moved-out people.

Probably the most important study developed shortly after the accident was to determine if the TMI nuclear accident has had any measurable impacts upon *pregnancy outcome* and *infant health* in the vicinity of the damaged nuclear reactor. The embryo, the fetus and the infant are highly sensitive to environmental insults, such as ionizing radiation and maternal psychological stress, depending upon the severity or intensity of the insults, the mode of exposure, and the gestational-postnatal age at exposure.

A carefully designed retrospective cohort study of pregnancy outcome was initiated in August, 1979. This study covered all pregnant women residing within a 10-mile radius of the TMI plant, who gave births between March 28, 1979 and March 27, 1980. This study cohort consisting of some 4,000 deliveries was compared with a control cohort of another 4,000 deliveries which took place during the immediately following one year period for women who also resided in the same 10-mile area communities.

Measures of adverse pregnancy outcome investigated were:*fetal deaths* (stillbirths with 16-week or more gestation including abortions after 16-week gestation), *neonatal deaths* (deaths within 28 days postpartum), *hebdomadal deaths* (deaths within seven days postpartum),*perinatal deaths* (combined measure of fetal and neonatal deaths), *prematurity* (gestation less than 37 weeks), *immaturity* (birth weight 2,500 grams or less), *congenital anomalies* (one or more developmental defects observed at birth), and *low Apgar score* (less than seven at one minute of delivery).

Since there are numerous factors other than radiation and stress that are known or suspected to influence the course of pregnancy and fetal outcome, it is important that the influences of such factors be considered. Detailed data on these factors have been collected, including maternal characteristics (sociodemographic, behavioral, and medical-obstetric histories), health care provider characteristics, and prenatal care attributes. The influences of all these factors were taken into account when maternal stress and/or radiation exposure were related to any of the various pregnancy outcome measures under study.

Maternal stress during and immediately following the TMI accident has been measured by overt personal statements of "anxiety-fear" as experienced and reported by individual pregnant women, and by actual stress-coping patterns described, such as taking extra medications (tranquilizers, sleeping pills, anti-hypertensive preparations, etc.) because of anxiety and fear.

Maternal radiation exposure during the 10-day crisis following the nuclear accident has been estimated by the Department of Radiation Health, University of Pittsburgh. For this purpose, already documented, reliable thermoluminescent dosemetry (TLD) and other source data including time-dependent dose-rate distribution compiled by government and non-government agencies were used to estimate *maximum possible* and *most likely* doses, to each individual pregnant woman, of whole-body gamma, thyroid doses to the mother and the fetus as well as combined gamma and beta doses to the skin. For estimating *maximum possible* doses the evacuation factor was not considered, but for determining *most likely* doses this factor was taken into account, i.e., those who evacuated during the accident were assigned smaller doses depending upon when and how long evacuation took place on an individual basis.

When pregnancy outcome measures were compared between the exposed study cohort and the unexposed control cohort, no significant differences were noted for any of the various outcome measures under study indicating that the impact of the TMI nuclear accident upon pregnancy outcome was negligible, if any. After adjusting for the influences of the many maternal and provider characteristics described earlier, the incidences of fetal and neonatal mortalities, congenital anomalies, prematurity, immaturity, and of low Apgar score within the study cohort were not significantly different from those within the control cohort.

A separate analysis of the comprehensive data by multivariate logistic analysis indicated that neither radiation exposure nor psychological stress as such was significantly correlated to the incidence of fetal-neonatal mortality, congenital anomalies, prematurity, immaturity or low Apgar score within the exposed study cohort.

It should be noted, however, that the excess medication taken by those pregnant women who were severely stressed during and/or shortly after the accident was significantly correlated to the incidence of low Apgar score which was measured at one minute postpartum, and to the incidence of immaturity, i.e., the risk of low birth weight. This was interpreted to mean that the one-minute Apgar scores among newborns were significantly influenced by maternal excess medication of tranquilizers, sedatives, and anti-hypertensives which was mediated through the accident-caused stress and anxiety. Our data also indicated that the low Apgar score at 5 minutes postpartum was not significantly correlated to the same maternal excess medication while prenant. This may suggest that the low Apgar score is a negative, but only a very short-term prognostic indicator with probably minimal clinical significance. However, the stress-mediated low birth weight can be a potentially significant long term health effect which requires special attention.

Apart from the above observations on pregnancy outcome, there was one other potentially important observation to be made particularly with respect to the effect of radioactive iodine upon thyroid function among newborn infants. Since State Health Department initiated a statewide screening program for congenital hypothyroidism in mis-1978, the available data were analyzed in relation to the March 28, 1979 nuclear accident.

During the March 28, 1979—March 27, 1980 pediod, only one case of *congenital hypothyroidism* was identified within a ten-mile radius of TMI among approximately 4,000 newborn infants. This incidence rate is well within a normal range of expectation.

An apparent clustering of seven cases of congenital hypothyroidism reported in Lancaster County during 1979 presented serious interests among epidemiologists and was subjected to a special in-depth analysis and investigation because of physical proximity of the county and timing of the TMI nuclear accident. From this investigation the following diagnostic and epidemiologic features emerged: (a) One of the seven cases identified was reported prior to the TMI accident, thus cannot be related to the nuclear accident. (b) One with severe multiple contral nervous system anomalies was born three months after the accident; this case is unlikely to have been associated with the TMI accident because of the late gestation period of the fetus when exposed to the accident, and also of coexisting other developmental anomalies which are unlikely to be related to radiation. (c) One case was of dysgenesis, representing one of discordant Amish twins, thus, non-supportive of the etiology secondary to radiation exposure. (d) One case of dyshormonogenesis from an Amish family where the condition (lack of enzyme to synthesize thyroxine) was inherited from the parents. (e) Another case of dysgenesis in whom the thyroid glands were displaced from the normal position. (f) For the remaining two cases thyroid scan was not conducted, thus, exact diagnostic entity remains unknown.

Having completed detailed diagnostic analysis and epidemiologic assessment of all cases reported in Lancaster during 1979, it was concluded that reported cases of congenital hypothyroidism were not related to the TMI nuclear accident, i.e., these types of anomalies are not expected to have resulted from direct or indirect exposure of the fetus to radioiodine. This conclusion was also supported by an independent Hypothyroidism Investigative Committee organized by the State Health Department, which included expertise in the fields of epidemiology, pediatric endocrinology, obstetrics, medical genetics, biostatistics, and radiation physics.

Apart from the incidence analysis described above, there was also an important biological consideration with respect to radiation in relation to congenital hypothyroidism.

First, after March 28 through December 31, 1979, no single case of congenital hypothyroidism was reported in Dauphin, Cumberland, Perry, Northumberland, Juaniata, Snyder, Mifflin, and Union Counties, the areas downwind (N, NW, NNW) from the Three Mile Island during the first 48 hours of the accident, when probably the largest amount of radioactive releases took place, thus the largest amount of contamination including I^{131}.

Second, the maximum combined (inhalation and ingestion) human thyroid dose of radioactive iodine in the vicinity of the TMI following the March 28, 1979 accident through April 1979 is estimated to be 7.5 mrad (Editorial: Annals of Internal Medicine, Vol. 91, No. 3, September 1979). At least 1,000 times greater thyroid doses (i.e., 7.5 rads) would be required to have significant acute damages to the thyroid glands; however, even at this dose level, many of the damaged cells may be repaired. Based on the experiences of the Marchallese exposed to fresh radioactive fallout and atomic bomb victims, it is considered likely that as much as 50 rads to 100 rads fetal thyroid doses would be necessary to cause irreversible tissue damages, such as congenital hypothyroidism and/or thyroid cancer. Acknowledging the fact that the fetal thyroid is much more sensitive to radioiodine than is the maternal thyroid (a conservative upper bound estimate is that the thyroid dose to a fetus may be as high as ten times the maternal thyroid dose), the maximum likely fetal thyroid dose of approximately 75 mrad and the maximum possible fetal thyroid dose of 190 mrad to 200 mrad in the vicinity of the damaged nuclear plant are still far too small to have caused congenital hypothyroidism.

In any epidemiological investigation of possible "cluster" of a disease or morbid condition, it is important to recognize the technical difficulty and methodological limitations associated with such investigations. It is the overall consistent pattern of observations that provides useful clues for conclusion, rather than a single isolated change or difference, which in most cases occurs without substantive epidmeiologic significance. This is particularly true when relatively small populations are being studied. One may or may not find a "statistically significant" change, difference, or clustering in morbid rates in an area depending upon how such population is delineated geographically and/or temporally. It is equally important that investigators carefully examine the observed relationships and determine if such relationships are consistent with the known biological theory or orientation, which is based on the previous studies and experiences. Out conclusions regarding congenital hypothyroidism around the TMI nuclear plant have been based on both the overall pattern of epidemiologic observations and in reference to existing scientific knowledge.

LONG-TERM EPIDEMIOLOGIC SURVEILLANCE

TMI Population Registry

Within three months after the March 1979 nuclear accident, a cross-sectional population census of some 36,000 persons living within 5 miles of the plant was undertaken jointly by state and federal governments.[31] The information collected through the census provided baseline data for future epidemiologic studies of possible health effects of the TMI accident. The data base, known as the *TMI Population Registry,* is comprised of demographic characteristics on each resident and a brief medical history of cancer diagnoses, thyroid disorders, prior radiation therapy and exposure to ionizing radiation on the job. Smoking histories were also included for teenagers and adults. In addition, each person's daily travel in and out of the 5-mile area during the 10-days after the accident was recorded so that TMI-related radiation doses could be estimated from the already documented time-place dependent radioactivity distribution in the area. After two months of data collection, the TMI Population Registry was considered to be 95 percent complete in coverge. For each resident included in the Registry, two radiation dose estimates (maximum possible and most likely) were given with respect to wholebody gamma and thyroid tissue respectively. Living status and whereabout of the registrants are updated annually for future contacts.

TMI Mother-Child Registry:

Within five months following the TMI accident, a carefully designed retrospective cohort study of pregnancy outcome was initiated.[32] This study included two separate cohorts, the exposed study group and the unexposed control group, all residing within 10 miles from the damaged nuclear plant. In each group there were approximately 4,000 mother-child pairs which constitute the *TMI Mother-Child Registry.* For each registered pair, detailed information regarding maternal characteristics and perinatal characteristics of the index infant were recorded. For the exposed study pairs estimated radiation doses (wholebody gamma and thyroid tissue) and the proxy measure of maternal stress during and shortly after the accident were documented on an individual basis, which can and will be related later to the various measures of possible long-term health effects. The TMI Mother-Child Registry includes 94% of all eligible cases of pregnancy in the area and provides the necessary baseline data for long-term epidemiologic studies. Living status and whereabout of all registrants are updated annually in preparation for such studies.

Objectives of Long-Term Studies

The aim of the TMI Health Effects Research Program is to provide factual information based on such studies which are epidemiologically sound and/or sociologically justified with respect to possible health effects of the TMI accident upon local residents. Based on the available TMI radiation exposure data and from the previously reported epidemiologic studies of low dose radiation, major adverse health effects from the TMI accident are not expected. Although this may provide assurance to many people at potential risk, the assurance is only as good as the radiation data itself, which has become a subject of debate. There is also a possibility that psychological stress from the accident and its aftermath, which has been well documented, will cause some adverse health effects among the TMI residents.

Although the effect of psychological stress is difficult to predict, these public health concerns should be addressed. We are taking a precautionary route by carefully documenting both the exposed population and its health experiences after the nuclear accident. The already established TMI Population Registry and the TMI Mother-Child Registry will provide reliable data bases for long-term followup studies of the health effects (physical, psychological and behavioral), if any, from the TMI nuclear accident for both the general population and for the special cohort of pregnant women and their in-utero exposed children. Causes of death and cancer diagnoses will be routinely ascertained by linkage to the State mortality and cancer incidence files. Data for other physical, psychological and behavioral health indices will be collected every five years, on the basis of a random sample through prospective followup surveys for both cohorts.

Regardless of the results of a variety of short-term and long-term studies undertaken, the primary mission of the TMI Health Research Program is to fullfill the need to respond to the much publicised, potentially important public health concerns. Because of the uniqueness of the TMI nuclear accident, thus its historical significance, as well as the scientific need to document health effects of very low dose radiation in humans, the rare opportunity presented by the TMI nuclear accident should not be lost in the pursuit of these important epidemiologic studies.

REFERENCES

1. Kemeny, J.G. (Chairman): Report of the President's Commission on the Accident at Three Mile Island. U.S. Gov. Printing Office, Washington, D.C., 1979.
2. Nuclear Regulatory Commission Special Inquiry Group: Three Mile Island, A Report to the Commissioners and to the Public. Vol. 1. U.S. Nuclear Regulatory Commission, Washington, D.C., 1980.
3. Woodward, K. Assessment of Offsite Radiation Doses from Three Mile island Unit 2 Accident. Pickard, Lowe, and Garrick (Consultants to Metropolitan Edison Corp.,), 1979 (TDR-TMI-116, revision O, August 31).
4. Ad Hoc Population Dose Assessment Group: Population Dose and Health Impact of the Accident at the Three Mile Island Nuclear Station. U.S. Govern. Printing Office, Washington, D.C., 1979.
5. Gur, D., et al. Radiation Dose Assignment to Individuals Residing Near the Three Mile Island Nuclear Station. Department of Radiation Health, University of Pittsburgh, PA 1983.
6. Committee on Federal Research in the Biological Effects of Ionizing Radiation: Followup Studies on Biological and Health Effects Resulting from the Three Mile Island Nuclear Power Plant Accident of March 28, 1979. NIH Pub. No. 79-2064. U.S. Dept. of Health, Education and Welfare, Washington, D.C., 1979.
7. Tokuhata, G.K. Three Mile Island Health Effects Research Program. Proceedings of the 56th Meeting, Pennsylvania Academy of Science, April, 1980.
8. Rahe, R., et al. Social Stress and Illness Onset. J. of Psychosomatic Research 8:35, 1964.
9. Zegans, L. Stress and the Development of Somatic Disorders. In: Goldberger and Breznitz (eds.). Handbook of Stress: Theoretical and Clinical Aspects. New York: Free Press, 1982.
10. Dunbar, H. Psychosomatic Diagnosis, New York: Hoeber, 1954.
11. Stein, M., Keller, S., and Schleifer, S. The Hypothalamus and the Immune Response. In: Weiner, Hofer, and Stunkard (eds.) Brain, Behavior and Bodily Disease. New York: Raven, 1981.

12. Lipton, M. Behavioral Effects of Hypothalamic Polypeptide Hormones in Animals and Man. In: Sachar (ed.) Hormones, Behavior and Psychopathology. New York: Raven, 1976.

13. Lisander, B. Somato-autonomic Reactions and Their Higher Control. In: Brooks, Koizumi, and Sato (eds.) Integrative Functions of the Autonomic Nervous System. New York: Elsevier, 1979.

14. Weitzman, E., Boyar, R., Kapen, S., and Hellman, L. The Relationship of Sleep and Sleep Stages to Neuroendocrine Secretion and Biological Rhythms in Man. Recent Progress Hormone Research 31:399, 1975.

15. Antelman, S. and Caggiula, A. Norephinephrine-dopamine Interactions and Behavior. Science 195:646, 1977.

16. Nuckolls, K.B. Psychological Assets, Life Crisis and the Prognosis of Pregnancy. Amer. J. Epid. 95:431, 1972.

17. Morishima, H.O. The Influence of Maternal Psychological Stress on the Fetus. Amer. J. Obs. and Gyn. 131:286, 1978.

18. Barlow, S.M. Delay of Postnatal Growth and Development of Offspring Produced by Maternal Restraint Stress during Pregnancy in the Rat. Teratology 18:211, 1978.

19. Smith, D.J. Modification of Prenatal Stress Effects in Rats by Adrenalectomy, Dexamethasone, and Chlorpromazine. Physiology and Behavior 15:461, 1975.

20. Dohrenwend, B.P., Dohrenwend, B.S., Kasl. S.V. and Warheit, G.J. "Technical Staff Analysis Report on Behavioral Effects" In: Report of the Public Health and Safety Task Force to the President's Commission on the Accident at Three Mile Island, U.S. Govern. Printing Office, Washington, D.C., 1979.

21. Houts, P.S., Miller, R.W., Tokuhata, G.K. et al. Health-Related Behavioral Impact of the Three Mile Island Nuclear Incidence. A Report Submitted to the TMI Advisory Panel on Health Research Studies, The Pennsylvania Department of Health. Part I (April 1980), II (Nov. 1980), and III (May 1981).

22. Bromet, E. and Dunn, L. Mental Health of Three Mile Island Residents. Psychiatric Epidemiology Program, Western Psychiatric Institute and Clinic, University of Pittsburgh, 1982.

23. Upton, A.C. Effects of Radiation on Man. Ann. Rev. Nucl. Sci. 18:495, 1968.

24. Upton, A.C. Radiation Injury: Effects, Principles, and Perspectives, University of Chicago Press, Chicago, Ill. 1969.

25. National Academy of Sciences Advisory Committee on the Biological Effects of Ionizing Radiation. National Academy of Sciences, National Research Council, Washington, D.C., 1972.

26. United Nations Scientific Committee on the Effects of Atomic Radiation. Sources and Effects of Ionizing Radiation. Report to the General Assembly. United Nations, New York, N.Y., 1977.

27. Battist, L., Buchanan, F., Congel, H., Peterson, C., Nelson, M., and Rosenstein, M. Population Dose and Health Impact of the Accident at Three Mile Island Nuclear Station. Preliminary Estimates for the Period March 28, 1979 through April 7, 1979. U.S. Nuclear Regulatory Commission, Washington, D.C., 1979.
28. Gerusky, T. Three Mile Island: Assessment of Radiation Exposures and Environmental Contamination. Ann. N.Y. Acad. Sci., 1981.
29. Upton, A. Radiobiological Effects of Low Doses: Implications for Radiological Protection. Radiat. Res. 71:51, 1977.
30. National Academy of Sciences Advisory Committee on the Biological Effects of Ionizing Radiation. National Academy of Sciences, Washington, D.C., 1980.
31. Goldhaber, M.K., et al. The Three Mile Island Population Registry. Public Health Reports 98(6):603, 1983.
32. Tokuhata, G.K. Pregnancy Outcome Around Three Mile Island. Pennsylvania Department of Health, Division of Epidemiology Research, Harrisburg, Pennsylvania, 1981.

Chapter Twenty-Six

ENVIRONMENTAL AND BIOLOGICAL EFFECTS OF IONIZING RADIATION

E. Willard Miller[1] and Shyamal K. Majumdar[2]

[1]Professor of Geography and Associate Dean
for Resident Instruction (Emeritus)
College of Earth and Mineral Sciences
The Pennsylvania State University
University Park, PA 16802
and
[2]Professor of Biology
Lafayette College
Easton, PA 18042

On July 16, 1945, on a desert in New Mexico, the United States exploded the world's first atomic bomb creating the nuclear age. Ten years later this nation launched the "atoms for peace" program designed to harness the awesome power of the atom for constructive purposes. As President Eisenhower stated at the time, "The atom stands ready to become man's obedient, tireless servant, if man will only allow it." Since then five nations have developed atomic weapons and nineteen nations have installed nuclear-electric power generating facilities.

It has been recognized from the beginning that nuclear weapons and nuclear power would generate sizeable quantities of ionizing radioactive wastes that could contaminate the environment. Nuclear wastes originate from a wide variety of sources. These include explosions of nuclear weapons, tailings from uranium mining, effluents from fuel enrichment and fabrication plants, the fission and transuranic byproducts of fission reactions, ordinary metals made radioactive in nuclear reactors by neutron bombardment, and the materials in all segments of the nuclear fuel cycle that became contaminated through contact with radioactive elements. The need to manage these wastes so that the environment will not be contaminated and health problems will not develop is one of the critical issues of modern times.

ATOMIC BOMB ENVIRONMENTAL CONTAMINATION

The testing of atomic bombs began in 1946 in order to determine their destructiveness. To acquire this knowledge the United States conducted 66 nuclear tests on Bikini and Enewetak. The sites chosen had to have numerous conditions. It had to be an area controlled by the United States in a climatic region free from storms and cold temperatures. It also had to have a sheltered area for anchoring target vessels and measuring the efforts of radiation. Further, it had to be an area with a small population that could be readily moved to another area. The search ended with the selection of Bikini and Enewetak. The following is limited to the Bikini experience (Weisgell, 1980).

The Bikinians were asked if they would be willing to sacrifice their island for the welfare of all men. After deliberations, the Bikinians agreed to move. Of three island choices given in the Marshall Islands, the Bikinians chose the uninhabited island of Rongerik. The removal of the people to their new home was done with extraordinary speed during the month of March 1946. The island of Rongerik was infertile and food shortages soon developed. By 1948 near starvation could no longer be ignored and new sites were explored. The Bikinians selected Kili, a fertile island 400 miles south of Bikini that had previously been used for copra plantations. This island has become the home of about 900 Bikinians. This drastic change from an atoll existence with abundant fish to an isolated island with no lagoon has taken a severe psychological and physical toll on the people. While Kili is more fertile than Bikini, the Bikinians are not skilled in intensive agriculture to make the island productive. In 1946 they were completely self-sufficient. Today they have lost the will to provide for themselves.

During the testing of bombs the U.S. Navy remained optimistic that as soon as testing was over the inhabitants would return. The dangers of radiation became evident in 1954. Winds carried the radioactivity from an explosion eastward to the islands of Rongelap and Utirik. Ninety percent of the Rongelapese people immediately suffered skin lesions and loss of hair. Today, 19 of the 21 Rongelapese who were under 12 years of age at the time of the test have developed thyroid tumors or other radiation-related illnesses.

In 1958 President Eisenhower declared a moratorium on U.S. atmospheric atomic testing in the Marshall Islands. In 1969 the Atomic Energy Commission declared that Bikini had virtually no radiation and there was no discernible effect on plant or animal life. The Bikinians were joyful that they could return to their island and resettlement began. The Department of Interior initially built 40 new homes and planted 50,000 new trees on the island. By 1973 the United States government indicated that construction was nearly complete, and all Bikinians could return permanently by Christmas.

Problems, however, were beginning to become evident (Weisgell, 1980). In 1972 the Atomic Energy Commission in a survey of Enewetak found that radiation levels were extremely high in certain places. In late 1974 the Secretary of the Interior, Roger C. B. Morton, alarmed at the findings of routine radiological surveys, halted construction on Bikini Island. In March of 1975 Defense Secretary James R. Schlesinger requested a thorough survey be conducted of the radioactivity on Bikini. Because of bureaucratic wrangling the surveys were delayed.

Meanwhile the Bikinians expressed a desire to build houses in the interior away from the lagoon (Trumbull, 1982). A routine survey in June 1975 indicated that the interior was too radioactive for housing and some wells were contaminated with radioactive plutonium. Further, the survey revealed that while coconuts were likely to be safe, breadfruit and pandanus, two basic food staples, contained unacceptably high levels of radiation.

The Bikinians, frustrated and confused, brought suit in federal court in October 1975 to force the U.S. to stop the resettlement program until a comprehensive radiological survey of the island would be made. The United States government agreed to make the survey and $2.6 million were appropriated in 1977, but again the Defense Department took no action saying that more money was needed. The squabble lasted three years. In the meantime tests in 1977 showed that levels of strontium 90 in well water on Bikini exceeded acceptable U.S. standards. Medical examinations revealed that the people in Bikini had absorbed cancer-causing radioactive elements such as strontium, plutonium and cesium.

In 1978 a medical team informed the 139 people living on Bikini that they could no longer eat locally grown food, but must import all food and water. It was shown that the population had acquired in a single year a 75 percent increase in body burdens of radioactive cesium 137. The people of Bikini may have ingested the largest amount of radiation of any known population. It was concluded that all people had to be removed from the island. In August 1978 the 139 people were removed, and no one has been allowed to live there since.

The Bikini experience has raised many questions and provided some answers. Certainly there is positive evidence that radioactive environmental contamination caused by atomic testing persists for many years. There is also strong evidence that the American government was less than responsible in its actions to the Bikinians. There remains the problem of where the Bikinians are to secure a permanent home. In the wider perspective, atomic testing on Bikini in the Marshall Islands illustrates the danger on a small scale of what a modern atomic holocaust could bring.

ENVIRONMENTAL IMPACT OF NUCLEAR FUEL CYCLE

The environmental impact of the nuclear energy industry has been of major importance since its origin (El-Hinnawi, 1980). Many studies now exist on each aspect of the nuclear fuel cycle beginning with the mining, milling, conversion, enrichment, fuel fabrication, power production, fuel reprocessing and, finally, waste management. The Atomic Energy Commission is a fundamental source of information (Atomic Energy Commission, 1972). The Environmental Protection Agency has analyzed the data in relation to human health (Environmental Protection Agency, 1973). When radioactivity causes a case of cancer, whether fatal or nonfatal, or when a deformity occurs due to a radiation-induced mutuation, it is counted as a single health effect (Hodges, 1977). A health effect, for example, of 0.1 from a nuclear facility means a 10 percent chance of one health effect.

The problem of protecting the public from harmful ionizing radiation from nuclear wastes has received not only national but international attention. The most direct health hazard from waste would be from gamma radiation. Gamma rays are particularly penetrating. Their effect is measured by units called rems, each of which is equal to the amount of radiation that is required to produce the same biological effect as one roentgen of X-radiation. The biological damage done by a gamma ray is essentially proportional to the ray's energy.

Tailings and Milling

Mill tailings are widely distributed in the world. They are found not only at the mining site but have been transported to many sites where they are processed. Because of their sheer volume the Environmental Protection Agency believes that mill tailings may present "the greatest environmental impact of all waste forms over the years." The extraction of only enough uranium to fuel a single 1,000 megawatt light-water reactor for a year requires the generation of at least 106 tons of tailings. The American operation of 73 commercial power plants, 145 military production and propulsion facilities, and 93 research reactors, plus the uranium exported, requires the mining of more than 20,000 tons of uranium ore every day.

There are approximately 29 open pits and 193 underground mines, shipping to 19 operating mills in the United States. The tailings are stored in piles reaching 100 feet high. There are at least 43 sites with tailing piles in the West and others in the East. The environmental problems of these sites can be illustrated by a report from the Sierra Club of a site near Canonsburg, Pennsylvania. In the area beneath the settling pond the levels of radium were 3,000 times greater than permitted by Federal guidelines. The water from this area drained into streams that flow into the Ohio River, a source of drinking water in western Pennsylvania.

Since uranium mining began in the West massive tailing piles have accumulated (Shapiro, 1981). In the early days of mining the environmental dangers from these tailing piles were not recognized. Many of the tailings provided the foundation for new settlements. They were frequently spread on roads to provide a base. Parking lots and shopping centers were built on uranium tailings.

It has now been found that these tailings have serious environmental consequences. In the uranium belt of western New Mexico, for example, the Environmental Protection Agency reported that shallow aquifers that are water sources have been highly contaminated with selenium attributable to excessive seeping from mill tailings (EPA, 1973). While selenium is not radioactive it has a toxicity comparable to arsenic. Radioactive contaminants above the maximum permissible levels have also been found in the region. In dry regions such as New Mexico and Utah there is also the danger that high winds will pick up the fine tailings and carry them into populated regions. The tailing piles are rarely stabilized to prevent wind erosion. In the Salt Lake Area of Utah Governor Scott Matheson has indicated that a 23 million ton pile of tailings is "the largest microwave oven in the West."

In addition to water runoff and wind contamination, tailing piles pose a hazard in their emissions of radioactive gas radon. In the process of changing uranium 238-to-radium 226, radon 222 is formed from the decay of radium 226, which has a half-life of only 3.8 days. However, in this short span these atoms have the capability of spreading alpha rays, a virulent form of radiation, wherever the wind spreads them. If the alpha radiation should enter someone's lungs, even in microscopic amounts, these particles could lodge in tissues and continue to radiate, resulting in fibrosis and lung cancer.

Conversion

In the milling process chemical methods are used to convert and purify the uranium ore into a semirefined uranium oxide (U_3O_8), known as yellowcake. The yellowcake must be further converted by additional chemical processes to produce UF_6. In this process there are small radioactive water and air emissions. The Environmental Protection Agency has indicated that conversion plant emissions produce an average dose of much less than 1 millirem/year to individuals within 50 miles of a plant (EPA, 1973). The five conversion plants are located at Metropolis, Illinois; Sequoyah, Oklahoma; Barnswell, South Carolina; Apollo, Pennsylvania; and West Valley, New York. This process has little possible environmental contamination.

Enrichment

The next step in the nuclear fuel cycle is the enrichment of the UF_6 (Hodges, 1977). This is accomplished by a gaseous diffusion process. The UF_6 is pumped through about 1,700 barriers in which the U-235 concentration is increased from the natural 0.7 percent to about 4 percent. In this process large quantities of heat, water and electricity are required (Glackin, 1976). In the enrichment of fuel for one year's operation of a 1,000-MWe LWR requires about 11 billion gallons of cooling water and 310 million kilowatt-hours of electricity. This is about 4 percent of the energy generated in the plant.

The large quantity of cooling water required is the major environmental impact in the enrichment process. Cooling towers are required to reduce the temperature of the water. Other environmental impacts are minimal. The radiation dose to an individual living at the plant boundary would be only about 2 mrem/year. The enrichment process occurs at three government plants located at Portsmouth, Ohio; Oak Ridge, Tennessee and Paducah, Kentucky. In the future private plants may be licensed to enrich UF_6.

Transportation

A massive amount of material must be transported from one place to another in the nuclear fuel cycle (Shapiro, 1981). It has ben estimated that to operate a single nuclear facility of 1,000 MWe capacity there are truck shipments totaling 16,800 miles of uranium ore on private land, 33,500 miles of shipments of uranium oxide and hexafluoride on public highways and 2,000 miles of rail shipment of fission products. In this transfer of radioactive material there is always potential danger to the environment. Especially designed containers have been built, and there have been no serious accidents to the present. A few minor accidents have released small amounts of radioactive substances. Radiation doses measured near trucks carrying unirradiated material (uranium not processed into fuel) are usually one mrem/hour or less, but varies from 3 to 55 mrem/hour within 15 feet of a rail car carrying high level wastes (AEC, 72).

Fuel Fabrication

The enriched UF_6 must now be converted to uranium dioxide, UO_2, and loaded into fuel rods which are then transported to the power facility. The Environmental Protection Agency has indicated that this process has little or no impact on the environment compared to the overall nuclear fuel cycle (EPA, 1973). At the plant boundary of a typical facility an individual might receive a maximum of 10 mrem/year in the lungs from normal breathing, but the average dose within 50 miles of the plant would be less than 0.1 mrem/year.

Power Production

The most controversial part of the nuclear fuel cycle has always been power production and the Three Mile Island accident has intensified this controversy. The power plant is the place in which new radioactive materials are produced. The following presents major concerns raised about power production at nuclear facilities.

In all nuclear power plants radioactive materials are produced in the nuclear reactor by several processes. The fissioning of the uranium and the neutron activation of the coolant produce many radioisotopes. These radioactive isotopes, particularly the gaseous ones, can escape from the fuel rods through pinhole defects to contaminate the coolant water. Technical specifications limit the amounts and rates of release of these radioactive materials into the environment. Constant monitoring of the facilities occurs. The Environmental Protection Agency has estimated that the health effect of one year's operation of a 1,000 MWe plant would be responsible for a total of only about 0.001

mrem/year, including all future health effects (EPA, 1973). It is thus evident that the environmental contamination of radioisotopes from an operating nuclear plant is minimal.

Another concern relates to the cooling system. If the coolant were to stop passing through the reactor core, the fission reactions would stop for the moderating influence of the coolant would end. At the same time the heat from the radioactive decay of the fission products would result in the melting of the fuel and if continued would lead to a break in the pressure vessel and containment system. At this point radioactive gases could escape into the environment. The hazards to the nearby population could be great. This could have occurred at the Three Mile Island accident, but was prevented. The fear of possible radioactive contamination was, however, great and has forced a reevaluation of the use of nuclear energy to produce electricity.

Fuel Reprocessing

The fuel rods in a nuclear plant must be replaced when they are spent. Because the spent rods are highly radioactive the normal practice is to store them underwater for a few months to reduce the radioactivity. The rods are then transported to a reprocessing plant where the wastes are removed and the usable uranium and plutonium recovered. The Environmental Protection Agency has attempted to predict the health effects at the reprocessing plant. It is estimated that a 5-metric-ton per day plant that would serve 45 nuclear plants of 1,000-MWe size would emit long-lived radioactive gases such as krypton 85 and tritium, at a rate that would produce about 2.5 health effects annually in the United States and about 100 health effects annually worldwide. If the processing plants were increased in number to service 1,000 nuclear power plants, the health effects would increase to about 210 annually worldwide, of which there would be 130 cancer deaths.

Nuclear Waste Management

The problems of disposing of radioactive wastes produced by nuclear power plants have attracted much attention and public fears have become so pervasive that the industry has been threatened. In the process of producing electricity for every metric ton (1,000 kilograms) of uranium in the initial load, 24 kilograms of uranium 238 and 25 kilograms of uranium 235 are consumed in a three-year period. During this period the enriched uranium 235 has been reduced from 3.3 percent to 0.8 percent of the total. This amount of uranium 235 has provided 800,000,000 kilowatt-hours of electrical energy.

In the process of producing electricity, 35 kilograms of radioactive waste products are produced. Of this amount 8.9 kilograms are various isotopes of plutonium, 4.6 kilograms of uranium 236, 0.5 kilogram of neptunium 237, 0.12 kilogram of Americium 243, and 0.04 kilogram of curium 244. Since only 25 kilograms of uranium 235 are consumed and a fifth of that amount is converted into uranium 236 and neptunium 237, it is calculated that only 60 percent of the energy produced comes from uranium 235. Thirty-one percent comes from plutonium 241 and 5 percent from the high energy neutrons in uranium 238.

The annual waste material from a 1,000-megawatt nuclear reactor is about two cubic meters. The simplest and most obvious way to dispose of these high-level wastes is to bury them permanently, deep underground (DeMarsily, 1977). This appears to be most reasonable since all rocks contain traces of naturally radioactive substances. It is estimated that the natural radioactivity in the substrata of the United States at depths of 600 meters is much greater than the radioactivity in the wastes that would be produced if the total electric supply of the nation were produced by nuclear fission. Of course, the nuclear facility radioactive wastes are much more concentrated and require special handling.

The means of disposal of radioactive wastes is still not definite but most recent research indicates the following type of procedure appears most promising. The wastes would be placed in a boro-silicate glass cylinder about 300 centimeters long and 30 centimeters in diameter. It would require 10 canisters each year to contain the waste from a single 1,000 megawatt nuclear power plant. Each of the canisters would be buried to a depth of at least 600 meters about 10 meters apart in order to dissipate the heat generated by the decay of the radioactive wastes. The problem of excessive heat can be controlled by delaying the burial by about one decade. It has been estimated that the United States would require about 400 1,000 megawatt nuclear plants to supply its needs. If this were to occur about one-half a square kilometer of space would be needed annually to store the radioactive wastes.

It is now thought that salt deposits would provide the best burial sites. Salt formations are firm with little possibility of fracture. Large deposits are located in areas with no tectonic or volcanic activity. In addition the waste heat would cause water to migrate toward the canisters. Typical salt deposits contain about 0.5 percent water trapped between the particles. The water reduces the temperature considerably. This is critical for glass devitrifies (crystallizes and becomes brittle) at temperatures higher than 700 degrees C. In time it is assumed that the stainless steel canisters will decay leaving the glass containers in contact with the salt.

Although complete risk cannot be eliminated in the buried radioactive wastes, the dangers must be placed in a proper perspective. A public fear is that the canisters will ultimately leak and the groundwater will contaminate food and drinking sources (Shapiro, 1981). Because the canisters will be placed in impervious layers of rock the typical rate of flow of water is less than 30 centimeters per day. At this rate it would take about 1,000 years for the water to migrate 100 kilometers (62.5 miles). During this period the decay of the wastes makes them far less dangerous. For example, cesium 137 decays to barium 137 in a period of 400 years (Cohen, 1977). During this period the total gamma ray hazard falls by more than four orders of magnitude.

There is also a fear that the waste materials would reach the surface due to erosion. This is an unfounded fear. When the waste is buried more than 600 meters deep, it would require millions of years to erode this material to the earth's surface (Cohen, 1977). The burial site must be in a stable tectonic area not subject to earthquakes or volcanic activity.

There is also a fear that social and political conditions are not sufficiently stable to guarantee safety. Monitoring of excessive ionizing radiation would require a minimum of personnel. Because of the radiational effects of the waste materials, radical groups will not disturb the permanent deposits.

Permanent nuclear waste disposal sites have not yet been chosen in the United States. In December 1984, the U.S. government proposed three sites. These were Hanford, Washington, Hereford, Texas and an isolated site in the desert of Nevada. The governors of Texas and Nevada are strongly opposed to having a nuclear waste site in their states. The least opposition comes from the proposed Hanford, Washington site. This is because Hanford was one of the first cities in 1943 to process nuclear material for the original atomic bombs. The final site selection will not be chosen until the late 1980s. In the meantime there will be much public debate and the courts of the United States may have to ultimately decide the disposal of nuclear radioactive wastes.

HEALTH EFFECTS AND EXPOSURE STANDARDS FOR NUCLEAR WASTES

The danger from nuclear wastes is very great and disposal must be absolutely safe. To illustrate, if an individual is within 10 miles of an unprotected radioactive waste container he would receive a dose of 500 rems (which is about a 50 percent chance of being fatal) within 10 minutes.

In order to assure safety, health standards for the measurement of radiation levels have been set by national and international commissions and agencies. In the United States the relevant organizations are the National Committee on Radiation Protection and Measurements (NCRP), the Nuclear Energy Commission (NEC) and the Environmental Protection Agency (EPA).

These standards include the following limitations to exposure. For individuals who are occupationally exposed the maximum permissible dose is 5 rems/year of whole-body irradiation. If a dose in a single year exceeds this, then in following years the dose must be reduced. The occupational dose to the skin in one year must not exceed 15 rems, to the hands not more than 75 rems, and to the forearms 30 rems. The maximum permissible exposure for the general public is set at one tenth the occupational limits, that is 0.5 rems/year excluding natural background and medical exposures (Stolwijk, 1983).

SHORT-TERM HEALTH EFFECTS OF IONIZING RADIATION

A major characteristic of nuclear wastes is its radioactivity. Wastes, however, are extremely complex and the radioactive isotopes may have a half-life of less than a second to many thousands of years. Further, each radioactive isotope emits a characteristic radiation when it decays. It can be electromagnetic such as X-rays, or it can be particles such as alpha, beta or neutron radiation. However, all radiation from radioisotopes has a similar ionizing effect when it touches matter, including biological tissue. In the ionizing process the tissue loses one or more electrons. The ionized atoms absorb a great deal of heat and in the process chemical changes occur in the biological tissue. These chemical and ultimately structural changes can have serious health consequences.

After extensive research it has been discovered that the health effect of a massive exposure in a short period is quite different than an exposure of about the same amount over a long period of time (Stolwijk, 1983). For example, if individuals are exposed to 1,000 rems over a short period, immediate severe illness will occur and almost all persons will die within one month. In contrast, if individuals are exposed to 10 times the amount of radiation in the natural background over several decades, none will suffer acute illness, and only a few will experience genetic defects causing illness and cancer.

In the use of radiation to treat some diseases the clinical effects vary with the size of the dose. When the radiation dose on the entire body is below 25 rems there is no evident chemical effect. In doses of 25 to 100 rems there is a decrease in the number of white blood cells. At doses between 100 and 200 rems there will be a continued reduction in white blood cells and leukopenia will develop accompanied by vomiting. At this level the human body will recover in several weeks. If the radiation dose is raised to between 200 and 600 rems vomiting will occur, accompanied by internal bleeding and complicating infections. Most hair will be lost within two weeks. Treatment will consist of blood transfusions and antibiotics extending over a period of one month to nearly a year. The mortality rate may be as high as 80 percent. When the dose is between 600 and 1,000 rems the treatment is the same, but the mortality rate is between 80 and 100 percent (Stolwijk, 1983).

LONG-TERM HEALTH EFFECTS OF IONIZING RADIATION

The exposure to ionizing radiation may not become evident for many years. The research on these delayed genetic and somatic effects is not as detailed or quantifiable as would be desired. Large human populations have not been exposed to doses of ionizing radiation over extended periods. While laboratory animals have been intensively studied, the extrapolation from animals to humans is exceedingly difficult. The long-term effects of radiation consist of three categories—genetic effects, human growth and development and the somatic effects (Stolwijk, 1983).

The genetic effects of ionizing radiation are most easily recognized in a single, dominant mutation associated with a disease or abnormality. About one percent of the population has some form of dominant mutation including anemia, dwarfism, extra fingers or toes. Recessive mutations require that two recessive genes are inherited, one from each parent. These produce such diseases as sickle cell anemia, cystic fibrosis, and Tay-Sachs disease. About 0.1 percent of all births are recessive diseases.

It must not be thought, however, that all mutations are caused by doses of ionizing radiation. Most mutations are a response to complex hereditary causes. In order to measure risk caused by ionizing radiation, a method is utilized to estimate the dose that will double the naturally occurring rate of spontaneous mutations. This is referred to as the "doubling dose." Scientific studies indicate that a doubling dose of chronic radiation is between 20 and 200 rems. Since the natural radiation background is only 3 to 5 rems in a reproductive lifetime, only a very small percentage of the mutations can be attributed to this effect. Even additional amounts of radiation will have only a small effect on increased mutations. The possible relationship between additional ionizing radiation and increased disease because of its mutational component is predicted to be between 0.5 and 5 percent in general illness for an additional dose of 5 rems per generation. This appears to be a figure that is much too high. Estimates indicate that the increase in the dose to which the population will be exposed by all nuclear power plants by the year 2,000 will be only 0.001 rems per year or 0.03 rem for a reproductive lifetime.

Evidence has now been accumulated to show that ionizing radiation has a major effect on a fetus in utero and on young children. From studies conducted on subjects of the Nagasaki, Hiroshima and Marshall Islands nuclear explosions, it was found that there was reduced growth rates, mental retardation, and microcephaly. Different types of doses do produce different health changes. When an individual was exposed within 16 weeks after conception the individual was more sensitive to health problems than if exposed later. Individuals exposed to 25 rems were mentally retarded and if the dose was 50 rems the mental retardation was profound. The individual could not care for himself or carry on a simple conversation. If the radiation occurs after birth, much larger doses are needed over a longer period of time to cause health problems. Evidence is limited, but chronic doses of 365 rems/year or less have not produced health effects.

With the exception of the massive doses of irradiation that produce immediate effects, the somatic effects of small doses over time may not become evident for years or even decades after exposure. The long-term effects may appear as a possible reduction of fertility but the most important effect will be the development of cancers such as leukemia or malignancies of the breast, thyroid, stomach or intestinal tract. As a result of long-term research, it has now been established that the increased risk of cancer is in proportion with the excess in radiation exposure. If a population is exposed to ionizing radiation in excess of the natural background it is estimated the total number of excess deaths per million exposed per rem would be 10 from thyroid, 50 from breast, 25 to 50 from lung, 20 to 50 from leukemia and 10 to 15 from gastrointestional cancer (Stolwijk, 1983). To place these figures in proper perspective, however, it must be remembered that cancer causes 200,000 deaths for each one million mortality. The increase in cancer mortality due to 10 excess rems in a population of one million would be 1,000 in addition to the expected 200,000.

CONCLUSIONS

The dangers of ionizing radiation have long been known. It is also recognized that great benefits can come from the use of radioactive materials. While radiation can cause disease, it can also be a major tool in treating many diseases. With the depletion of the nonrenewable fossil fuels, nuclear energy can be a major source of energy. The use of radioactive materials requires eternal vigilance. A basic question is, can society meet this challenge?

SELECTED REFERENCES

Brodine, Virginia, ed., 1975, *Radioactive Contamination,* New York: Harcourt Brace, Jovanovitch, 190 pp.

Cohen, Bernard, 1977, "The Disposal of Radioactive Wastes from Fission Reactors," *Scientific American,* 236 (June): 21-31.

__________, 1974, *Nuclear Science and Society,* Garden City, NY: Anchor Press, 268 pp.

Colglazier, E. William, Jr., ed., 1982, *The Politics of Nuclear Waste,* New York: Pergamon, 264 pp.

DeMarsily, G., E. Ledoux, A. Barbreau and J. Margat, 1977, "Nuclear Waste Disposal: Can the Geologist Guarantee Isolation?" *Science,* 197: 519-27.

Edsall, John T., 1974, "Hazards of Nuclear Fission Power and the Choice of Alternatives," *Environmental Conservation,* 1:21-30.

Eichholz, Geoffrey G., 1976, *Environmental Aspects of Nuclear Power,* Ann Arbor, MI: Ann Arbor Science Publishers, 683 pp.

El-Hinnawi, E. E., ed., 1980, *Nuclear Energy and the Environment,* Elmsford, NY: Pergamon, 300 pp.

Fetter, S. A. and K. Tsipis, 1981, "Catastrophic Releases of Radioactivity," *Scientific American,* 244 (April): 41-47.

Flax, Susan Jo, 1981, "Radioactive Waste Management," *Harvard Environmental Law Review,* 5 (2): 259-295.

Glackin, James J., 1976, "The Dangerous Drift in Uranium Enrichment," *Bulletin of Atomic Scientists,* 32 (February): 22-29.

Hodges, Laurent, 1977, *Environmental Pollution,* 2nd ed., New York: Holt, Rinehart and Winston, 496 pp.

Kittel, J. H., 1984, "Nuclear Waste Management - Issues and Progress," *Journal of Environmental Sciences,* 27 (March/April): 34-41.

Klingsberg, C. and J. Duguid, 1982, "Isolating Radioactive Waste," *American Scientist,* 70 (March/April): 182-190.

Lee, K. N., 1980, "Federalist Strategy for Nuclear Waste Management," *Science,* 208 (May 16): 679-684.

Lester, Richard and David Rose, 1977, "The Nuclear Wastes at West Valley, New York," *Technology Review,* 79: 20-29.

Lipschutz, Ronnie O., 1980, *Radioactive Waste: Politics, Technology, and Risk,* Cambridge, MA: Ballinger, 247 pp.

Lovins, Amory B., 1976, "Energy Strategy: The Road Not Taken," *Foreign Affairs,* 55 (October): 65-96.

McCarthy, Gregory J., ed., 1979-1980, 2 Vols., *Scientific Basis for Nuclear Waste Management,* New York: Plenum.

Moore, John G., ed., 1981, Vol. 3, *Scientific Basis for Nuclear Waste Management,* New York: Plenum, 650 pp.

Murray, Raymond I., 1974, *Nuclear Energy,* New York: Pergamon Press, 278 pp.

Northrup, Clyde J. Jr., et al., eds., 1980, *Scientific Basis for Nuclear Waste Management,* New York: Plenum, 925 pp.

Olds, F. C., 1981, "Nuclear Waste Management," *Power Engineering,* 85 (May): 48-56.

Raudenbush, M. H., 1983, "Looking at Waste Management Worldwide," *Nuclear Engineering International,* 28 (August): 30-33.

Richter, D. and S. Fareeduddin, 1982, "Developing Guidelines in Managing Radioactive Waste," *IAEA Bulletin,* 24 (June): 6-12.

Sagan, L. A., ed., 1974, *Human and Ecologic Effects of Nuclear Power Plants,* Springfield, IL: Thomas, 536 pp.

Shrader-Frechette, K. S., 1980, *Nuclear Power and Public Policy: The Social and Ethical Problems of Fission Technology,* Boston, MA: D. Reidel, 176 pp.

Shapiro, Fred C., 1981, *Radwaste,* New York: Random House, 288 pp.

Stolwijk, Jan A. J., "Nuclear Waste Management and Risks to Human Health," in Walker, Charles A., Leroy C. Gould and Edward J. Woodhouse, eds., 1983, *Too Hot to Handle? Social and Policy Issues in the Management of Radioactive Wastes,* New Haven, CT: Yale University Press, pp. 75-93.

Temples, James R., 1980, "The Politics of Nuclear Power: A Subgovernment in Transition," *Political Science Quarterly,* 95 (Summer): 239-60.

Trumbull, Robert, 1981, "An Island People Still Exiled by Nuclear Age: Natives of Bikini Want Only to Return to Their Pacific Paradise: But Contamination from U.S. Atomic Tests Puts Their Dream Out of Reach," *U.S. News, 93 (October 18): 48-50.*

Vaughan, Burton E., 1983, "Critical Considerations in the Assessment of Health and Environmental Risks - What We Have Learned from the Nuclear Experience," *Science of the Total Environment,* 28 (June): 505-514.

Walker, Charles A., Leroy C. Gould and Edward J. Woodhouse, eds., 1983, *Too Hot To Handle? Social and Policy Issues in the Management of Radioactive Wastes,* New Haven, CT: Yale University Press, 209 pp.

Weisgall, Jonathan M., 1980, "The Nuclear Nomads of Bikini," *Foreign Policy,* (Summer): 74-98.

Woodhouse, Edward J., 1982, "Managing Nuclear Wastes: Let the Public Speak," *Technology Review,* (October): 12-13.

Environmental Protection Agency, 1973, *Environmental Analysis of the Uranium Fuel Cycle,* 3 Vols., Washington, D.C.: Environmental Protection Agency.

Nuclear Energy Policy Study Group (NEPSG), 1977, *Nuclear Power, Issues and Choices,* Cambridge, MA: Ballinger, 418 pp.

Union of Concerned Scientists, 1975, *The Nuclear Fuel Cycle,* Cambridge, MA: MIT Press, 291 pp.

United Nations, Scientific Committee on the Effects of Atomic Radiation (UNSCEAR), 1977, *Sources and Effects of Ionizing Radiation,* New York: United Nations, 725 pp.

U.S. Atomic Energy Commission, 1972, *Environmental Survey of the Nuclear Fuel Cycle,* Washington, D.C.: U.S. Atomic Energy Commission, 339 pp.

Appendix
A. Nuclear Power: The Next Generation
B. The Great State of Uncertainty in Low-Level Waste Disposal

A. Nuclear Power: The Next Generation*

A national program is under way to accelerate the evolution of nuclear technology by feeding 25 years of experience into the design of an advanced, simplified light water reactor. The goal is to emerge within five years with the basics of a standardized plant that is safer, lower in cost, and approved by NRC.

Originally, the light water reactor (LWR) was chosen for use in electric power generation because of its potential for having a simple design and for being easily operated and maintained. Although today's commercial nuclear power plants work extremely well, unpredictable regulatory decisions have forced many new components to be added on rather than incorporated into the design. "Treated that way," explains Karl Stahlkopf, director of EPRI's Nuclear Systems and Materials Department, "a nuclear plant can become hung with retrofits that don't necessarily work well together, making reactors more complex but not necessarily safer."

Now a major new program involving cooperative work by electric utilities, EPRI, reactor manufacturers, architect-engineers, and the Nuclear Regulatory Commission (NRC) has been launched to ensure the availability of the nuclear option in the 1990s. The program is based on an emerging consensus that if nuclear power is to remain a viable power generating option for the next decade, when orders for new plants are expected to pick up, an all-out effort must be made now to simplify and standardize plant design. A fresh approach will also

---

*Reprinted with permission from EPRI Journal, Volume 10, Number 2, March 1985, Brent Barker, Editor in Chief.

be taken to address fundamental institutional issues, such as uncertain licensing requirements and the difficulty of funding new plants. In addition, the program will explore possibilities for building relatively small (400-600 MW) nuclear plants to meet further demand growth after the turn of the century.

This unique, integrated approach to developing a new generation of reactors has come in response to urgent utility requests. Recent surveys conducted by EPRI show that utilities generally want to retain the nuclear option in order not to become solely dependent on coal, and they feel the LWR remains the best near-term candidate for meeting this need. In particular, utilities report that they require detailed specifications for nuclear plants that will offer minimal lifetime cost and assurance of NRC approval.

In response, the five-year, $20 million Advanced Standardized LWR Industry Program has been established in EPRI's Nuclear Power Division. The program will produce a detailed requirements document specifying the design of simplified, standardized LWR plants. Each portion of the requirements document will be reviewed by the Utility Steering Committee and an industry review group consisting of nuclear steam supply system vendors and architect-engineers. These sections will then be passed to NRC for approval. Eventually, utilities can use the document to order a nuclear plant with assurance that it will meet accepted industry standards and all current licensing requirements and that it can be constructed and operated reliably and economically. In addition, conceptual plans for a small nuclear plant will also be produced.

"The current hiatus in nuclear power in the United States is characterized by uncertainties. Our best judgment is that the next generation of nuclear plants must therefore be based on the widest and best experience we have—that is, the light water reactor," comments Solomon Burstein, vice chairman of the board of Wisconsin Electric Power Co. and chairman of the new program's Utility Steering Committee. "Our history, together with the experiences of other nations, tells us that opportunities exist to simplify, modularize, and standardize light water reactor plants."

Path of evolution

Over the years LWRs have become complicated indeed. For example, a typical 1000-MW nuclear plant today contains between 30,000 and 40,000 valves, compared with only about 4000 valves in a commensurate fossil fuel plant. Many of these valves are themselves complex subsystems requiring a motor, position indicator, controls, remote readouts, and other equipment. Elimination of a single, large motorized valve from a plant can save tens of thousands of dollars in capital costs alone, plus further savings in maintenance, inspection, and testing.

A recent review of several pressurized water reactor (PWR) facilities indicated that as many as 40% of their valves could be eliminated by taking advantage of changes in knowledge and functional requirements that have occurred since the systems were originally designed. Even more valves could be eliminated if an integrated approach is taken to simplify the overall design of new plants.

The cumulative effect of such redundancies is staggering. From 1971 to 1982, the per kilowatt requirements of new PWR plants have increased by a factor of 2.75 for cubic feet of concrete, 2.4 for feet of cable, 6.2 for work hours of craft labor, and 7.1 for work hours of nonmanual field and engineering service. Over the same period, total plant costs have roughly quadrupled. Yet one of the strongest indications that room for improvement does exist comes from the fact that similar plants with the same projected date of service may now have estimated costs that differ by a factor of 4.

What can be done? Much publicity has been given recently to proposals that represent a radical departure from the LWR design that has dominated nuclear power generation in the United States for more than two decades. In particular, ASEA -- Atom of Sweden has proposed the PIUS (process inherent ultimately safe) reactor that is entirely immersed in a pool of water, which would automatically flood the core in case of a primary system break. Alternatively, proponents of the high-temperature gas-cooled reactor (HTGR) cite its ability to withstand loss of coolant for many hours before core damage occurs. Although both of these designs might eventually offer potential advantages, neither has been built at the 1000-MW scale, and they would thus require construction of a prototype before any commercial orders could be expected.

The delay involved in building and testing a prototype essentially eliminates such novel designs from consideration in the present attempt to create improved reactors ready for order in the 1990s. Further, little or no government support can now be expected for such a prototype. Meanwhile, faced with a diminishing backlog of orders for new plants, the American nuclear industry is slowly winding down its operations. Major suppliers are shifting their emphasis from plant construction to nuclear service contracts, and second-tier suppliers are casting about for new lines of work.

It was against this background that a new, industrywide consensus emerged on the need to continue reactor development along an evolutionary path. On the one hand, if new nuclear plants are to be ordered before the turn of the century, they will have to be based on proven LWR technology, without the need for a prototype and capable of being built by the manufacturing infrastructure already in place. On the other hand, present LWR designs leave room for improvement, especially if an integrated approach is taken to simplify and standardize the plants and to rationalize regulatory requirements.

The program

Several specific characteristics utilities want to see before ordering new nuclear plants were gathered from the recent industry surveys. The broadest requirements are that such plants will have to be able to compete with comparable coalfired power stations both in cost per kilowatt and in terms of firm schedules for construction and licensing. A target construction cost of $1600/kW has been tentatively adopted, together with a required construction time of 6 years or less. Utilities also want to be able to order plants that will be highly reliable, so tentative design requirements include an average annual availability rate of 80% and an expected lifetime of up to 60 years.

To meet these requirements, design emphasis will have to be placed on minimizing full life-cycle costs, not just on achieving lowest possible initial cost, as in the past. The design will be based on proven LWR technology to take advantage of the experience gained with these systems over the past two decades. Improved features and increased design margins will be provided so that plant operations are simplified and there is more time for the operators to deal with transient and upset conditions, thus keeping minor incidents and operator errors from cascading into major accidents. Provisions will also have to be made at the outset for handling radioactive wastes.

"The guiding philosophy in choosing particular systems," says Stahlkopf, "will be to take advantage of the best off-the-shelf technology." Thus some mechanical systems may be changed very little, while other systems, like instrumentation and control, may take advantage of such state-of-the-art technology as fiber optics.

Already much progress has been made toward ensuring the licensability of nuclear plants designed under the new program. Working with industry representatives, NRC has reduced its list of unresolved safety and licensing issues that will apply to the new plants from 638 to 101, and these, in turn, have been grouped into 34 topics for resolution. In addition, NRC has agreed to review at its own expense the various sections of the program's requirements document for approval. These developments mean that by the time utilities order plants based on the requirements document, each component system will have undergone complete NRC review and approval.

The requirements document will contain 14 system packages, covering such items as safety systems, plant controls, radioactive waste processing, primary coolant systems, and so forth. When designing one of these component systems, a contractor will prepare a package for the requirements document and then submit it to two review groups: one composed of U.S. utility representatives; the other, representatives from nuclear system suppliers and architect-engineers. After approval by these two groups, each package will pass to the Utility Steering Committee for further review and submittal to NRC.

For the small-plant evaluation portion of the program, conceptual designs will be produced for small units of both boiling water reactors (BWRs) and PWRs. If utility interest in small reactors follows, such conceptual designs could be used by nuclear suppliers to produce detailed plans for their own commer-

Growth in Material and Labor Requirements

Many individual factors have contributed to rising construction costs of nuclear plants. Since 1971 the amounts of concrete and cable required to build a plant have more than doubled, while requirements for pipe and raceway (trays that hold cables) have more than quadrupled. Similarly, costs of craft labor and nonmanual field and engineering services have increased by a factor of from 6 to 7. By standardizing plant design and shop-fabricating key components, new nuclear plants can be built with sharply reduced requirements for both material and labor.

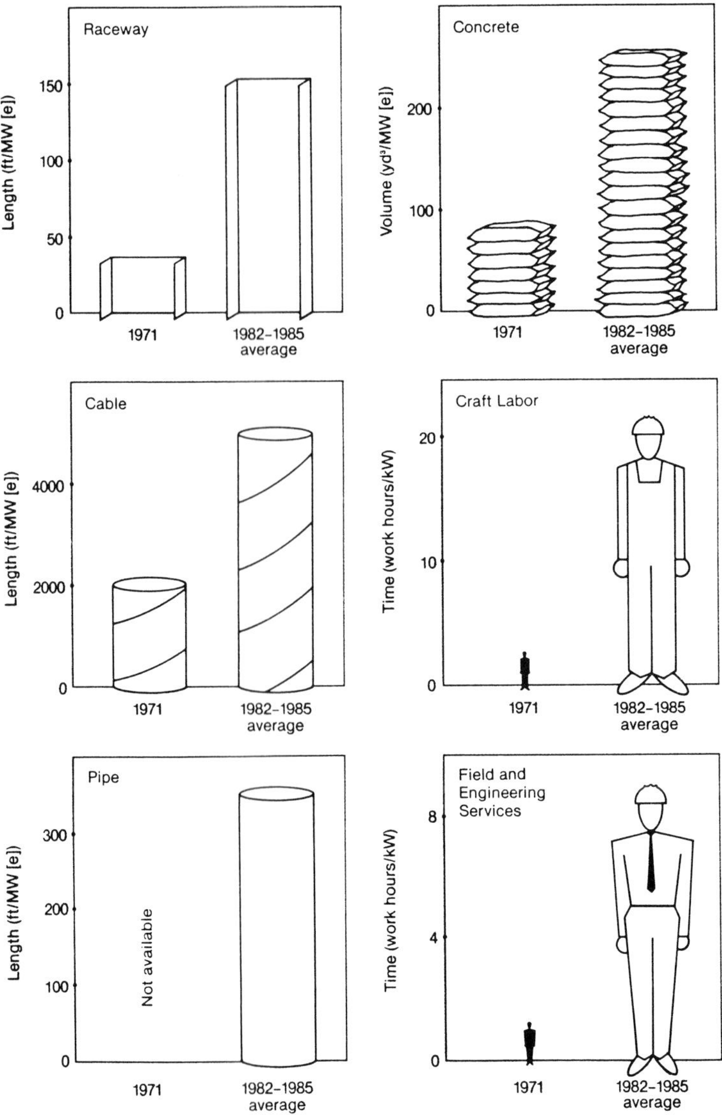

cial plants, which would also satisfy the requirements document approved by NRC.

"We believe that utilities should stipulate in detail their requirements for a nuclear plant on the basis of their own extensive construction and operating experience rather than relying primarily on the suppliers," explains John Taylor, EPRI vice president, Nuclear Power Division. "The aim of the current program is to ensure that a balanced set of requirements is available to assure lifetime safety, reliability, and economy. That's why from the beginning we are involving the reactor manufacturers and architect-engineers who have the state-of-the-art design and fabrication technology and are putting them to work directly with utility people who have extensive plant experience."

Optimizing design

Many of the specific steps involved in simplifying and optimizing LWR plants are already well recognized. The requirements document will particularly address the need to eliminate unnecessary plant complexities, reevaluate design margins, and identify ways of making new plants easier to construct, maintain, and operate.

In some cases, simple configuration changes could be made that would increase plant safety. One example involves preparations for a small-break LOCA in PWRs. In current designs, a small (6-in-diam [15-cm] or less) break in the cold-leg pipe carrying water into the reactor requires automatic actuation of water injection equipment or operator actions to ensure that the water level does not drop below the top of the core. In the unlikely event that such actions are not taken, the core rapidly overheats and can be extensively damaged. This potential problem can be avoided, however, by raising the elevation of various pipes and nozzles in the system so as to maintain water levels above the top of the core, while letting steam escape through the break.

Eliminating needless redundancy is another approach to simplification. An important example of the sort of redundant equipment that might be eliminated by an overall plant redesign is the containment spray system. Of course, detailed studies must be completed to assess whether the removal of this system can be accomplished without compromising plant safety. Originally intended to reduce the spread of radioactive iodine to the atmosphere after a loss-of-coolant accident (LOCA), the containment spray system delivers water containing sodium hydroxide to spray heads near the top of the containment structure. The sodium hydroxide is added to increase the pH of the water and thus enhance its ability to retain elemental iodine, but recent studies have shown that most iodine released during the LOCA is in the form of highly soluble cesium-iodide. As a result of these studies, it now appears that the entire containment spray system would be a candidate for elimination because cesium-iodide could easily

Containing the Small-Break LOCA

One important design change in new nuclear plants that can help prevent a relatively minor incident from causing expensive core damage involves increasing design margins for small-break loss-of-coolant accidents. When a small break occurs in the cold leg (flow toward the core) of a cooling system, coolant flows out of the break until the water level inside the reactor vessel lowers to that in the loop seal between the steam generator and the reactor coolant pump. When the levels equalize, steam escapes through the break and the system depressurizes, allowing the high-flow-rate, low-pressure injection to initiate and ensure core coverage. In today's reactors, unless appropriate actions are taken, the water levels may not equalize until the core is uncovered enough to potentially cause clad damage. By raising both the cold-leg pipe and the loop seal, core uncovering can be prevented.

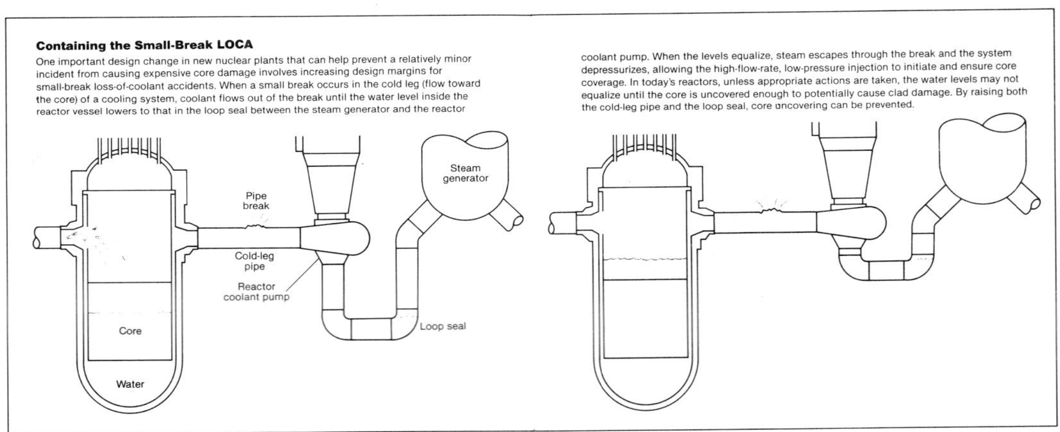

Improved Response to Transients Through Larger Pressurizers

The pressurizer attached to the hot leg (flow away from the core) of a reactor cooling system is designed to keep the system pressure from rising or falling excessively during a transient. Increasing the volume of the pressurizer by as much as 50% in new plants would allow pressure to be maintained by relatively smaller changes in water level during a transient, giving the operator more time to react and lowering the likelihood of having to bleed off pressure by opening relief valves.

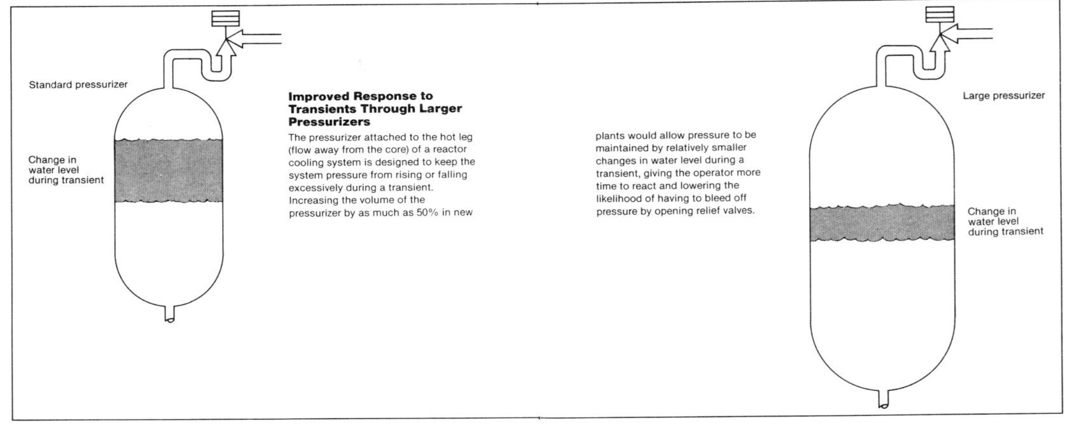

be captured by water already in the reactor coolant system or by steam condensing in the containment atmosphere.

Eliminating this one system could save $8-$10 million in the cost of building a plant, because this system contains about 60 valves, a large tank for water and a small one for sodium hydroxide, and two trains for delivering the mixed solution. By avoiding the possible corrosive effects of sodium hydroxide, plant designers could also consider using aluminum in the containment structure for the first time. In addition, removal of the redundant system would also eliminate a large number of technical specifications and surveillance tests now required.

Utilities have also requested that new plants have better response during transients. In response to this request an increase in design margin for the coolant pressurizer in PWRs is being proposed. This unit is essentially a tank containing both water and steam that is inserted along the hot leg of the coolant system to maintain the coolant pressure at a level high enough to prevent boiling in the core. The problem with current pressurizers is that they are not large enough, in themselves, to maintain pressure during transients. When a turbine trips, for example, operators must quickly open power-operated relief valves to limit pressure rise. Conversely, when the power removed from the system temporarily exceeds that being produced in the core, coolant pressure can drop far enough for steam bubbles to form in the reactor vessel despite the presence of the pressurizer.

Both types of problems could be addressed by increasing the size of the pressurizer—perhaps doubling its volume, compared with present models. Although such enlarged pressurizers would cost about 50% more to build, their presence would greatly reduce the chances that a minor transient could get out of hand and cause damage to the primary cooling system. In particular, operators would have more time to react to a transient and less need to use power-operated relief valves. (It was the mechanical malfunction of one such valve, combined with operator error, that caused the accident at Three Mile Island.)

"The type of design changes being proposed for study in the new program should substantially reduce the financial risk to utilities of building new nuclear power plants," asserts Taylor. "The chance of a LOCA causing the type of damage we saw at Three Mile Island will be greatly diminished, which means that plant safety will also be enhanced."

Longer life and shorter construction

Several design changes will also be required to provide an extended life of up to 60 years for the new, optimized LWRs. Recent experiments show that radiation damage to the reactor vessel itself can be kept low enough for extended life by careful choices of vessel materials and by providing a larger ring of water between the core and the vessel walls. Some internal reactor structures may have to be designed for remote replacement in case they are weakened by irradiation, corrosion, or thermal fatigue. Improved decontamination methods should reduce the sort of internal contamination that leads to corrosion, however—a step that will also make maintenance easier and reduce radiation exposure to maintenance personnel.

To increase fuel reliability, designers will also consider lowering the amount of heat generated along each foot of a fuel rod. Currently this linear heat rate is about 14 kW/ft (4.2kW/m) and studies will be made to determine the effects of lowering this rate to 9-10 kW/ft (2.7-3kW/m). Such a reduction is expected not only to raise fuel reliability by placing less stress on the rods but also to lower requirements for the emergency core cooling system. By having lower power density, a reactor can be controlled with slower reaction times during transients. Because lower heat rates will require more fuel rods in a core, the new design is likely to cost more, and various other trade-offs will have to be investigated.

A recurrent theme running through many of the design changes now being considered is modularization. If new nuclear plants can be constructed with numerous shop-fabricated modules, the economy of scale once brought to large plants by conventional construction practices may be broken. Such a development would make a significant contribution both to large standardized plants through reduced costs and better quality control and to small plants by reducing their economic disadvantage of size.

Off-site construction of modular units should help reduce the lead time and the total capital cost of nuclear plants, as well as reduce the risk to a utility from construction-related delays. Quality assurance is generally easier to maintain if major modules are shop-fabricated because of the more closely controlled environment of a factory and the learning curve improvements that come with experience in building many similar units. Modular construction can also facilitate standardization of plant components and improve overall construction control.

Two possible scenarios for building a modular plant might go something like this:

Large components are built at separate factories and shipped by rail for final on-site assembly, or a complete reactor and most of its safety-related systems are constructed on a barge at a shipyard. The barge would then be shipped by coastal and inland waterways to the installation site and floated into position over a ready-made foundation (reactor base mat). In either case, most of the shield building would already have been constructed, and final plant preparation could proceed relatively quickly.

Toward smaller plants

The portion of the advanced LWR program devoted to small reactors will

Milestones for Advanced LWRs

The national program will draw together the expertise of utilities, EPRI, NRC, reactor vendors, architect-engineers, major laboratories, and major universities to produce a detailed requirements document by 1990 that will specify the design of simplified, standardized LWR plants. Utilities will then be able to use the document to order a nuclear plant with assurance that it will meet industry standards and all current licensing requirements. In addition, conceptual plans for a small nuclear plant will be produced.

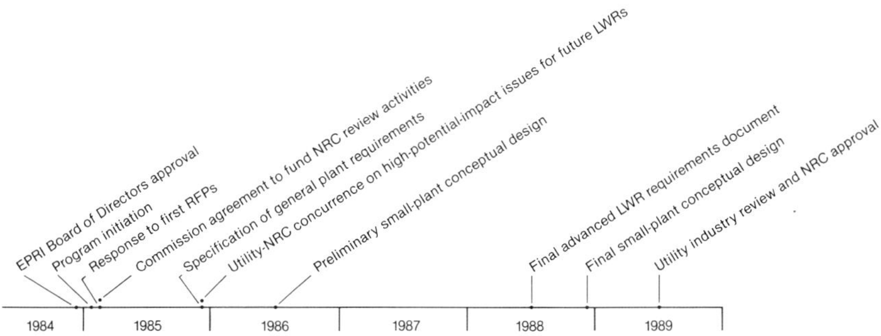

essentially try to determine whether modularization can be carried far enough to make nuclear plants economically viable in units of 600 MW or less. Such plants would offer utilities the option of adding new nuclear capacity in relatively small units, which may be better suited to uncertain growth projections and better able to compete with the small coal plants expected to be available in the 1990s. In addition, construction of small plants would require shorter lead times and less capital raised in a single bond issue.

The choice of whether to build large or small plants is becoming more complex. One the one hand, large plants have traditionally been cheaper to build on a per kilowatt basis. On the other hand, bringing such plants on-line often creates substantial periods of too much or too little generating capacity. During the period before a major plant is finished, for example, a utility may have to purchase power at high rates to meet rising demand. Then, suddenly, when the plant is finished, the utility may have excess capacity it cannot fully use. In addition, the construction delays that have increasingly plagued very large new plants can substantially add to the ultimate cost of the electricity produced.

Several unique features of small plants will be evaluated during the current program. The potential that small reactors have for increased natural circulation will receive particular attention, because this would reduce the amount of power needed for pumping and provide an extra margin against LOCAs. A small BWR might be able to operate entirely on natural circulation, which would eliminate the need for external recirculation loops. In a small PWR, enhanced natural circulation would reduce the potential for core damage during a LOCA in which pumped circulation is lost. (Ways are also being sought to increase the amount of natural circulation in larger plants up to 30% of total circulation, which could mean that smaller pumps might be used.)

By reducing the size of PWRs the use of soluble boron as a core poison might be eliminated. Boron carried by coolant in large PWRs absorbs neutrons and thus evens out the temperature distribution in a core, eliminating hot spots. The problem is that boron injection and removal is complex and expensive. In a small reactor the extensive fluid systems related to boron use may be eliminated if additional gray control rods are used to produce even temperature distribution. (Rods in a conventional PWR are black—that is, they absorb vitrually all the neutrons that encounter them. Gray rods would absorb some predetermined fraction of the neutrons passing through them.)

For both types of smaller reactors, however, several trade-offs will have to be made between size and other considerations. The purpose of the current program is to establish an optimal level for such items as power density, pressurized volume, number of loops, vessel size, and so forth. Designers of improved small plants will then attempt to incorporate features unique to the small size, but they will not introduce unproven technology that would delay design introduction. The design will have to satisfy the requirements document, which is size-independent. Economic analyses will also be carried out to assess the effect of various trade-offs and to determine relative advantages of small plants. Preliminary economic studies sponsored by EPRI indicate that small plants might be able to cost as much as 20% more than large plants (in terms of overnight capital expenditures for construction, which ignore the time it takes to build a plant) and still be able to produce electricity at the same busbar price as a large plant.

Cooperation for a turnaround

The new program began in January and final delivery of the simplified LWR requirements document and conceptual small-reactor designs is scheduled for 1989. During their meeting in December, EPRI's Board of Directors committed $20 million to the program. In addition, the monetary value of review efforts donated by NRC, nuclear suppliers, utilities, and the architect-engineers is expected to be at least as high at EPRI's direct cost.

Requests for proposals (RFPs) from contractors bidding to work on the requirements document were sent out during December and contractor selection was scheduled for the end of February. RFPs for the small-reactor portion of the program were to be released in January, with contractor selection scheduled to take place in March.

Overall guidance for the advanced LWR program is being provided by the Utility Steering Committee. Oversight of the program's budget comes from the Nuclear Power Division's Utility Advisory Committee and its Advanced Concepts Subcommittee. Various departments of the Nuclear Power Division are also providing assistance to the program through a management matrix—a cooperative effort involving shared management responsibilities within the division.

"I've never seen more cooperation between NRC and industry than in preparing for this program," states James A. Coffey, site director of TVA's Browns Ferry nuclear plant and chairman of the Nuclear Power Division's Advanced Concepts Subcommittee. "I believe this program can bring about a major turnaround in the future of commercial nuclear power plants."

This article was written by John Douglas, science writer. Technical background information was provided by Karl Stahlkopf and Warren Bilanin, Nuclear Power Division.

B. The Great State of Uncertainty in Low-Level Waste Disposal*

Technical options for safe disposal of low-level radwaste have been developed, but their application has been stalled by tough institutional problems.

When the clock strikes midnight next New Year's Eve, most utilities—as well as thousands of hospitals, research laboratories, and industrial concerns—may confront the reality of having nowhere to dispose of low-level radioactive waste. That's the deadline for each of the 50 states to have assumed responsibility for waste generated within its borders; January 1, 1986, is also when states now operating low-level radwaste disposal sites may begin refusing to accept the nuclear waste of others.

Such an apparent impasse most certainly was not what Congress intended when it enacted the Low-Level Radioactive Waste Policy Act in 1980. The law was designed to encourage states to form regional compacts that would cooperatively plan and operate disposal sites, thereby more equitably allocating the economic and political costs that are now borne by three states with active sites: Nevada, South Carolina, and Washington.

More than 34 states have since joined multistate compacts, yet to date no new site for low-level waste (LLW) has been selected by the newly formed alliances. Several states have launched extensive efforts to evaluate potential sites and establish waste management programs, but the political anathema embodied by the "not in my backyard" response of the public has, in most cases, kept progress to a crawl.

Meantime, the more than 20,000 medical and academic institutions, laboratories, government agencies, and industrial companies—including utilities operating nuclear power plants—that generate LLW face an increasingly uncertain future for disposal options. These sources now produce an annual volume of about 77,000 m^3 of LLW containing around 500,000 curies of radioactivity; a DOE study predicts annual volume could more than double by 1990.

The situation follows a decade in which the cost of shallow land burial of a 55-gal drum of LLW has increased 10-fold. For an example of particular interest to utilities, the cost of burying a drum of various boiling water reactor (BWR) waste streams at the Barnwell, South Carolina, commercial site increased 227-295% between 1980 and 1983 alone.

---

*Reprinted with permission from EPRI Journal, Volume 10, Number 2, March 1985, Brent Barker, Editor in Chief.

"Low-Level waste disposal is in a state of transition," says Michael Naughton, a project manager for radioactive waste and coolant technology in EPRI's Nuclear Power Division. "As recently as 1975 the disposal of such wastes was viewed as a rather straightforward proposition. Costs were in line with other operating expenses and burial site availability was assumed.

"Now, however, utilities, as well as thousands of other generators of low-level waste, are having to make long-term disposal plans in the face of substantial uncertainties regarding many of the variables controlling cost and site availability," Naughton adds.

In response to the changes of the last 10 years, EPRI is sponsoring a broad research effort on LLW disposal, ranging from more-accurate and less-expensive waste assay methods to studies of volume-reduction systems and other waste minimization techniques, solidification criteria, and evaluations of disposal economics. The work is helping provide utilities with the information they need to chart economically and technically sound strategies for disposing of wastes, whatever the effects of unpredictable political and institutional outcomes.

LLW and disposal experience

As defined in the 1980 law, low-level radioactive waste is any material not classified as spent reactor fuel, high-level waste (from reprocessing reactor fuel), transuranic waste (containing significant amounts of such long-lived isotopes as those of americium or plutonium), or uranium or thorium by-product material, such as mill tailings. The Nuclear Regulatory Commission (NRC) carries the definition a step further, specifying that LLW should be acceptable for underground disposal and defining subclassifications according to the degree of care required for disposal.

Nuclear power plants operated by utilities in 25 states accounted for about 66% of the waste volume shipped to commercial disposal sites in 1982; the utility sector's share of the total radioactivity of waste disposed of in that year was a slightly higher percentage. LLW from nuclear reactors includes dry trash, used equipment, and solidified liquids and sludges. Items range from spent resins from ion-exchange processes, filter materials, lubricating oils, contaminated tools, clothing, and packaging (which have relatively low levels of radioactivity) to irradiated reactor components, such as in-core instrumentation and control rods (which typically contain higher levels of radioactivity).

Nearly all the remaining LLW comes from a variety of institutional and industrial sources, including hospitals, universities, pharmaceutical companies, nuclear fuel fabricators, and other manufacturers spread over all 50 states. These sources produce a range of waste types, as do nuclear reactors, but the isotopic content of industrial and institutional wastes can be quite different, depending on the specific application.

Faced with a 1986 deadline for finalizing their low-level radioactive waste disposal plans, individual states are forming regional compacts to share responsibility. Four compacts involving 24 states have been ratified and submitted to Congress for approval, while five other compacts have yet to receive complete state ratification.
State-ratified compacts submitted to Congress
Proposed compacts awaiting state ratification
Undetermined/uncertain

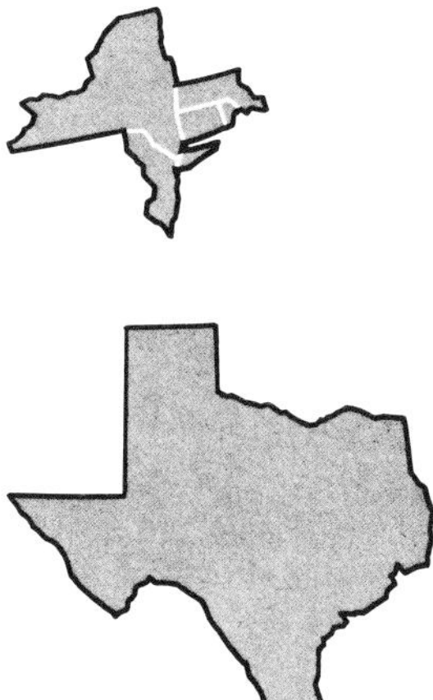

Connecticut, Massachusetts, New Jersey, New York, and Rhode Island have not yet arranged interstate alignment, and their plans for LLW disposal remain uncertain. Texas chose to plan for LLW disposal on its own and has begun a site selection process.

Until the 1960s most low-level wastes were disposed of at sea in steel drums—an expensive option that later gave way to shallow land burial as wastes were shipped to federally owned disposal sites. But as the volume of commercial waste grew with the increased use of nuclear energy for power production, commercial burial sites were licensed by individual states and became the principal facilities for LLW disposal, including some government defense-related LLW.

During the late 1960s and early 1970s the nation had six commercial disposal facilities—operating in Illinois, Kentucky, Nevada, New York, and South Carolina and at the Hanford federal reservation in Washington.

For a time, these sites served the country's needs for shallow land burial capacity. Long trenches, each about 15 m deep, were dug, filled with drums or other waste packages, backfilled with earth, and covered with clay caps to minimize the infiltration of water that could cause leaching of radioactivity. Yet it was problems of precisely this nature that in 1975 led to the closing of the West Valley, New York, site and, two years later, of the site at Maxey Flats, Kentucky. Both sites now are closely monitored for radioactivity migration from the trenches.

In 1978, after protracted negotiations between the state and the operator over expanding burial capacity, The Sheffield, Illinois, site was also closed. Water infiltration of trenches has since become a concern there as well.

With three of the six commercial sites closed, the governors of Nevada, South Carolina, and Washington became concerned about their states' accepting all of the nation's LLW. They urged Congress to develop a plan to make each state responsible for disposing of LLW generated within its borders. Moreover, the governors of Nevada and Washington temporarily closed the Beatty and Richland sites, respectively, in 1979 in protest against packaging and transportation rule violations in some waste shipments. And the following year South

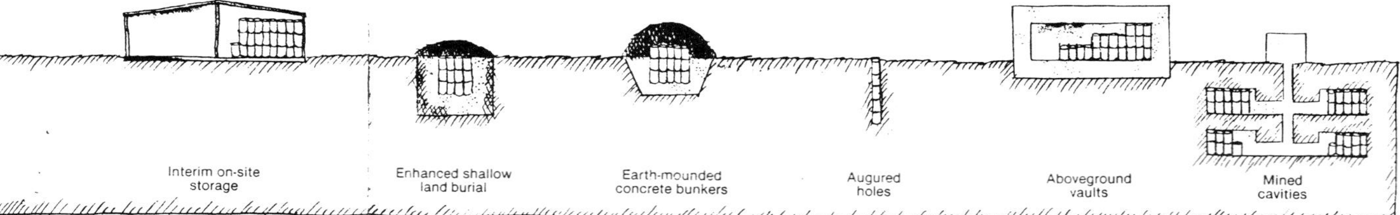

Engineered Storage and Disposal Options

With a potential shortage of shallow land burial capacity looming, some utilities are planning for interim on-site storage of LLW until new burial sites are developed. Meanwhile, many states are considering engineered alternatives to shallow land burial in the face of current political uncertainties. Engineered approaches range from earth-mounded concrete bunkers and augured holes for concrete containers to above- and below-ground vaults and mined geologic cavities. Such approaches are under development in other countries as well. France, for example, is pursuing the concrete bunker approach, while Sweden proposes to excavate a coastal granite formation beneath the sea. All the engineered methods, however, entail substantially higher cost than shallow land burial and, EPRI believes, offer only marginal added safety. For the same reason, Japan plans to use traditional shallow land burial in its LLW disposal program.

Carolina announced it would increasingly limit the amount and type of waste acceptable at the Barnwell site.

As matters now stand, only the Richland and Barnwell sites are accepting significant amounts of commercial LLW. Third-party inspection requirements and associated additional charges at the Beatty site have reduced the volume shipped there to about 2% of the LLW disposed of commercially; the site reportedly will soon be closed. Washington State continues to express a desire to limit shipments to the Richland site to those wastes generated in the Northwest.

States forming compacts

With a serious shortage of LLW disposal capacity looming by the middle of the decade, a combination of state and federal government initiatives led to enactment of the Low-Level Radioactive Waste Policy Act in late 1980. Declaring LLW disposal to be the responsiblity of the states, the act recommended the formation of regional compacts that would provide for disposal in accordance with NRC licensing criteria. Member states in a compact could exclude other states from using their disposal facilities after January 1, 1986. Congressional approval was required for a compact to take effect.

Soon after passage of the 1980 law, NRC issued the first comprehensive set of criteria for regulating LLW disposal facilities. Contained in Title 10 of the Code of Federal Regulations, Part 61 (10 CFR 61), the criteria cover procedures and performance objectives to guide states in licensing LLW facilities.

The NRC's requirements, based on insights gained from past practices and current experience at existing sites, are significantly more stringent than those that applied in the 1960s and 1970s. Despite problems at some disposal sites in the past, most experts believe shallow land burial continues to be a safe and viable approach to LLW disposal if sites are chosen with careful consideration of underlying geology and operated according to strict procedures. Political developments, however, are forcing serious consideration of enhanced burial techniques (incorporating engineered structures and barriers) and alternatives to shallow land burial.

Although the licensing guidelines contained in 10 CFR 61 apply to any land disposal method, specific technical requirements assumed shallow land burial. NRC is now assessing the need for technical criteria for other disposal or interim storage techniques, such as concrete bunkers or engineered structures above ground.

At least 31 states have ratified regional compacts, but the lack of a single new disposal site in the firm planning stage one year before the exclusion deadline reflects the continuing negotiations among compact states and nonmembers alike over who will join the host state ranks. According to experts, it takes from five to six years to select, license, and begin full operation of a disposal site.

Four regional compacts are awaiting congressional approval; three of these—covering the Northwest, Southeast, and Rocky Mountains—would use the existing commercial sites in Washington, South Carolina, and Nevada, respectively. But the Southeast compact provides for continued use of the Barnwell site only through 1992, and a search is under way for a site in Colorado to eventually replace the Beatty, Nevada, facility.

The fourth compact, among five states in the central United States, has begun a site selection process.

Five other regions—including part of the Northeast, the Midwest, Central Midwest, and Western states—have draft agreements in various stages of state approval. Pennsylvania has agreed to host a disposal site in a compact agreement with West Virginia, Maryland, and Delaware. A New York State proposal to establish an interim, monitored storage facility in New York awaits legislative and executive approval. But in Massachusetts, a major LLW-producing state, voters have retained a veto right over possible disposal sites, clouding efforts to make agreements with other states with continued uncertainty.

Elsewhere, California and Arizona have proposed to join in a two-state Western compact, but the agreement, which calls for California to host an LLW disposal site, also awaits approval by that state's legislature. Texas, perhaps further along the road to a new LLW site than any other state or compact group, decided early on to establish its own LLW disposal authority and is currently narrowing the search for a suitable site.

Based on available information, it appears that states accounting for more than half of the volume and nearly three-quarters of the radioactivity of the nation's LLW may not be ready to assume their responsibilities by next January's exclusion deadline. There are indications that these states may push for a congressional amendment to the 1980 act that contains a mechanism for providing disposal capacity beyond the 1986 deadline while compacts proceed with establishing their own facilities.

Implications for utilities

Although the problems of providing new sites for disposal are being worked out on the state and federal level, most producers of LLW find themselves in the role of observers of the unfolding institutional drama. For the most part, utilities are not actively involved in the compact negotiations or in the site selection process, although the outcome of these negotiations will directly affect future disposal costs and related technologic choices.

Despite the uncertainty and political volatility of the site availability issue, utilities and other significant generators of LLW must nonetheless continue planning for management and ultimate disposal of waste. For waste producers, the cost of available options is the driving force behind strategic planning for LLW.

Volume Reduction Options

In response to rising burial and transportation costs, most utilities are pursuing a combination of
LLW packaging options that reduce waste volume to the minimum practical level. A wide range of
volume reduction (VR) equipment is now available for this purpose. Leading candidates appear to
be supercompaction systems for dry waste and evaporation/solidification systems for liquid
wastes. Incineration is another option for dry waste; although it has been little used commercially
in this country because of political and regulatory concerns, extensive experience with
incineration has been gained at many of the national laboratories.

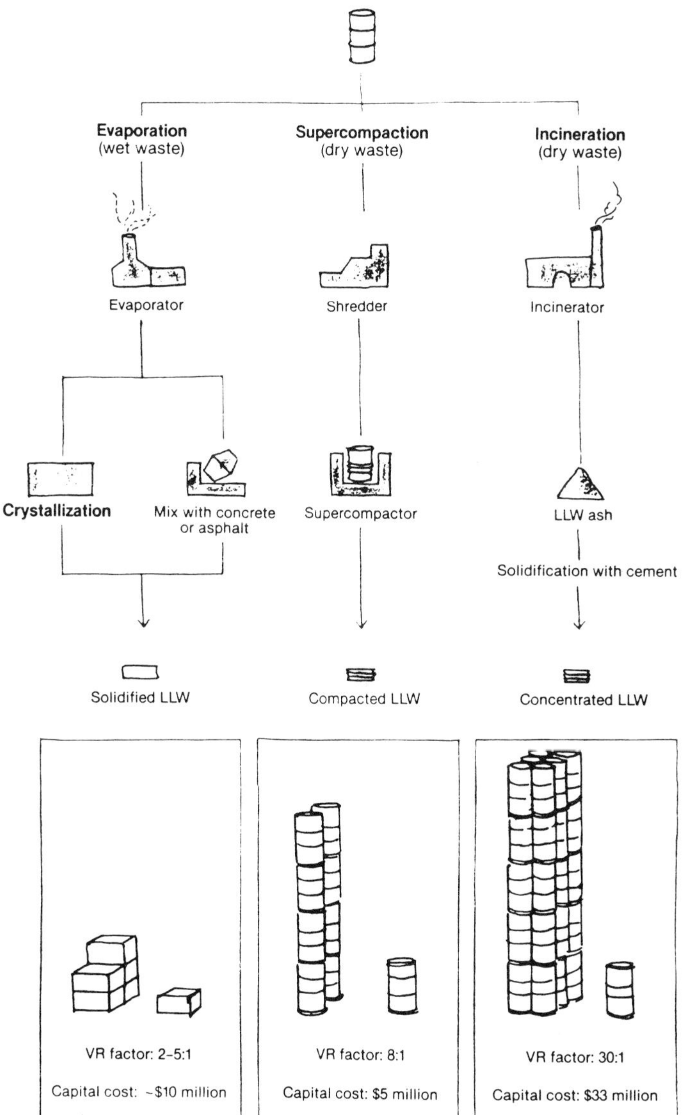

Thus, regardless of where the waste drums are eventually shipped for disposal or stored in the meantime, utilities have for many years sought to minimize the volume of LLW requiring disposal, both through operational practices that limit the amount produced and techniques such as compaction and incineration that further reduce waste volume. In addition, most utilities are now planning to add on-site storage facilities at nuclear plants to handle LLW until adequate commercial burial capacity is available.

EPRI has assisted the industry with surveys of utility experience with LLW volume reduction systems and plan for on-site storage. A comprehensive data base has been established for characterizing in detail the quantity and type of waste from various nuclear plant sources to aid utilities in managing LLW programs.

The Institute has developed a computer code for assessing a wide range of volume reduction equipment under various economic conditions and scenarios. EPRI studies have also provided utilities with technical performance criteria for evaluating LLW solidification technologies and containers. And new techniques are being developed for accurately determining the radionuclide content of LLW.

Volume reduction

In response to rapidly increasing disposal costs and uncertain availability of commercial disposal capacity, many nuclear utilities are considering upgrading radwaste volume reduction (VR) systems and on-site storage facilities. Under EPRI sponsorship, Burns and Roe, Inc., surveyed nuclear utilities around the country to determine the influence of site-specific and regional factors on utility plans in this regard. Results characterize typical nuclear utility planning as an aid to other utilities that are about to commence similar planning. A report on the project includes design details of 22 onsite storage facilities and sketches of 14 such structures as well as drawings of advanced VR facilities.

Of 77 nuclear plants originally targeted for the survey, information was subsequently gathered from 54. Results indicated that as of late 1983, 36 storage facilities were planned or under construction. Reinforced concrete structures were the favored design for on-site storage, although some facilities feature simple metal-sided warehouse structures. The majority of the facilities were sized to provide storage capacity for five years at an average LLW generation rate of 12,000 ft^3 (340 m^3) per plant over the period for pressurized water reactors (PWRs) and 23,000 ft^3 (650 m^3) per unit for BWR plants.

As for VR systems, the survey indicated 27 nuclear plants expressed interest in planning for some type of advanced VR equipment, such as super-compaction or incineration; about 35 nuclear plants had already ordered or installed advanced VR systems as retrofits or new facilities.

"Although most nuclear plants can be expected to go ahead with plans to build on-site storage facilities, their future actions are not so apparent with respect to VR installations," notes Naughton.

"VR facilities can be an expensive alternative, with costs ranging $10-$50 million per plant. Utilities will most likely pursue low-cost VR options in the near term. But if burial costs are dramatically increased by state compact decisions, maximum VR capability could become a necessity for all plants," he adds. Naughton points out that recent pricing changes at current commercial disposal sites that include a curie surcharge on the radioactive content of LLW tend to reduce the economic benefit of volume reduction.

Yet there are significant commitments to advanced VR systems: Consumers Power Co. has installed a bitumen (asphalt) solidification system made by Waste Chem Corp. at its Palisades plant; similar systems have been ordered for at least five other plants. Commonwealth Edison Co. has installed Aerojet Energy Conversion Corp's fluidized-bed incinerator-dryer system at its recently completed Byron plant. Duke Power Co. is building a new VR facility at its Oconee nuclear station that includes the Aerojet incinerator system, a crystallizer for liquid wastes made by HPD, Inc., and Stock Equipment Co's cement-based solidification system.

"These are strong statements in support of advanced volume reduction," adds Naughton, "especially considering that the Oconee plant is in South Carolina." Under a contract with Sargent & Lundy, EPRI plans to document installation and initial operation of the VR incinerator system at Byron, now in preoperational testing.

Incineration of LLW is a VR option for dry material, ion exchange resins, and contaminated oil. It has been widely and successfully used at government and industrial facilities in this country and at many nuclear plants in Europe and Japan. At least eight Aerojet incinerator-dryer systems have been ordered for U.S. nuclear plants, but most utilities are awaiting a clearer picture of the long-range economics of incinerator systems before making firm commitments.

An EPRI-funded survey performed by Gilbert Associates, Inc., gathered detailed information on the design and operation of 26 LLW incinerator facilities and development prototypes in Canada, Europe, Japan, and the United States. The project assessed the licensability of incinerator types most likely to be used at U.S. plants.

The study found LLW incinerator technology to be well established on the basis of 40 years' experience in which some 4000 tons of material have been processed. Incinerators, which reduce bulk material by a factor of 30 or more of its original volume to produce a concentrated LLW ash product, can be safely and reliably operated; high-efficiency offgas filtration systems keep gaseous radioactive effluent releases well below regulatory limits, the survey indicated.

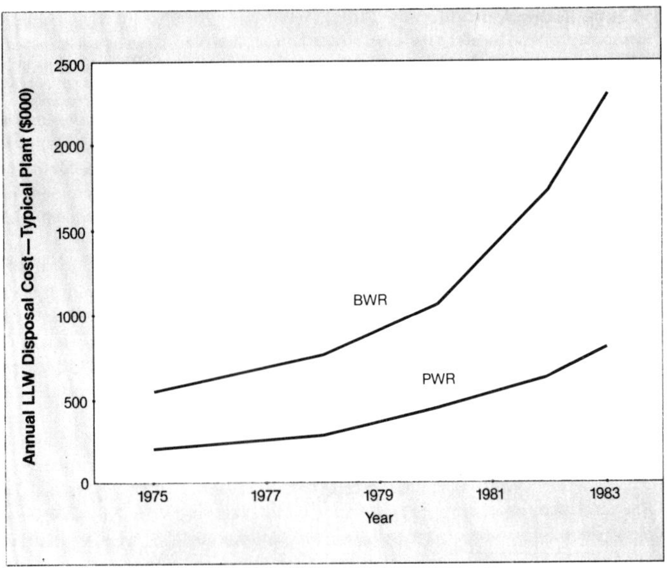

Cost of LLW disposal has risen dramatically in the last decade, primarily because of higher burial costs resulting from increased charges at disposal sites. The overall increase has been particularly acute for BWR plants, which produce an average 50% more LLW than do PWRs.

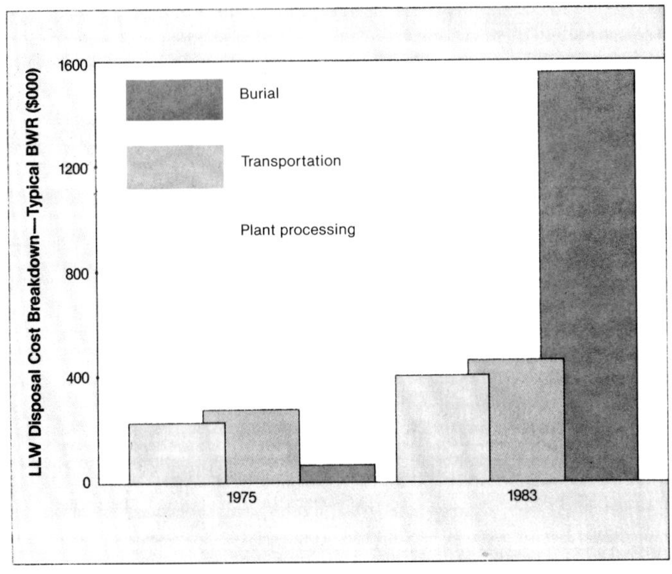

Choosing among options

Perhaps one of the most difficult tasks in utility LLW management planning involves assessing the comparative economics of various VR equipment and storage options under widely varying assumptions. Thorough assessment of a single VR option may require hundreds of inputs to complex, tedious calculations.

To aid in these assessments, EPRI funded a project by Burns and Roe and The Analytic Sciences Corp. to compare the long-range economics of various VR technologies and to develop a model that could easily and quickly analyze a wide range of technologies and economic factors. The resulting computer code, VRTECH, has emerged as a powerful tool for such evaluations.

VRTECH, available through the Electric Power Software Center, has calculated the disposal costs of over 3000 separate cases and configurations in only three minutes of computer time. In a generic analysis of VR systems, the code was used to explore economic sensitivities by calculating costs under 40,000 different combinations of equipment, waste generation rates, storage options, burial prices, and other parameters.

Highlights of project results include the conclusions that the most important factors in VR economics are the rate of LLW generation and the rate of escalation in burial prices. VR is generally most cost-effective at BWR plants—which produce an average of 50% more LLW than PWRs—and can result in significant savings in transportation, operation, and storage costs. Calculations indicate that VR can yield a net saving today despite higher waste activity charges at commercial sites because the reduction in volume charges exceeds the increase in curie surcharges.

One utility that has already used VRTECH to realize a substantial saving in LLW management is the Tennessee Valley Authority. The federal utility had plans dating from 1980 that called for onsite storage and incineration of radwaste at its Browns Ferry, Sequoyah, and Watts Bar nuclear plants, and it was approaching a critical decision point on the purchase of VR equipment.

Faced with an urgent need to reanalyze its options given recent changes in costs, technologies, and legislation, TVA worked with Burns and Roe and Analytic Sciences to use VRTECH for evaluating VR alternatives including incinerators, supercompactors, and solidification systems. With VRTECH, TVA was able to organize a tremendous amount of site-specific information and compare it with individual and collective plant waste streams as a basis for decision making.

As a result TVA changed its approach to LLW handling; VRTECH calculations identified an optimal combination of VR equipment requiring minimal capital costs that still provided significant waste volume reduction. "Using the EPRI approach and computer program, we identified a radwaste management program with a potential net saving of $137 million over the remaining life of our nuclear plants," says TVA's Douglas Michlink.

Other approaches

In addition to VR equipment for minimizing LLW disposal volume, specific operating practices at nuclear plants can help utilities limit the amount of LLW produced. Under EPRI contract, Gilbert Associates used an extensive data base on various LLW categories from two-thirds of the nation's operating nuclear plants to analyze waste generation experience for insights on key operating factors that influence the amount of waste generated.

From a wealth of information, it was found, for example, that BWRs can reduce wet waste volume by about 40% simply by disposing of condensate demineralizer resins rather than regenerating this material. In addition, PWRs that convert from concentrator to demineralizer processing of wet waste can reduce this source of LLW by as much as 75%. Overall, some 100 waste-minimization techniques were documented. Many plants have successfully implemented dry waste minimization programs, resulting in LLW reductions of 20-60%, the study found.

Utilities can also reduce burial costs of radwaste shipments through more-accurate assay techniques for identifying the amounts of specific isotopes contained in waste drums. Recent radioactivity surcharges at commercial disposal sites are an incentive for utilities to define as precisely as possible the isotopic content of shipments.

One approach to waste assay under development employs direct gamma scanning of the waste container. The presence and amount of gamma-emitting isotopes can be correlated with other important radionuclides that would otherwise require direct sampling of waste to quantify; the technique thus provides an accurate means of determining the full spectrum of radioactivity in a waste container. Virginia Electric Power Co.. successfully applied this technique on LLW resin from its Surry nulcear plant. EPRI is continuing development of gamma-scanning technology.

The institutional challenge

Technology for safe, effective treatment and disposal of low-level radioactive waste has been in hand for many years. The problem of LLW is more of a political and institutional nature. Utilities and other waste producers are already under strong economic incentives to minimize waste and to reduce the volume of material requiring disposal. Now, in addition, they must plan long-range strategies for LLW management despite continuing, major uncertainties over the availability of disposal site capacity.

The prospects of these uncertainties being neatly resolved in the near-term are probably not favorable. Until they are resolved, however, there will continue to be compelling cost incentives to reduce other areas of uncertainty as much as possible. EPRI has made significant contributions in this regard and will continue to provide utilities with technical and planning assistance for informed decision making.

Further reading

Long-Term Low-Level Radwaste Volume Reduction Strategies, Vols. 1-5. Final report for RP 1557-13, prepared by The Analytic Sciences Corp., November 1984. EPRI NP-3763.

On-Site Storage of Low-Level Radwaste: A Survey. Final report for TPS 82-657, prepared by Burns and Roe, Inc., July 1984. EPRI NP-3617.

U.S. Nuclear Regulatory Commission. Proceedings of the State Workshop on Shallow Land Burial and Alternative Disposal Concepts. Held at Bethesda, Maryland, May 2-3, 1984. NUREG/CP-0055.

Radwaste Incinerator Experience. Final report for RP1557-4, prepared by Gilbert Associates, Inc., October 1983. EPRI NP-3250.

U.S. Department of Energy. Understanding Low-Level Radioactive Waste. Prepared by EG&G Idaho, Inc. October 1983. DOE/LLW-2.

This article was written by Taylor Moore. Technical background information was provided by Michael Naughton, Nuclear Power Division.

Subject Index

Management of Radioactive Wastes: Issues and Progress

Editors: Dr. Shyamal K. Majumdar,
>   Professor of Biology, Lafayette College
>   Easton, Pennsylvania 18042

>   Dr. E. Willard Miller,
>   Professor of Geography and Associate Dean for Resident
>   Instruction (Emeritus), The Pennsylvania State University,
>   University Park, Pennsylvania 16802

EDITORIAL BOARD

Dr. Robert F. Schmalz, Department of Geoscience, 536 Deike Bldg., The Pennsylvania State University, University Park, PA 16802

Dr. George C. Shoffstall, President, Pennsylvania Academy of Science

Arthur A. Socolow, Director and State Geologist, Pennsylvania Geological Survey, P.O. Box 2357, Harrisburg, PA 17120

Dr. George K. Tokuhata, Director, Division of Epidemiological Research, Pennsylvania Department of Health, P.O. Box 90, Harrisburg, PA 17120

Charles G. VanNess, Geologist, Pennsylvania Power and Light Company, Two North Ninth St., Allentown, PA 18101

Dr. Warren F. Witzig, Head, Department of Nuclear Engineering, 231 Sackett Bldg., The Pennsylvania State University, University Park, PA 16802

Prof. Stanley J. Zagorski, PAS President, Associate Dean, College of Science and Engineering, Gannon University, Perry Square, Erie, PA 16541

ADVISORY BOARD

Honorable Justice John P. Flaherty, Supreme Court of Pennsylvania, Six Gateway Center, Pittsburgh, PA 15222. Chairman of the Advisory Board, and the Honorary Executive Board of the Pennsylvania Academy of Science, Pittsburgh, PA

Honorable Richard Caliguiri, Honorary Executive Board of the Pennsylvania Academy of Science, Mayor of the City of Pittsburgh, PA

Buddy A Beach, Vice President of the Environmental Quality Control, Consolidation Coal Company, Consol Plaza, Pittsburgh, PA 15241

Dr. J. Benoit Bundock, Scientific Advisor to the Minister of Environment, 2360 Chemin Ste-Foy, Shinte-Foy, Quebec, G1V 1H8, Canada

Karin Carter, Attorney, Assistant Counsel, Pennsylvania Dept. of Environmental Resources, Bureau of Regulatory Counsel, P.O. Box 2637, 101 South 2nd St., Harrisburg, PA 17120

Dr. Al P. Dufour, Chief Bacteriology Section, Microbiology Branch, HERL, Environmental Protection Agency, 26 West St., Clair St., Cincinnati, Ohio 45268

Sister M. Gabrielle, PAS Past-President, Grove and McRobert Road, Pittsburgh, PA 15234

Donald A. Lazarchik, P.E. Director, Bureau of Solid Waste Management, Pennsylvania Department of Environmental Resources, P.O. Box 2063, Harrisburg, PA 17120

Mark M. McClellan, Executive Director, Citizen Advisory Council, PA. DER, 8th Floor, Executive House Apartment, 2nd and Chestnut Streets, Harrisburg, PA 17120

Dr. Bruce D. Martin, PAS Past-President, Associate Vice President for Academic Affairs, Duquesne University, Pittsburgh, PA 15219

Dr. Heinz G. Pfeiffer, Manager, Technology & Energy Assessment, Pennsylvania Power and Light Co., Two North Ninth St., Allentown, PA 18101

Dr. Arun K. Sharma, Professor, Calcutta University, President, Indian National Science Academy, New Delhi, India.

Daniel E. Wiley, Manager, R/D, PPG Industries, Inc., Industrial Chemical Division, Pittsburgh, PA 15222

Dr. J.B. Yasinsky, General Manager, Advanced Power Systems Divisions, Nuclear Center, Westinghouse Electric Corporation, P.O. Box 355, Pittsburgh, PA 14230

Prof. Stanley J. Zagorski, PAS President, Associate Dean, College of Science and Engineering, Gannon University, Perry Square, Erie, PA 16541

OFFICERS OF THE PENNSYLVANIA ACADEMY OF SCIENCE

HAZARDOUS AND TOXIC WASTES:
Technology, Management and Health Effects

Table of Contents

CONTRIBUTORS OF
Hazardous and Toxic Wastes book

D.G. Ackerman Jr., (Chapter 4). TRW Inc., California

R.F. Anderson, (Chapter 11). Boston University, Massachusetts

R.L. Bittle, (Chapter 15). Pennsylvania Department of Environmental Resources, Pennsylvania.

K. Caputo, (Chapter 15). Pennsylvania Department of Environmental Resource, Pennsylvania

R. Carnes, (Chapter 2). U.S. Environmental Protection Agency, Arkansas

J.J. Cudahy, (Chapter 3). I.T. Enviroscience, Tennessee

R.G. Dinsmore, (Chapter 19). Western Electric Company, Pennsylvania

A. Fisher, (Chapter 21). U.S. Environmental Protection Agency, Washington, D.C.

C.L. Fraust, (Chapter 19). Western Electric Company, Pennsylvania

G.R. Galida, (Chapter 16). Pennsylvania Department of Environmental Resources, Pennsylvania

T.M. Grabowski, (Chapter 12). Sun Refining and Marketing Company, Pennsylvania

M.R. Greenberg, (Chapter 11). Rutgers, The State University, New Jersey

C.W. Heath, (Chapter 26). Emory University, Georgia

Teh-Wei-Hu, (Chapter 22). The Pennsylvania State University, Pennsylvania

R.A. Hunter, (Chapter 23). RAH Consulting, Pennsylvania

D.L. Kaltrieder, (Chapter 22). The Pennsylvania State University, Pennsylvania

P.M. King, (Chapter 20). P.P.G. Industries, Inc., Pennsylvania

D. Lesak, (Chapter 14). Hazard Management Associates, Pennsylvania

T. Longosky, (Chapter 5). Envirosafe Service, Inc., Pennsylvania

G. Lumb, (Chapter 27). Hahnemann Medical University, Pennsylvania

J.E. McClure, (Chapter 18). GRW Engineers, Inc., Kentucky

C.H. McPherson, Jr. (Chapter 13). U.S. Environmental Protection Agency, Georgia

M.S. Monmonier, (Chapter 8). Syracuse University, New York

E.T. Oppelt, (Chapter 2). U.S. Environmental Protection Agency, Ohio

J.W. Osheka, (Chapter 20). P.P.G. Industries, Inc., Pennsylvania

R.R. Parizek, (Chapter 10). The Pennsylvania State University, Pennsylvania

D. Pitts, (Chapter 3). I.T. Enviroscience, Tennessee

A.J. Raymond, (Chapter 12). Sun Refining and Marketing Company, Pennsylvania

R.W. Regan, (Chapter 1). The Pennsylvania State University, Pennsylvania

H.G. Rückel, (Chapter 6). Zweckverband Sondermullplatze Mittlefranken, West Germany

B.M. Russell, (Chapter 19). Western Electric Company, Pennsylvania

A.J. Smith, (Chapter 13). U.S. Environmental Protection Agency, Georgia

C.L. Smith, (Chapter 5). Envirosafe Service, Inc., Pennsylvania

A. Scharsach, (Chapter 7). Von Roll Ltd., Switzerland

R.F. Schmalz, (Chapter 9). The Pennsylvania State University, Pennsylvania

G.A. Schnell, (Chapter 8). State University of New York, New York

G.A. Tokuhata, (Chapter 24). The University of Pittsburgh, Pennsylvania

G. VanderVelde, (Chapter 4). Chemical Waste Management, Illinois

R.R. Van Stockum, Jr., (Chapter 17). Louisville, Kentucky

J.F. Villaume, (Chapter 25). Pennsylvania Power and Light Company, Pennsylvania

J.A. Weaver, (Chapter 22). The Pennsylvania State University, Pennsylvania

E. Webber, (Chapter 23). V.F.L. Technology Corporation, Pennsylvania

SOLID AND LIQUID WASTES:
Management, Methods and Socioeconomic Considerations

Table of Contents

CONTRIBUTORS OF
Solid and Liquid Wastes book

R. P. Albertson, (Chapter 21), 3315 Simmons Street, Oakland, CA 94619

Dale E. Baker, (Chapters 13 and 19), Professor of Soil Chemistry, 221 Tyson Bldg., The Pennsylvania State Univ., University Park, PA 16802.

John A. Bartone, (Chapter 22), President, Organic Processing Systems, Inc., 409 East 26th Street, Erie, PA 16504.

R. D. Bartusiak, (Chapter 9), Engineer, Homer Laboratories, Bethlehem Steel Corporation, Bethlehem, PA 18016.

Robert C. Bealer, (Chapter 29), Professor of Sociology, Weaver Bldg., The Pennsylvania State University, University Park, PA 16802.

Patricia T. Bradt, (Chapter 18), Adjunct Prof. of Biology, Lehigh University, Bethlehem, PA 18015.

J.B. Bundock, (Chapter 17), Scientific Advisor for the Minister of Environment for Quebec, 2549 Carre Pijart, Sainte-Foy, Quebec, Canada, G1V1#8.

Karin W. Carter, Esq., (Chapter 26), Asst. Counsel, Pennsylvania Dept. Environmental Resources, 505 Executive House, P.O. Box 2357, 1015 Second Street, Harrisburg, PA 17120.

Donald Crider, (Chapter 29), Professor of Sociology, Weaver Bldg., The Pennsylvania State University, University Park, PA 16802.

Robert F. Denoncourt, (Chapter 15), Professor of Biology, York College of Pennsylvania, York, PA 17405.

Roger H. Downing, (Chapter 28), Research Assistant, Institute for Research on Land and Water Resources, The Pennsylvania State University, University Park, PA 16802.

Judith Dudley, (Chapter 18), Research Technician, Lehigh University, Bethlehem, PA 18015.

Al Dufour, (Chapter 16), Chief, Health Effects Research Division, Cincinnati, OH 45268.

William A. Eberhardt, (Chapter 3), Environmental Manager, Procter and Gamble Paper Products Company, P.O. Box 32, Mehoopany, PA 18629.

J.S. Evans, (Chapter 10), Senior Project Engineer, Woodward-Clyde Consultants, Plymouth Meeting, PA 19462.

H.Y. Fang, (Chapter 10), Professor, Department of Civil Engineering, Fritz Engineering Lab 13, Lehigh University, Bethlehem, PA 18015.

Hays B. Gamble, (Chapter 28), Associate Director, Institute for Research on Land and Water Resources, The Pennsylvania State University, University Park, PA 16802.

Richard A. Gammon, (Chapter 16), Prof. of Microbiology, Gannon Univ., Erie, PA 16541.

J.R. Graham, (Chapter 17), Member of the Minnesota Bar and Scientific Consultant to the Ministry of the Environment, Province of Quebec, Canada.

David C. Hubinger, (Chapter 8), Marketing and Training Manager, Chemicals and Pigment Treatment, Technical Service Laboratory, Chestnut Run, E.I. Dupont De Nemours & Co., Wilmington, DE 19898.

Teh-Wei-Hu, (Chapter 23), Prof. of Economics, The Pennsylvania State University, University Park, PA 16802.

Prem Shankar Jha, (Chapter 25), Senior Asst. Editor, The Times of India, 7, Bahadurshah Zafar Marg, New Delhi - 110002, India.

Gerald A. Kraus, (Chapter 16), Assoc. Prof. of Microbiology, Gannon Univ., Erie, PA 16541.

I.J. Kugelman, (Chapter 10), Professor of Civil Engineering, Lehigh University, Bethlehem, PA 18015.

Donald A. Lazarchik, (Chapter 5), Director, Bureau of Solid Waste Management, Pennsylvania Dept. of Environmental Resources, Harrisburg, PA 17120.

Robert Lewis, (Chapter 2), Assistant Director of Planning for the Environmental Services Department, City of Pittsburgh, Pittsburgh, PA 15201.

William C. Livingood, (Chapter 27), Professor of Health Education, Health Department, East Stroudsburg University, East Stroudsburg, PA 18301.

Brian K. Mathias, (Chapter 4), Adminstrative Resident, The Western Pennsylvania Hospital, 4800 Friendship Avenue, Pittsburgh, PA 15224.

James J. McKeown, (Chapter 3), Regional Manager, NCASI, Tufts University, Medford, MA

John R. McNamara, (Chapter 24), Professor of Economics, Lehigh University, Bethlehem, PA 18015.

E. Willard Miller, (Chapter 14), Professor of Geography (Emeritus), 318 Walker Bldg., The Pennsylvania State University, University Park, PA 16802.

P.J. Morin, (Chapter 17), Scientific Consultant to the Ministry of the Environment, Province of Quebec, Canada.

Patrick F. Mutch, (Chapter 4), Asst. Executive Driector, The Western Pennsylvania Hospital, 4800 Friendship Avenue, Pittsburgh, PA 15224.

Allen R. O'Dell, (Chapter 6), Joint Planning Commission, Lehigh-Northampton Counties, ABE Airport Govt., Build, Allentown, PA 18103.

Michael R. Overcash, (Chapters 11 and 12), Professor of Chemical Engineering, North Carolina State University, Raleigh, NC 27650-5035.

Raymond Regan, (Chapter 1), Associate Professor of Civil Engineering, The Pennsylvania State University, University Park, PA 16802.

Robert F. Schmalz, (Chapter 20), Professor of Geoscience, 503 Deike Bldg., The Pennsylvania State University, University Park, PA 16802.

A.K. Sharma, (Chapter 7), President, Indian National Academy Science, New Delhi, and Professor, Department of Botany, University of Calcutta, Calcutta, 700019, India.

William R. Sierks, Esq., (Chapter 26), Asst. Counsel, Pennsylvania Dept. of Environmental Resouces, 505 Executive House, P.O. Box 2357, 1015 Second Street, Harrisburg, PA 17120.

Olev Taremäe, (Chapter 6), Joint Planning Commission, Lehigh-Northampton Counties, ABE Airport Govt. Build, Allentown, PA 18103.

James Walker, (Chapter 2), Director of Environmental Services, City of Pittsburgh, Pittsburgh, PA 15201.

Ann M. Wolf, (Chapters 13 and 19), Research Asst. Dept. of Agronomy, 221 Tyson Bldg., The Pennsylvania State University, University Park, PA 16802.

James A. Weaver, (Chapter 23), Manager, Economics Studies, Engineering Science, Inc., Fairfax, Virginia.

Stanley J. Zagorski, (Chapter 16), Prof. of Biology, Gannon Univ., Erie, PA 16541.